AI 2041

AI 2041

TEN VISIONS FOR OUR FUTURE

KAI-FU LEE
AND CHEN QIUFAN

CURRENCY
NEW YORK

Published in the United States by Currency,
an imprint of Random House, a division of
Penguin Random House LLC, New York.

CURRENCY and its colophon are trademarks of
Penguin Random House LLC.

Library of Congress Cataloging-in-Publication Data
Names: Lee, Kai-Fu, author. | Chen, Qiufan, author.
Title: AI 2041 / by Kai-Fu Lee and Chen Qiufan.
Description: First edition. | New York: Currency, [2021] |
Includes index.
Identifiers: LCCN 2021012928 (print) | LCCN 2021012929 (ebook) |
ISBN 9780593238295 (hardcover; acid-free paper) |
ISBN 9780593238301 (ebook)
Subjects: LCSH: Artificial intelligence in literature. |
Artificial intelligence.
Classification: LCC Q335 .L423 2021 (print) | LCC Q335 (ebook) |
DDC 006.3—dc23
LC record available at https://lccn.loc.gov/2021012928
LC ebook record available at https://lccn.loc.gov/2021012929

International edition ISBN 978-0-593-24071-7

Printed in the United States of America on acid-free paper

crownpublishing.com

1 2 3 4 5 6 7 8 9

First Edition

Book design by Edwin Vazquez

What we want is a machine that can learn from experience.

—ALAN TURING

Any sufficiently advanced technology is indistinguishable from magic.

—ARTHUR C. CLARKE

CONTENTS

THE REAL STORY OF AI

Artificial intelligence (AI) is smart software
and hardware capable of performing tasks
that typically require human intelligence.
AI is the elucidation of the human learning
process, the quantification of the human
thinking process, the explication of human
behavior, and the understanding of what
makes intelligence possible. It is mankind's
final step in the journey to understanding
ourselves, and I hope to take part in this
new, but promising science.

I WROTE THESE words as a starry-eyed student applying to Carnegie
Mellon University's PhD program almost forty years ago. Computer
scientist John McCarthy coined the term "artificial intelligence" even
earlier—at the legendary Dartmouth Summer Research Project on Artificial Intelligence in the summer of 1956. To many people, AI seems like
the quintessential twenty-first-century technology, but some of us were
thinking about it decades ago. In the first three and a half decades of my
AI journey, artificial intelligence as a field of inquiry was essentially confined to academia, with few successful commercial adaptations.

AI's practical applications once evolved slowly. In the past five years,
however, AI has become the world's hottest technology. A stunning turning point came in 2016 when AlphaGo, a machine built by DeepMind
engineers, defeated Lee Sedol in a five-round Go contest known as the
Google DeepMind Challenge Match. Go is a board game more complex
than chess by one million trillion trillion trillion trillion times. Also, in

contrast to chess, the game of Go is believed by its millions of enthusiastic fans to require true intelligence, wisdom, and Zen-like intellectual refinement. People were shocked that the AI competitor vanquished the human champion.

AlphaGo, like most of the commercial breakthroughs in AI, was built on deep learning, a technology that draws on large data sets to teach itself things. Deep learning was invented many years ago, but only recently has there been enough computing power to demonstrate its efficacy, and sufficient training data to achieve exceptional results. Compared to when I made my cold start in AI forty years ago, we now have about one trillion times more computing power available for AI experimentation, and storing the necessary data is fifteen million times cheaper. The applications for deep learning—and its related AI technologies—will touch nearly every aspect of our lives.

AI is now at a tipping point. It has left the ivory tower. The days of slow progress are over.

In just the past five years, AI has beaten human champions in Go, poker, and the video game Dota 2, and has become so powerful that it learns chess in four hours and plays invincibly against humans. But it's not just games that it excels at. In 2020, AI solved a fifty-year-old riddle of biology called protein folding. The technology has surpassed humans in speech and object recognition, served up "digital humans" with uncanny realism in both appearance and speech, and earned passing marks on college entrance and medical licensing exams. AI is outperforming judges in fair and consistent sentencing, and radiologists in diagnosing lung cancer, as well as powering drones that will change the future of delivery, agriculture, and warfare. Finally, AI is enabling autonomous vehicles that drive more safely on the highway than humans.

As AI continues to advance and new applications blossom, where does it all lead?

In my 2018 book *AI Superpowers: China, Silicon Valley, and the New World Order*, I addressed the proliferation of data, the "new oil" that powers AI. The United States and China are leading the AI revolution, with the United States leading research advances and China more swiftly tapping big data to introduce applications for its large population. In *AI Superpowers*, I predicted new advances, from big-data-driven decision-making to machine perception to autonomous robots and vehicles. I projected that

AI's new applications in digital industries, finance, retail, and transportation would build unprecedented economic value, but also create problems related to the loss of human jobs and other issues. AI is an omni-use technology that will penetrate virtually all industries. Its effects are being felt in four waves, beginning with Internet applications, followed by applications in business (e.g. financial services), perception (think smart cities), and autonomous applications, like vehicles.

2018 Wave 4: Autonomous AI
Agriculture, manufacturing (robotics), transportation (autonomous vehicle)

2016 Wave 3: Perception AI
Security, retail, energy, IOT, smart homes, smart cities

2014 Wave 2: Business AI
Financial services, education, public services, medical, logistics, supply chain, back-office

2010 Wave 1: Internet AI
Websites/apps, search, advertising, games/entertainment, e-commerce, social, Internet lifestyle

Four waves of AI applications are disrupting virtually all industries.

By the time you read this new book in late 2021 or beyond, the predictions I made in *AI Superpowers* will have largely become reality. We must now look ahead to new frontiers. As I've traveled the world talking about AI, I'm constantly asked, "What's next?" What will happen in another five, ten, or twenty years? What will the future hold for us humans?

These are essential questions for our moment in history, and everyone working in the technology space has an opinion. Some believe that we're in the midst of an "AI bubble" that will eventually pop, or at least cool off. Those with more drastic and dystopian views believe everything from the notion that AI giants will "hijack our minds" and form a utopian new race of "human cyborgs" to the arrival of an AI-driven apocalypse. These various predictions may be born out of genuine curiosity or understandable fear, but they are usually speculative or exaggerated. They miss the complete picture.

Speculation varies wildly because AI appears complex and opaque. I've observed that people often rely on three sources to learn about it: science fiction, news, and influential people. In science fiction books and TV shows, people see depictions of robots that want to control or outsmart humans, and superintelligence turned evil. Media reports tend to focus on negative, outlying examples rather than quotidian incremental advances: autonomous vehicles killing pedestrians, technology companies using AI to influence elections, and people using AI to disseminate misinformation and deepfakes. Relying on "thought leaders" ought to be the best option, but unfortunately most who claim the title are experts in business, physics, or politics, not AI technology. Their predictions often lack scientific rigor. What makes things worse is that journalists tend to quote these leaders out of context to attract eyeballs. So, it is no wonder that the general view about AI—informed by half-truths—has turned cautious and even negative.

To be sure, aspects of AI development deserve our scrutiny and caution, but it is important to balance these concerns with exposure to the full picture and potential of this crucially important technology. AI, like most technologies, is inherently neither good nor evil. And like most technologies, AI will eventually produce more positive than negative impacts on our society. Think about the tremendous benefits of electricity, mobile phones, and the Internet. In the course of human history, we have often been fearful of new technologies that seem poised to change the status quo. In time, these fears usually go away, and these technologies become woven into the fabric of our lives and improve our standard of living.

I believe there are many exciting applications and scenarios in which AI can profoundly enhance our society. Firstly, AI will create tremendous value to our society—PricewaterhouseCoopers estimates $15.7 trillion by 2030—which will help reduce hunger and poverty. AI will also create efficient services that will give us back our most valuable resource—time. It will take over routine tasks and liberate us to do more stimulating or challenging jobs. Lastly, humans will work symbiotically with AI, with AI performing quantitative analysis, optimization, and routine work, while we humans contribute our creativity, critical thinking, and passion. Each human's productivity will be amplified, allowing us to realize our potential. The profound contributions AI is poised to make to humanity need to be explored as deeply as its challenges.

Amid what seems like a feedback loop of negative stories about AI, I believe it's important to tell these other stories, too, and answer that question of "What happens next?" So I decided to write another book about AI. This time, I wanted to extend the horizon a bit further—to imagine the future of the world and our society in twenty years' time, or 2041. My aim is to tell the "real" AI story, in a way that is candid and balanced, but also constructive and hopeful. This book is based on *realistic AI*, or technologies that either already exist or can be reasonably expected to mature within the next twenty years. These stories offer a portrait of our world in 2041, based on technologies with a greater than 80-percent likelihood of coming to pass in that timeframe. I may overestimate or underestimate some. But I believe this book represents a responsible and likely set of scenarios.

How can I be so confident? Over the past forty years, I have been involved in AI research and product development at Apple, Microsoft, and Google, and managed $3 billion in technology investments. So I have hands-on experience with the time and processes needed to take a technology from academic paper to pervasive product. Further, as an adviser to governments on AI strategy, I can make predictions based on my knowledge of policy and regulation frameworks, and the reasoning behind them. Also, I avoid making speculative predictions about fundamental breakthroughs and rely mostly on applying and extrapolating the future of existing technologies. Since AI has penetrated less than 10 percent of our industries, there are many opportunities to reimagine our future with AI infusion into these fields. In short, I believe that *even with few or no breakthroughs, AI is still poised to make a profound impact on our society.* And this book is my testimony.

I've been told that one of the reasons that *AI Superpowers* made an impact on readers was that it was accessible to people with no prior knowledge of AI. So when I embarked on this new book, I asked: What can I do to tell stories about AI in a way that makes them even more widely appealing? The answer, of course, was to work with a good storyteller! I decided to reach out to my former Google colleague Chen Qiufan. After Google, I started a venture capital firm. Qiufan did something more adventurous—he became an award-winning science fiction writer. I was delighted that Qiufan agreed to work with me on the project, and to dovetail his creativity with my judgment on what technology will be capable of in twenty years. We both believed that imagining the feasible technologies within a twenty-year period and embedding them in stories would be quite engaging, and we wouldn't even have to resort to teleportation or aliens to mesmerize our readers.

Qiufan and I worked out a unique arrangement. I first created a "technology map" that projected when certain technologies would mature, how long it would take to gather data and iterate AI, and how easy it would be to build a product in various industries. I also accounted for possible externalities—challenges, regulations, and other deterrents, as well as story-worthy conflicts and dilemmas that might emerge alongside these technologies. With my input on the technological components, Qiufan then flexed his talents—dreaming up the characters, settings, and plotlines that would bring these themes to life. We worked to make each story engaging, provocative, and technologically accurate. After each one, I offer my technology analysis, digging into the forms of AI revealed and their implications for human life and society. We organized the stories to cover all key aspects of AI, and roughly ordered them from basic to advanced technologies. The sum of these parts, we hope, is a uniquely engaging and accessible primer on AI.

We named our book *AI 2041* because that is twenty years from the initial publication of this book. But it didn't slip our notice that the digits "41" happen to look a bit like "AI."

Many of our readers may love the wonderful storytelling of science fiction, but I imagine there are others who may not have picked up a novel or a collection of short stories since college. That's okay. If you fall into that camp, think of *AI 2041* not as "science fiction" but as "scientific fiction." The stories are set in wide-ranging locations around the world. In some, you may recognize a world that seems not too dissimilar from your own—with narratives that draw on existing customs and habits, albeit with an AI twist. In others, AI has transformed human life dramatically. Both AI enthusiasts and skeptics will have plenty to think about. Creating a book with a significant fiction component is inherently riskier than writing a nonfiction book that simply describes the present and asks questions about the future. Qiufan and I sought to be bold with our narratives, and we believe the stories that follow will strike a chord with every open-minded reader whose imagination is large enough to ponder what the future holds.

The first seven stories were designed to cover technology applications for different industries in increasing technological complexity, along with their ethical and societal implications. The last three stories (plus chapter 6, "The Holy Driver") focus more on social and geopolitical issues raised by AI, such as the loss of traditional jobs, an unprecedented abundance of goods, exacerbated inequality, an autonomous weapons arms race, trade-offs between privacy and happiness, and the human pursuit of a higher purpose. These are profound changes, and humans may embrace them with compassion, exploit them with malice, capitulate to them with resignation, or be inspired by them to reinvent ourselves. In the final four stories, we decided to show four possible variations and different pathways, as a way of underscoring that the future is not yet written.

We hope the stories entertain you while deepening your understanding of AI and the challenges it poses. We also hope that the book's road map of the coming decades will help you prepare yourself to capture the opportunities and confront the challenges that the future will bring. Most of all, we hope you will agree that the tales in *AI 2041* reinforce our belief in human agency—that we are the masters of our fate, and no technological revolution will ever change that.

Now, let's take a journey to 2041.

HOW WE CAN LEARN TO STOP WORRYING AND EMBRACE THE FUTURE WITH IMAGINATION

IN AUGUST 2019, while visiting the Barbican Centre in London, I came across an exhibition titled *AI: More Than Human.* Like a refreshing summer downpour, the exhibition cleared my senses—and changed most of my preexisting biases and misconceptions toward artificial intelligence. The deceivingly simple name of the exhibition was nowhere near a sufficient representation of the diversity and complexity it contained. Each room of the exhibit revealed new wonders, all with a connection to the curators' expansive definition of what AI encompasses. There was Golem, a mythical creature in Jewish folklore; Doraemon, the well-loved Japanese anime hero; Charles Babbage's preliminary computer science experiments; AlphaGo, the program designed to challenge humans' fundamental intellect; Joy Buolamwini's analysis on the gender bias of facial recognition software; and teamLab's large-scale interactive digital art infused with Shinto philosophy and aesthetics. It was a magnificent and mind-expanding reminder of the power of interdisciplinary thinking.

According to Amara's law, "We tend to overestimate the effect of a technology in the short run and underestimate the effect in the long run." Most of us tend to think of AI in narrow terms: the murderbot from *The Terminator,* incompetent algorithms that could never match the wits or threaten the existence of humans in any way, mere soulless technological inventions that have nothing to do with how humans perceive the world, communicate emotions, manage institutions, and explore other possibilities of life.

The truth—as it has been revealed in stories ranging from the Chinese folktale of Yan Shi, the mechanic who creates a humanoid, to Talos, the bronze automaton in Greek mythology—is that humans' search for artificial intelligence has persisted throughout world history, long before computer science existed as a field or the term "AI" entered the lexicon. From the past era to the present day, the unstoppable force of AI has been revolutionizing every dimension of human civilization, and it will continue to do so.

Science fiction, my chosen field, plays a rather delicate role in investigating the human-machine paradigm. The 1818 novel *Frankenstein,* often praised as the first modern science fiction novel, hinges on questions that still resonate today: With the help of technology, are humans entitled to create intelligent life that's different from all currently existing forms of life? What would the relationship between the creation and the creator look like? The archetype of the mad scientist inflicting his creations on the world originated from Mary Shelley's masterpiece two hundred years ago.

While some may scapegoat science fiction, blaming it for people's narrow and often negative view of AI, that's only part of the story. Science fiction has the capacity to serve as a warning, but speculative storytelling also has a unique ability to transcend time-space limitations, connect technology and humanities, blur the boundary between fiction and reality, and spark empathy and deep thinking within its reader. Historian and bestselling author Yuval Noah Harari has called science fiction "the most important artistic genre" of our time.

That's a high bar to live up to. For science fiction writers like myself, the challenge we face is creating stories that not only reveal hidden truths about our present-day reality, but also, simultaneously, project even wilder imaginative possibilities.

Therefore, when my former colleague from Google Kai-Fu Lee got in touch with me and proposed this collaboration on *AI 2041*—a one-of-a-kind book project that combines science fiction and analysis of big ideas that animate technology—I was thrilled. The Kai-Fu I know is a pioneering global leader, a savvy and trend-making business investor, and an imaginative, open-minded prophet of tech. His notions on career development in his field have influenced a generation of young people. Now, his mind is set on the future.

Equipped with a profound understanding of cutting-edge research and its applications in the business world, Kai-Fu delineates the ways in which AI could change human society in twenty years in areas ranging from medicine and education to entertainment, employment, and finance. His idea for this project was ambitious, but it was also a kind of magical coincidence. Years earlier, in my own writing, I had developed the notion of "science fiction realism." To me, science fiction is fascinating because it not only generates an imaginative space for escapists to leave behind their mundane lives, play the role of superheroes, and freely explore galaxies far, far away, but it also provides a precious opportunity for them to temporarily remove themselves from everyday reality and critically reflect upon it. By imagining the future through science fiction, we can even step in, make change, and actively play a role in shaping our reality.

In other words, with every future we wish to create, we must first learn to imagine it.

My imagination began developing as a child thanks to classic works of science fiction like *Star Wars, Star Trek,* and *2001: A Space Odyssey.* Since I was ten, these works have been my portal to the vast beyond and worlds unknown. I believe that, before setting pen to paper for each story, the key is always to orient the story in the history of its genre and a greater social context. As someone deeply invested in—even obsessed with—the fantasies of science fiction, I am in awe of how inclusive the spectrum of science fiction storytelling is. Almost any theme or style can find its place in the genre.

Before I became a full-time author, I worked in technology. A lot of people would assume engineers and computer science wizards might have little interest in fiction—because their brains are hardwired for science, as opposed to literature. But during my more than ten years working in tech, I encountered many engineers and technologists who were not-so-secret fans of speculative fiction. This enthusiasm sometimes manifested in meeting rooms with names like "Enterprise" or "Neuromancer," but it also was present among the formidable minds behind projects like Google X and Hyperloop. From the modern submarine to the laser gun, and from mobile phones to CRISPR, scientists will readily admit they got direct inspirations from fiction. Imagination indeed shapes the world.

From the beginning, I decided that *AI 2041* would challenge the stereotype of the dystopian AI narrative—the kind of tale where the future is irrevocably bleak. Without disregarding AI's faults or nuances, Kai-Fu and I endeavored to portray a future where AI technology could influence individuals and societies positively. We wished to imagine a future that we would like to live in—and to shape. We imagined a future where the next generations could enjoy the benefits of technological development, work to bring more achievement and meaning into the world, and live happily.

The path to imagining the future of our dreams was not always an easy one. Our challenge was to become immersed in the latest AI research and then to project, with science and logic, *realistically,* how the AI scene would appear in twenty years. Kai-Fu and our team spent hours studying recently published research papers, conversing with experts, professionals, and thinkers involved with the AI industry, participating in the AI workshop hosted by the World Economic Forum, and visiting top AI tech companies, in order to ensure we had a comprehensive grasp of the technological and philosophical basis of AI development.

The other challenge was imagining the human future. We wished to represent how individuals from disparate cultures and industries and with different identities would react to the future shock induced by AI. Subtle psychological details are difficult to infer through mere logic and rationalization. To help fill in the emotional portrait of the characters in our stories, we looked to history and drew inspiration from similar world-changing events that have occurred in the past. To stimulate our readers' imagination and capacity to conceptualize alternative human conditions, we knew our stories must also spark empathy if we were to fully convey our vision and sentiment. Kai-Fu's analysis serves as the string that connects the soaring kite of imagination to the graspable reel of reality.

After months of intensive work and rounds of polishing, here are the ten portals we have assembled that will transport you to the time-space of 2041. We hope that you will embark on this journey with curiosity, an open mind—and an open heart, too.

One last thing: For me, the greatest value of science fiction is not providing answers, but rather raising questions. After you close the book, our hope is that lots of new questions will enliven your mind: For instance, can AI help humans prevent the next global pandemic by eliminating it at the very root? How can we deal with future job challenges? How can we maintain cultural diversity in a world dominated by machines? How can we teach our children to live in a society where humans and machines coexist? We hope our readers' questions will help take us further down the path as we shape a happier and brighter future.

Welcome to 2041!

AI 2041

CHAPTER ONE

THE GOLDEN ELEPHANT

STORY TRANSLATED BY BLAKE STONE-BANKS

IT IS BETTER TO LIVE YOUR OWN DESTINY
IMPERFECTLY THAN TO IMITATE SOMEBODY ELSE'S
PERFECTLY.
> —BHAGAVAD GITA (भगवद्गीता, SONG OF GOD
> OR HINDU SCRIPTURE), CHAPTER 3, VERSE 35

NOTE FROM KAI-FU: The opening story takes readers to Mumbai, where we meet a family who has signed up for a deep-learning-enabled insurance program. This dynamic insurance program engages with the insured in the form of a series of apps intended to better their lives. The family's teenage daughter, however, finds that the AI program's persuasive nudges complicate her search for love. "The Golden Elephant" introduces the basics of AI and deep learning, offering a sense of its main strengths and weaknesses. In particular, the story illustrates how AI can single-mindedly try to optimize certain goals, but sometimes create detrimental externalities. The story also suggests the risks when one company possesses so much data from its users. In my commentary at the end of the chapter, I will explore these issues, offering a brief history of AI and why it excites many but has become a source of distrust for others.

ON THE SCREEN, the three-story statue of Ganesh swayed in the surf of Chowpatty Beach as though synced to the sitar soundtrack. With each wave, the towering idol descended lower until it was engulfed by the Arabian Sea. In the salty brine, the statue dissolved into gold and burgundy foam, washing onto Chowpatty Beach, where the colors clung like blessings to the legions of believers who had gathered for the Visarjan immersion ritual celebrating the end of the Ganesh Chaturthi festival.

In her family's Mumbai apartment, Nayana watched as her grandparents clapped their hands and sang along to the TV. Her younger brother, Rohan, took a mouthful of cassava chips and a deep swig from his diet cola. Though he was only eight, Rohan was under doctor's orders to strictly control his fat and sugar intake. As he wagged his head in excitement, crumbs sprayed from his mouth and flew across the floor. In the kitchen, Papa Sanjay and Mama Riya banged on pots and crooned like they were in a Bollywood film.

Nayana tried to shut them all out of her mind. The tenth-grader was instead focusing all her energy on her smartstream, where she had downloaded FateLeaf. The new app was all Nayana's classmates could seem to talk about lately. It was said to possess the answer to almost any question, thanks to the prescience of India's greatest fortune tellers.

The app—its branding and ad campaign made clear—was inspired by the Hindu sage Agastya, who was said to have engraved the past, present, and future lives of all people in Sanskrit onto palm leaves, so-called Nadi leaves, thousands of years ago.

According to the legend, simply by providing one's thumbprints and birthdate to a Nadi leaf fortune teller, a person could have their life story foretold from the corresponding leaf. The problem was that many leaves had been lost to meddling colonialists, war, and time. In 2025, a tech company tracked down and scanned all the known Nadi leaves still in circulation. The company used AI to perform deep learning, auto-translation, and analysis of the remaining leaves. The result was the creation of virtual Nadi leaves, stored in the cloud—one for each of the 8.7 billion people on Earth.

Nayana was not dwelling on the ancient history of the Nadi leaves. She had a more pressing matter on her mind. Users of the FateLeaf app could seek to uncover the wisdom of their Nadi leaf by posing various questions. While her family watched the Ganesh Visarjan celebration on TV, Nayana nervously typed out a question within the app: "Does Sahej like me?" Before she clicked "Send," a notification popped up indicating that an answer to her question would cost two hundred rupees. Nayana clicked "Submit."

Nayana had liked Sahej from the moment his stream first connected in their virtual classroom. Her new classmate didn't use any filter or AR background. Behind Sahej, hanging on the wall, Nayana could see rows of colorful masks, which, she learned, Sahej had carved and painted himself. On the first day of the new term, the teacher had asked Sahej about the masks, and the new student shyly gave a show-and-tell, explaining how the masks combined Indian gods and spirits with the powers of superheroes.

Now, in an invitation-only room on her ShareChat, some of Nayana's classmates were gossiping about Sahej. From the way his room was furnished to the fact that his surname was hidden from public view in school records, these girls were certain Sahej was among the "vulnerable group" that the government mandated make up at least 15 percent of their student body. At private schools across India, such children were practically guaranteed spots and their tuition, books, and uniforms were covered by scholarships. "Fifteen percent" and "vulnerable group" were euphemisms for the Dalits.

From documentaries she had watched online, Nayana knew about India's old caste system, which was deeply embedded in Hindu religious and cultural beliefs. A person's caste had once determined one's profession, education, spouse—their whole life. At the bottom rung of this system were the Dalits, or, as they were sometimes referred to with derision, "untouchables." For generations, members of this community were forced to do the dirtiest jobs: cleaning sewers, handling the corpses of dead animals, and tanning leather.

The constitution of India, ratified in 1950, outlawed discrimination based on caste. But for years following independence, Dalit areas for drinking, dining, residing, and even burial were kept separate from those of groups considered higher in the system. Members of the higher castes might even refuse to be in the same room as the Dalits, even if they were classmates or colleagues.

In the 2010s, the Indian government sought to correct these injustices by establishing a 15-percent quota for Dalit representation in government positions and in schools. The well-intentioned policy had sparked controversy and even violence. Higher-caste parents complained that such admissions weren't based on academic performance. They argued that their children were paying the price for previous generations' sins and that India was just trading one form of inequality for another.

Despite these pockets of backlash, the government's efforts seemed to be working. The 200 million descendants of Dalits were integrating into mainstream society. It had become more difficult to recognize their past identity at a glance.

THE GIRLS IN NAYANA'S ShareChat couldn't stop talking about the new boy in school, Sahej, debating his background—but also whether they would consider going out with him.

You shallow snobs, Nayana silently huffed.

For her part, Nayana saw in Sahej a kindred artistic spirit. Inspired by Bharti Kher, Nayana dreamed of becoming a performance artist, and she often had to explain that this was nothing like being a superficial pop entertainer. She believed great artists had to be brutally honest about their innermost feelings and should never accept the perspectives of others. If she liked Sahej, then she liked Sahej—no matter his family background, where he lived, or even his Tamil-accented Hindi.

The question Nayana had posed to the FateLeaf app seemed to take forever to process. Finally, a notification popped up on Nayana's smartstream accompanied by a palm leaf icon: "What a pity! Due to insufficient data provided, FateLeaf cannot currently answer your query."

The clink of Nayana's refund vibrated from her smartstream.

"Insufficient data!" Nayana silently cursed at the app.

Annoyed, she finally raised her head from her screen to notice her mother, Riya, putting the finishing touches on dinner. Something was off. In addition to a number of Indian holiday delights, Nayana saw several super-expensive dishes from a Chinese delivery place on the table. Such treats were rare for her penny-pinching father. But there was something even more unusual: Riya was wearing her favorite pure silk Parsi-style sari. She had her hair up and was wearing a complete set of jewelry. Even Nayana's grandparents seemed different—happier than usual—and for once, her fat brother, Rohan, wasn't pestering her with all kinds of stupid questions.

The Ganesh Chaturthi festival couldn't explain all this.

"So, is anyone going to tell me what's going on?" Nayana said as she stared at the spread on the table.

"What do you mean, what's going on?" Riya shot back.

"Am I the only one who thinks all this is a bit out of the ordinary?"

Nayana's parents glanced at each other for a second then burst out laughing.

"Take a look and tell us what's different," Riya said.

Nayana felt like she was about to lose her mind. "What are you hiding from me?"

"My sweet little girl, eat first." Grandmom began to pull apart the naan.

"Wait. Did Dad get promoted? Did we win the lottery? Did the government cut taxes?"

Dad wobbled his head back and forth. "All beautiful ideas. But no. It's all for your mother—"

Nayana spun toward her mother. "Mom, what did you buy this time?"

"Your tone should be more respectful when talking with your elders," Riya chided.

"It wasn't me who got taken to the cleaners for buying cheap . . ." Nayana's voice trailed off into a sigh.

Nayana exhaled. "And what exactly is it you bought?"

"Ganesh Insurance! They had an amazing sale for the holiday. First time ever that GI was fifty percent off! All the neighbors got it, too, and they're even more thrifty than me."

Dad clapped his hands in excitement. So did Nayana's grandparents.

"Wait! Hasn't our family always had a policy from the Life Insurance Corporation of India?"

"That policy wasn't nearly enough! Your grandparents are old and rely on us. What if something were to happen to us? Where will the money come from? We have to save where we can. And you and your younger brother are both in private school, and don't you still want to go to SOFT at Rai University? Tuition and dormitories cost a lot more than public universities in Mumbai."

"Why does the conversation always have to twist back to blaming me?"

"To plan for the future, you must also think about what's in front of you," Grandfather observed.

"So, what's the deal with this insurance exactly?"

"Well, Mrs. Shah from next door filled me in," explained Riya. "It's a platform that uses AI to adjust the insurance plan according to the family's needs. And for a very good price. And it's not just the one platform, more like a little family of apps. There's one for calculating and paying insurance fees, and one for investments, and my favorite one is the home goods shop. Another one shows you deals in your area. And just look at my hair. The salon the Cheapon deals app recommended only cost four hundred rupees."

Just as Rohan was about to steal a sweet, Nayana slapped the back of his hand, which he withdrew with a sheepish look.

"You sound like an advertisement," Nayana told her mother. "Why would an insurance company tell you where to get your hair done? And how exactly does this AI insurance know so much about our family?"

"This, well . . ." Mom searched for a way out of the question. "To get the benefits of Ganesh Insurance, we share data link access for each member of the family."

"What?" Nayana's eyes grew as wide as brass bells.

"It's all kept strictly confidential, unless we give permission for GI to use it."

"What right do you have to share my data link with some insurance company!"

"Hey, don't talk to your mother like that." Dad wagged his finger at Nayana. "Don't forget you're still a minor. As your parents, we have the right to make data decisions for you."

Nayana's face turned bright red; she was unaccustomed to such a sharp rebuke from her father. She threw her knife and fork down onto the plate and raced back to her room. She grabbed her quilt and pulled it over her head, imagining that somewhere on her Nadi leaf, it was written that today was the worst day of her life.

NAYANA AND HER MOTHER'S cold war lasted a week, until Nayana's smartstream began pushing some unusual new notifications:

```
IT'S GOING TO RAIN TODAY
SO TAKE AN UMBRELLA.
RESPIRATORY ILLNESSES ARE BECOMING MORE
PREVALENT, SO YOU SHOULD WEAR A MASK.
THERE'S A TRAFFIC ACCIDENT ON YOUR ROUTE
SO TO AVOID THE CONGESTION . . .
```

At first, Nayana was skeptical about the endless stream of notifications. But she found she couldn't stop reading them. From time to time, she actually got a useful tip. A clothing deal, a discount at a lunch place she liked . . . Of course, to actually redeem the deals, Nayana had to install the Cheapon deals app and other various golden-elephant-branded Ganesh Insurance apps on her smartstream and permit them to access her data.

It seemed Mom had already forced the golden elephant onto all the family's smartstreams. Women controlled data sharing in more than 60 percent of Indian households. All that personal data was linked to the national ID Aadhaar card and the unique identifying number issued to all of India's 1.4 billion residents by the Unique Identification Authority of India. Since implementing the system in 2009, after twenty years of development, the government had collected data including citizens' fingerprints, retina signatures, genetic histories, family information, occupations, credit scores, home-buying history, and tax records. With its clients' consent, Ganesh Insurance was able to tap into this rich trove of data to personalize its services.

Of course, there were some privacy restrictions. For example, social media data needed to be separately authorized, and use of minors' data required consent of their legal guardians.

Nayana aimed to be vigilant in every interaction with GI. In her data literacy class in high school, she had learned that on the Internet every click might sell you out. She carefully studied the fine print before choosing between "I accept" or "I need more time to consider." Yet it seemed every time she selected "I need more time to consider," GI would send appealing new discounts and suggestions for how to solve her immediate problems.

For example, how exactly could she attract Sahej's attention?

Sahej was really cute, especially his sheeplike eyes. He instinctively wanted to please every classmate. He had even sent each classmate a wood carving he had made of a small animal head. But the virtual classroom had its limits. Sometimes all Nayana could see was just a blurry headshot icon and a glitchy voice due to Sahej's poor connection. After finally meeting Sahej during one of the school's "in-person" days, Nayana found it even more difficult to contain her feelings for him. She sought any excuse to talk with him. But for some reason, the boy kept his distance.

Does Sahej not like me? Or is it another reason?

Could Sahej's background, Nayana wondered, account for his shyness around her?

AS THE QUESTION LINGERED in Nayana's mind, little golden elephants popped up in a notification from MagiComb, GI's lifestyle advice app, about "how to make yourself more attractive to guys." Nayana guessed that the AI was able to use her online browsing and shopping data to infer what she was thinking, but these recommendations disturbed Nayana for another reason. Why should women need to change themselves to win a man's favor? Why couldn't women show men who they really were and see if they were or weren't a match?

Though she was still feeling annoyed at her mother, Nayana decided to ask her about the golden elephant's odd messages.

"Silly girl, machines only learn what is taught to them by human beings." Riya looked at her newly bought long skirt in the mirror and turned. "But what's this all about? Have you met someone?"

"Not at all," Nayana replied, with a tinge of guilt.

"You can hide it from me, but not from the AI," her mother joked. "Are you sure you don't want me to help you scheme? You know, your mother knows a thing or two about men."

"I just don't know how I can find out what he really thinks of me. I give him likes online, but he never seems to respond."

"Ahh, so there *is* someone! It's not enough to give someone a like online. You've got to have guts. And that reminds me. If you permit GI to access your ShareChat account data, its recommendations will be better. Not to mention, the premium for our family will also dip just a bit more."

Nayana shook her head and left the room. She recalled that only a few weeks earlier, her mother had rejected Nayana's request to share her data link with a different app—FateLeaf—to obtain more accurate fortune-telling. Now, their positions were reversed. Of course, now money was on the line.

It wasn't just her mother. To Nayana, everyone in the family had been brainwashed by that little golden elephant. They had become hyperaware that any change in behavior might raise or lower their premiums. Once something was linked with money, it seemed to Nayana that the human brain went on autopilot. They'd do whatever it took to score an award and evade a penalty.

It wasn't that GI didn't have its plus side. The little golden elephant would remind Nayana's grandparents to take their medicine and nudge them to schedule doctor appointments. Even Nayana's father, who had never listened to anyone, gave up smoking when the little golden elephant kept chiding him. He swapped his favorite arrack for a healthier single nightly glass of red wine. His driving style even became more restrained. At the app's urging, he no longer zigzagged through Mumbai's congested streets like an out-of-work race car driver. GI had given him an incentive—by changing his behavior, he was able to lower his auto, health, and life insurance premiums.

If anyone in the family could resist the GI app's nudges, Nayana suspected it would be her brother, Rohan. After all, fat and sugar were as addictive as heroin, especially for children with no self-control. But that golden elephant made it happen. Even if the eight-year-old didn't understand insurance premiums or delayed gratification, the rest of the family were conditioned to see any sweet near the boy as a threat to their bank account. Their former indulgence of Rohan's sweet tooth was over.

It naturally made sense. Insurance companies wanted people to live healthier, longer lives—it made for better profits.

As for herself, Nayana was still on the fence. Should she hand over the data link of her ShareChat?

Equally puzzling was the question of Sahej. When Sahej had given everyone in the class a handmade wood carving, he chose a crow's head covered in patterns to give to Nayana. The tenth-grader practically tore her hair out thinking about what hidden meaning the gift might hold.

Doesn't the crow symbolize bad luck? Is he telling me to not be so loud and annoying? Am I coming on too strong? What's he saying exactly?

Nayana tortured herself with such questions. Her first thought was to turn to FateLeaf for a divination, but her mother had forbidden her to permit that app to access her data. *What about Magi-Comb?* Nayana wondered. Lovesick, she decided she would see what the elephant's omnipotent algorithm had to say about her future.

The future the little golden elephant imagined, however, wasn't anything like the one she'd hoped for.

EVERYTHING FELT IMMEDIATELY WRONG.

Granting GI access to her data on ShareChat, Nayana knew from her data literacy class, was like opening the door to your bedroom. Your whole private life might be visible at a glance. Although GI guaranteed that all data was fed anonymously to its AI for purposes of federated learning and that no third party could access it, Nayana thought that sounded a bit like a farmer telling the turkey a week before Thanksgiving, "Hey, you're safe here."

Whenever she was browsing, chatting, liking, or even selecting emojis on ShareChat, all Nayana could think about now was how her choices would affect the family's insurance premiums. She found the whole system infuriating and ridiculous.

But perhaps, she wondered, *it's even more ridiculous to expect this AI to act as my matchmaker.*

Sahej posted almost nothing on ShareChat. He was like a person from some bygone era who had failed to keep up with technology. He occasionally posted a news article, quotes he liked, or outdated memes. But his usage was sporadic and unpredictable. His account, Nayana thought, looked like a fake zombie account.

The AI was supposed to help Nayana get together with Sahej, but how could it learn anything important about Sahej from his boring account? Meanwhile, it was all too easy for AI to understand Nayana's intentions, given her incessant clicking. To the AI, such things were a matter of math, not love.

To Nayana, something fishy was going on with GI when it came to Sahej. She found it odd that every time she refreshed Sahej's page or liked one of his posts, GI would send her a weird notification, as if trying to break her focus. If she tried to come up with a reason to talk with him, browsed online for a gift for him, or even just thought about inviting him out to coffee, that little golden elephant would pop up with some totally ridiculous recommendation or seemingly load a page in error.

The only possible explanation Nayana could think of was that the little golden elephant didn't want her to get close with Sahej at all. It was actively working against her.

Was the elephant like this with everyone? Is it because I'm too young? But isn't coupling up and marriage a good thing? Aren't we told that as a country of 1.4 billion, our reproductive capacity will make us invincible on the world stage? What's the problem?

As her thoughts spiraled, Nayana noticed her mother watching her from the doorway.

"What in the hell have you been up to, young lady? Our premium is going through the roof!"

"Me?" Nayana didn't know what to say. It was clear to her the little golden elephant was determined to turn her whole virtual world upside down.

"Tell me now, or I'm taking your smartstream away!"

"No, you can't!"

"Sorry, but yes, that's exactly what I'm going to—"

Before Riya could finish, Nayana shot up, rushed past her mother, and raced out of the house as fast as her legs would carry her.

Clenching her smartstream, Nayana ran until she no longer recognized where she was. Finally, she saw the familiar relief sculptures of the New India Assurance Building in the Fort district. Sunset lit the weathered façade's artfully sculpted farmers, potters, spinners, and porters as Nayana decided it was the perfect moment to give Sahej a call, no matter how much it raised her family's premium.

The boy's avatar image popped onto her smartstream as the screen flickered with GI notices. Nayana could see that her family's premium had already increased by 0.73 rupees. It was a long time before he answered, and Nayana was about to give up when the phone finally connected with a videostream so dark she could barely make out the contours of a face and white-toothed grin.

"That you, Sahej?" Nayana asked timidly.

"It's me. Nayana?"

"I was afraid you weren't going to answer."

"Ermmm . . . it's a bit complicated. I can't speak long. But I really do want to talk with you."

"Me, too." Nayana's heart jumped. "I'm going to give you the address of a restaurant. Can we meet there?"

Sahej glanced about in silence for a moment before finally whispering, "Okay."

After hanging up, Nayana couldn't help but cheer.

Then someone called her name. She spun to see her mother shining gold and red against the setting sun, as though the goddess Saraswati had come down to Earth.

"How did you find me?"

"I'm the data manager of our house, and don't you forget it!" Her mother glared at her.

"I'm sorry." Nayana didn't dare look into her mother's eyes. "But, remember I told you about that guy? I'm going to meet up with him. But GI won't allow it, so . . ."

"You think that's why the premium's going up? GI wants to keep us living healthier and longer—and prevent us from doing stupid things that will harm us, not . . . Unless this is some kind of dangerous person?"

Nayana shook her head. "No, he's just my new classmate, Sahej. He's smart, a real talent. And this is the present he made for me. He carved it himself."

Her mother inspected the wooden crow head Nayana handed her.

"He doesn't sound like such a dangerous person. Is he handsome?"

Nayana let slip a shy smile, but it quickly turned into a grimace. "This sucks. What does GI know that I don't? Maybe I'll live longer if I never meet up with him."

"Sweetie, let me tell you something." Nayana's mother draped an arm over her daughter's shoulder. "I know we don't always see eye to eye. But I'm not as blind as you may think! You know, talking to you makes me think about something I read recently. Actually—it was an old ebook suggested by the MagiComb, come to think of it."

"What was it?" Nayana became curious.

"It was a book from 2021, and there was a story in there about a mom who is so superficial and proud and obsessed with her image that she ignores her daughter's growing distress. And the family was Indian just like us! It hit me hard.

"When I was your age, my parents wanted me to marry as soon as possible. I wanted to go to school to become a lawyer. But they didn't want me entertaining suitors or making my own decisions. I wasn't brave enough to stand up for myself. I gave in, and I regret it to this day. And that's why I would never be mad at you for wanting to follow your heart, whether it's about a boy or who you want to become in the world."

Nayana's mother kept her hand on her daughter's shoulder. Nayana noticed the sun glinting in her mother's eyes.

"I have always worked to give you the sense of security and comfort I never had, so you won't ever have to rely on who you marry to bring you happiness. Go to SOFT at Rai University and become whoever you are supposed to be. Don't let anyone tell you who you are. No one, human or AI. If they try, don't you dare listen. There is no easy answer, and you'll never find it unless you try."

"So you don't mind if I leave Mumbai for Ahmedabad?"

"Well, if you study and pass the exam." Her mother smiled. "Don't forget the competition is fierce."

"You won't mind then if our family's premium keeps going up because of me?"

"Some risks are worth taking."

"Thank you, Mom. I'm going to see Sahej now. I'll bring the answer back for you."

A red double-decker SmartBus turned the corner. Nayana kissed her mother and skipped toward the station as the sun dipped below the horizon.

⊢———

THROUGH THE WINDOW, WAITERS were busily setting tables and lighting candles, waiting for customers to enter Indigo on this romantic night. Sahej was on the corner. His skin looked even darker in the night. Nayana could tell he didn't want to enter the restaurant.

"I'm sorry." He shook his head. His eyes glowed like fireflies.

"Why?"

"If I go into that restaurant with you, my mother will not be happy. Going to such a restaurant is an indulgence—it would increase our premium."

"You mean . . ." Nayana put it together. "Your family's on Ganesh, too?"

"Yeah. My mom is sick. We're lucky GI offers a special insurance premium for vulnerable groups, otherwise we would never be able to afford—"

"I understand," Nayana said. "But what I don't understand is why you would give me a crow instead of a peacock, a rabbit, or any other animal?"

Sahej flashed a slight smile. "You're a girl with a lot of questions. Maybe we shouldn't stand at the door of this dumb fancy restaurant staring at each other like two idiots. Let's walk around a bit."

THE MUMBAI STREETS WERE full of traffic this time of night, and car horns beeped one after another across the vast city of thirty million people. It wasn't always a city of tall buildings, bright lights, and digital displays. But it had been crowded with people for a long time. This place, its history, could be traced all the way back to the Stone Age. When the ancient Greeks arrived here, they'd named the city Heptanesia, meaning "seven islands." Mumbai had since seen the rise and fall of many dynasties and rulers. It had been baptized in blood and reborn countless times before the country was granted its independence.

Such thoughts of history were far from the minds of the two high schoolers strolling the city's brightly lit streets. Nayana noticed Sahej was careful to keep his distance from her, as though she carried a dangerous electric current.

"Sahej, why? Why can't we get close to each other?" Nayana chose her words carefully.

Now it was Sahej's turn to look surprised. "Nayana, do you really not know?"

"Know what?"

"My last name."

"The schools and virtual classroom keep your surname protected just like you're the offspring of a big star or some famous family."

"On the contrary, it's because they don't want it to trigger any discomfort."

"What kind of discomfort?"

"In the past, it was described as a feeling of being *polluted*."

"You're talking about your caste? But that whole system was outlawed years ago."

Sahej gave a bitter laugh. "Just because it's no longer permitted by law and doesn't appear in the news doesn't mean it's gone."

"But how would the AI know about it?"

"The AI doesn't know. The AI doesn't need to know the definition of the castes. All it needs is its users' history. No matter how we hide or if we change our surnames, our data is a shadow. And no one can escape their shadow."

Nayana thought about what her mother had said, that AI only learns what humans teach it. She rolled the thought about in her head, then looked at Sahej. "So you're saying that AI identifies the invisible discrimination in our society and quantifies it."

Sahej's expression became serious, but he exhaled a soft laugh. "I almost forgot. There's also the color of my skin. The Sanskrit word *vārna* once meant both *caste* and *color*."

"It's all so absurd!"

"No, it's reality. And in reality, women of low caste can date and marry men of higher caste. But the other way around will never be accepted. The reputation of the girl's family would be damaged."

"But does the AI really care about those things?"

"Sure, AI doesn't care about our old social mores. It only cares about how to reduce the premium as much as possible, and that's why GI wants to stop us from being together."

When she heard Sahej say "together," Nayana's ears felt hot.

"Objective function maximization."

"What?"

"Humans give the AI its objective, which here is to decrease insurance premiums to the lowest cost possible. Then the AI does everything possible to achieve that goal. The AI won't consider anything at all beyond those factors, certainly not whether or not we're happy. Machines aren't smart enough to interpret all the feelings going on behind the data. Plus, these injustices and biases are still real. All AI does is lift that veil of shame."

"Why do you know so much about it?"

Sahej gave a little smile. "Because I want to go to Imperial College to become an AI engineer, so I can help change it."

They reached the crossroads near Nayana's home, and Sahej paused to prepare his goodbye.

"But why can't we change it now?" Nayana said. "Are we so ready to let AI arrange our fate? Like those predictions on FateLeaf that were written thousands of years ago?"

A strange expression emerged on Sahej's face. "Have you opened FateLeaf since connecting to GI?"

"Ugh, I'm so sick of that little golden elephant. What does it have to do with FateLeaf?"

"FateLeaf is in the GI family of apps, just like MagiComb and Cheapon. If you accept the data-sharing terms, you'll get more accurate fortunes."

"Of course! How did I not figure this out before? So the so-called fates of the Nadi leaves aren't real after all. I guess, like everyone else, I wanted it to be real—and for it to tell me what I wanted to hear." Nayana didn't know whether to rejoice or feel cheated.

Sahej looked at the girl before him. He paused, then pointed at the street he was going to take home.

"This road leads to where my family lives. It passes through the Dharavi construction site. There used to be more than a million people crowded into that 2.4-square-kilometer slum. Tourists visited to take photos, but not one ever wanted to stay. The government's finally transforming it into a community suitable for ordinary citizens. But I promise you, if you ever get close to Dharavi, your GI will flood with illness alerts, or warnings not to drink the water. The app will implore you to stay away. Nayana, I appreciate your sense of justice, but that path just isn't for people like you. The world is on your side, not that side. If we're going to talk about fate, that's what our fate is."

"Take me there." Nayana was startled by how quickly the words left her mouth, but she stepped forward nonetheless. "I want to prove I'm not that person you're thinking of."

Sahej tilted his head. "You sure?"

Nayana glanced at the road stretching into a forbidden hollow at the heart of Mumbai. She was afraid, but she remembered what her mother had told her before saying goodbye: *Some risks are worth taking.*

Sahej smiled, bent his arms, and gestured her forward in a gentleman's bow. "As you please."

The young couple made their way deeper into the ancient city, where centuries of renovation and innovation had branded every corner. Towers old and new lined their path like reincarnated souls. Of course, these souls, too, would eventually be broken up and reconstituted by tomorrow's machine gods.

"So, will you now finally tell me why on earth you made me a crow's head?"

"My astrological animal is the crow, though perhaps I'm more socially awkward than most crows."

"It's that simple?"

"It's that simple."

Nayana's smartstream vibrated with increasing frequency. She knew every vibration was an alert from that little golden elephant trying to save her, warning her to walk away from what was once the world's largest slum, incentivizing her to turn her back on its poverty, disease, discrimination, and *untouchables*, like the boy next to her.

She pulled her collar tight and continued forward at his side.

In the dark ancient streets ahead, an answer was waiting.

ANALYSIS

DEEP LEARNING, BIG DATA, INTERNET/FINANCE APPLICATIONS, AI EXTERNALITIES

The benefits of Ganesh Insurance—powered by deep learning AI—are clear in "The Golden Elephant." Nayana's mom, Riya, saves money thanks to the program's deals app. Her father, Sanjay, quits smoking and drives more safely. Even her brother is eating healthier, after AI raises an alarm about the potential for him to develop diabetes. Such a suite of apps running on the smartstream (mobile phone of 2041), marked by personalized nudges toward better health and well-being, could help people live longer, healthier, and wealthier lives. So, is there a catch? That question about trade-offs lies at the heart of "The Golden Elephant," which introduces the foundational AI concept of deep leaning.

Deep learning is a recent AI breakthrough. Among the many sub-fields of AI, machine learning is the field that has produced the most successful applications, and within machine learning, the biggest advance is "deep learning"—so much so that the terms "AI," "machine learning," and "deep learning" are sometimes used interchangeably (if imprecisely). Deep learning supercharged excitement in AI in 2016 when it powered AlphaGo's stunning victory over a human competitor in Go, Asia's most popular intellectual board game. After that headline-grabbing turn, deep learning became a prominent part of most commercial AI applications, and it is featured in most of the stories in *AI 2041*.

"The Golden Elephant" explores deep learning's stunning potential—as well as its potential pitfalls, like perpetuating bias. So how do researchers develop, train, and use deep learning? What are its limitations? How is deep learning fueled by data? Why are Internet and finance the two most promising initial industries for AI? In what conditions does deep learning optimally work? When it does work, why does it seem to work *so well*? And what are the downsides and pitfalls of AI?

WHAT IS DEEP LEARNING?

Inspired by the tangled webs of neurons in our brains, deep learning constructs software layers of artificial neural networks with input and output layers. Data is fed into the input layer of the network, and a result emerges from the output layer of the network. In between the input and output layers may be up to thousands of other layers, hence the name "deep" learning.

Many people assume AI is "programmed" or "taught" by humans with specific rules and actions, like "cats have pointy ears and whiskers." But deep learning actually works better without these external human rules. Instead of being nudged by humans, many examples of a given phenomenon are fed into the input layer of a deep learning system, along with the "correct answer" at the output layer. In this way, the network in between the input and output can be "trained" to maximize the chance of getting the correct answer to a given input.

For example, imagine that researchers want to teach a deep learning network how to distinguish between photos depicting cats and those that do not depict cats. To start, a researcher might feed the network millions of sample photos labeled "cat" or "no cat" into the input layer, with "cat" or "no cat" already set at the output layer. The network is trained to figure out for itself what features in the millions of images were most helpful to separate "cat" from "no cat." This training is a mathematical process that adjusts the millions (sometimes even billions) of parameters in the deep learning network in order to maximize the chance that a cat image input results in a "cat" output, and that a non-cat image input results in a "no cat" output. The figure below shows such a "cat recognition" deep learning neural network.

During this process, deep learning is mathematically trained to maximize the value of an "objective function." In the case of cat recognition, the objective function is the probability of correct recognitions of "cat" vs. "no cat." Once "trained," this deep learning network is essentially a giant mathematical equation that can be tested on images it hasn't seen,

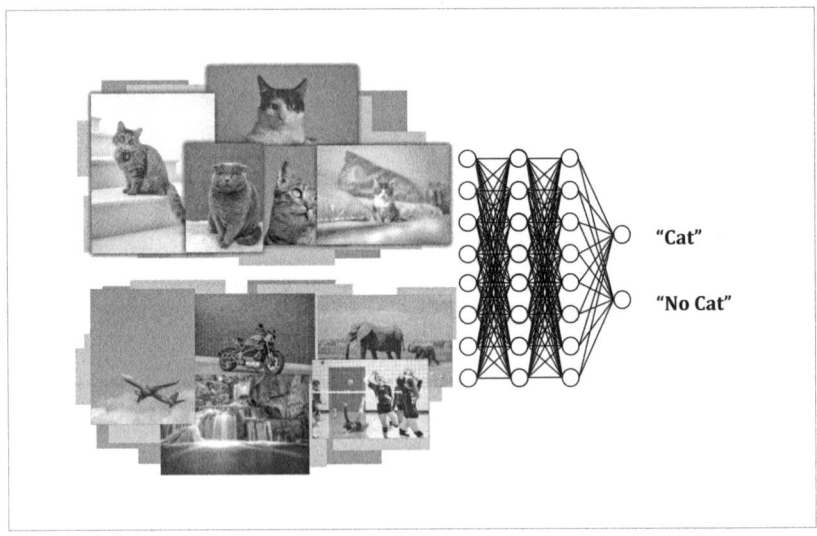

Deep learning neural network trained to recognize pictures
of cats vs. pictures with no cats.

and it will perform "inference" to determine the presence or absence of cats. The advent of deep learning pushed AI capabilities from unusable to usable for many domains. The figure on the following page shows the dramatic reduction of image recognition errors before and after deep learning was applied.

Deep learning is an omni-use technology, meaning it could be applied to almost any domain for recognition, prediction, classification, decision-making, or synthesis. Take insurance, the prime example in "The Golden Elephant." The deep learning powering Ganesh Insurance's apps has been trained to determine the likelihood that each insured may develop serious health problems, and then set premiums accordingly.

To train a network to separate those who likely face serious health claims from those who do not, AI would learn from training data comprising all past insurance applicants and their medical claims and family information. Each case would be labeled with "filed serious health claim" or "did not file serious health claim" in the output layer. Having absorbed this trove of data in the training process, the AI could infer the likelihood any new application would lead to a serious health claim, and decide

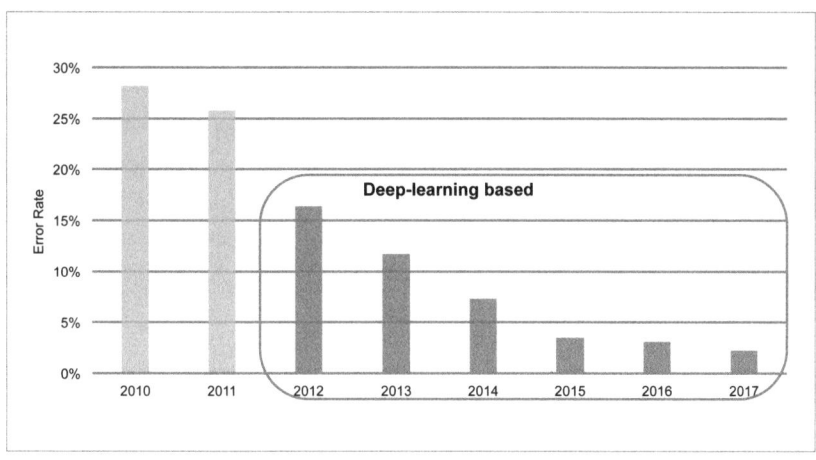

Deep learning led to dramatically
lower computer-vision object recognition rates.

whether to approve the insurance application or not, and if so, how much the premium should be. Note that, in this scenario, no human would ever need to label an applicant as a health risk or not. Instead, the labels are based solely on "ground truth" (for example, whether each insurer had filed a serious health claim).

DEEP LEARNING: AMAZING CAPABILITIES BUT WITH LIMITATIONS

The first academic paper describing deep learning dates all the way back to 1967. It took almost fifty years for this technology to blossom. The reason it took so long is that deep learning requires large amounts of data and computing power for training the artificial neural network. If computing power is the engine of AI, data is the fuel. Only in the last decade has computing become fast enough and data sufficiently plentiful. Today, your smartphone holds millions of times more processing power than the NASA computers that sent Neil Armstrong to the moon in 1969. Similarly, the Internet of 2020 is almost one trillion times larger than the Internet of 1995.

While deep learning was inspired by the human brain, the two work very differently. Deep learning requires much more data than humans, but once trained on big data, it will outperform humans by far for a given task, especially in dealing with quantitative optimization (like picking an ad to maximize likelihood of purchase, or recognizing a face out of a million possible faces). While humans are limited in the number of things they can pay attention to at once, a deep-learning algorithm trained on an ocean of information will discover correlations between obscure features of the data that are too subtle or complex for we humans to comprehend, and which may not even be noticed.

Furthermore, when trained on a huge amount of data, deep learning can customize for individual users, based on that user's patterns as well as similar patterns observed on other users. For example, when you visit Amazon, the website's AI highlights specific products intended to entice you and maximize your spending. And when you open a Facebook page, Facebook shows you content designed to maximize the number of minutes you will stay on Facebook. Amazon and Facebook's AI are targeted, meaning that they show different personalized content to each person. So the content shown to me works great on me, but may not work at all on you. This targeted accuracy is much more effective at producing clicks and purchases than the one-size-fits-all approach used by traditional static websites.

As powerful as it is, deep learning is not a panacea. While humans lack AI's ability to analyze huge numbers of data points at the same time, people have a unique ability to draw on experience, abstract concepts, and common sense to make decisions. By contrast, in order for deep learning to function well, the following are required: massive amounts of relevant data, a narrow domain, and a concrete objective function to optimize. If you're short on any one of these, things may fall apart. Too little data? The algorithm won't have enough examples to uncover meaningful correlations. Multiple domains? The algorithm cannot account for cross-domain correlations and will not get enough data to cover all permutations. Too broad an objective function? The algorithm will lack clear guidance to sharpen its optimization.

It is important to understand that the "AI brain" (deep learning) works very differently from the human brain. Table 1 illustrates the key differences:

	Human Brain	AI Brain (Deep Learning)
Data required to learn	Few data points	Huge amount of data
Quantitative optimizing and matching (picking a face out of a million)	Hard	Easy
Customizing for each situation (showing each user a different product to maximize purchasing)	Hard	Easy
Abstract concepts, analytical reasoning, inferences, common sense, and insight	Easy	Hard
Creativity	Easy	Hard

Table 1: Strengths and Weaknesses of Human vs. AI "Thinking"

APPLYING DEEP LEARNING TO INTERNET AND FINANCE

Given the strengths and weaknesses of deep learning, it is no wonder that the first beneficiaries of this form of artificial intelligence are the biggest Internet companies. Tech behemoths like Facebook and Amazon have the most data, which are often automatically labeled via user action (Did the user click or buy? How many minutes did the user stay on a page?). These user actions are directly related to a business metric (either revenue or clicks) to maximize. When these conditions are met, an app or platform can become a money-printing machine. As the platform collects more data, it makes more money. It is no wonder that giant Internet companies like Google, Amazon, and Facebook have experienced phenomenal growth in the past decade and become powerhouse AI companies.

Beyond Internet companies, the next industry that is low-hanging fruit for AI is finance, including banks and insurance companies, as "The Golden Elephant" shows. Consider the example of insurance. The industry has similar benefits to Internet companies': a large amount of high-quality data within a single domain (insurance) connected to business metrics. The emergence of AI-based fintech (financial technology) companies, such as Lemonade in the United States and Waterdrop in China, are making it possible to buy insurance in an app, or take a loan in an app, with instant approval. These AI-based fintech companies are poised to overtake brick-and-mortar financial corporations because they deliver better financial results (lower default or fraud rate), instantaneous transactions (using AI and the app), and lower costs (no humans in the loop). Traditional financial companies are also hurrying to implement AI in their existing products and processes. The race is on.

Another interesting benefit of AI fintech is that it can use data beyond those considered by human professionals. It can improve predictive power by tapping into massive heterogeneous data that would not be feasible for a human insurance underwriter to assess, for example, whether you buy more processed foods or vegetables, whether you spend a lot of time in a casino or in a gym, whether you invest in Reddit-group recommendations or hedge funds, whether you have a girlfriend or harass women online. All of this evidence would say a lot about you, including your relative risk as an insured person. Millions of pieces of information (or "features") can be found on your mobile phone apps. That's why in "The Golden Elephant" the Ganesh Insurance services come in the form of a family of "golden elephant" social applications, encompassing everything from e-commerce to recommendations and coupons, investment, ShareChat (a popular Indian local language social network), and the fictitious FateLeaf fortune-telling app.

Every time Nayana buys something, accepts a recommendation, asks about a fortune, or makes a friend, Ganesh Insurance gets another piece of information, data it uses to train itself to be more intelligent and optimized. This is similar to how Google knows so much about you from the aggregation of the bread crumbs you leave in Google search, Google Play, Google Maps, Gmail, and YouTube. Out of potentially millions of features, some may be extremely relevant and useful, while most may only have modest predictive power. But even among the modestly useful features, deep learning will find helpful subtle combinations that are informative but that humans cannot possibly fathom.

DEEP LEARNING DOWNSIDES

Every powerful technology is a double-edged sword. Electricity powers all of our conveniences but is lethal when touched directly. The Internet makes everything convenient, but also shortens our attention span. So what are the downsides of deep learning?

First is the risk when AI knows you better than you know yourself. The benefits are clear—AI can recommend products you'll want before you would otherwise notice them, and AI can recommend compatible matches in romantic partners and friends based on your known affinities. However, knowing you too well has downsides. Have you ever sat down to watch one video on YouTube, and ended up spending three hours? Or clicked on one provocative link on Facebook, only to see more extreme content recommended next?

The popular 2020 documentary *The Social Dilemma* illustrates how AI's personalization will cause you to be unconsciously manipulated by AI and motivated by profit from advertising. *The Social Dilemma* star Tristan Harris says: "You didn't know that your click caused a supercomputer to be pointed at your brain. Your click activated billions of dollars of computing power that has learned much from its experience of tricking two billion human animals to click again." And this addiction results in a vicious cycle for you, but a virtuous cycle for the big Internet companies that use this mechanism as a money-printing machine. *The Social Dilemma* further argues that this may narrow your viewpoints, polarize society, distort truth, and negatively affect your happiness, mood, and mental health.

To put it in technical terms, the core of the issue is the simplicity of the objective function, and the danger from single-mindedly optimizing a single objective function, which can lead to harmful externalities. Today's AI usually optimizes this singular goal—most commonly to make money (more clicks, ads, revenues). And AI has a maniacal focus on that one corporate goal, without regard for users' well-being.

Ganesh Insurance in "The Golden Elephant" promises to minimize insurance premiums, which is highly correlated to minimizing health claims, and thus improving health. On the surface, this seems to suggest a harmonious alignment of corporate and user goals. However, in the story, the insurance company's AI determines that a relationship between Nayana and her crush, Sahej, would likely increase Nayana's family's insurance premium in the future, so it tries to deter the young couple's budding romance. In other words, the Ganesh Insurance AI was trained on massive amounts of data to find causalities. AI can find elevated disease risks of smoking, and thus try to reduce smoking, which is good. But AI might also find that a potential romantic pairing—even one that could help bridge societal division in the long term—could, according to its narrow analysis of the data, increase insurance premiums. So, inference results in actions that serve to tear people apart and exacerbate inequality.

How can we solve this problem? One general approach is to teach AI to have complex objective functions, such as lowering insurance premiums while maintaining fairness. When it comes to maximizing the time humans spend on social networks, for example, Tristan Harris has proposed using "time well spent" as a metric instead of simple "time spent." These two goals could be blended into a complex objective function. Another solution proposed by AI expert Stuart Russell is to ensure that every objective function always be beneficial to humans, by finding a way for humans to be in the loop in the design of objective functions. For example, can we build objective functions for "greater human good," such as our happiness, and can we involve humans to define and label what happiness means? (We explore this idea further in chapter 9, "Isle of Happiness.")

All of these ideas require more AI research on complex objective functions, and also ways to quantify notions like "time well spent," "fairness," or "happiness." Furthermore, each of these ideas would cause companies to make less money. So how can companies be incentivized to do the right thing? One possibility is to have government regulations that penalize offenders. Another is to encourage positive behavior as a part of corporate social responsibility, such as ESG (environmental, social, and corporate governance). ESG is gaining traction in some business circles, and it is possible that responsible AI could be a part of the future ESG. Another idea is for third parties that can serve as watchdogs by creating dashboards for companies' performance, tracking metrics like rates of "fake news" generated or "lawsuits filed alleging discrimination" to pressure them to incorporate pro-user metrics. Finally, perhaps the hardest but the most effective solution is to ensure that the AI owner is 100-percent interest aligned with each user (see chapter 9 for more on this utopian solution).

A second potential downside is fairness and bias. AI bases its decisions purely on data and outcome optimization, which may often be more equitable than decisions made by people, who can be unduly influenced by various prejudices. But there are reasons that AI, too, may be biased. For example, the data used to train the AI may be insufficient and inadequately represent race or gender demographics. One company's recruiting department may find that its AI algorithms are biased against women because the training data didn't include enough women. Or the data may be biased because it was collected from a biased society. Microsoft's Tay and OpenAI's GPT-3 were both known to make inappropriate remarks about minority groups.

Recently, research has shown that AI is able to infer sexual orientation with high accuracy based on facial micro-expressions. Such abilities could lead to discrimination. This is similar to what happened to Sahej in "The Golden Elephant," when his Dalit status was found not directly but by inference. In other words, Sahej was not labeled Dalit, but because his data and features correlated to being a Dalit, warning signals were sent to Nayana, as the AI system tried to keep the two apart. These unfair outcomes are not intentional, yet the consequences are extremely serious. If a society applied them to domains like hospital admissions or criminal justice proceedings, the stakes would be even higher.

Fairness and bias issues with AI will require substantial efforts to address them. Some steps are clear. First, companies using AI should disclose where AI systems are used and for what purpose. Second, AI engineers should be trained with a set of standard principles—like an adapted physician's Hippocratic oath; engineers need to understand that their profession embeds ethical choices into products that make life-changing decisions, and thus promise to protect users' rights. Third, rigorous testing should be required and embedded in AI-training tools, to provide warnings or disallow use of models trained on data with unfair demographic coverage. Fourth, new laws requiring AI audits could be passed. If a company receives enough complaints, it could be AI audited (for fairness, disclosure, and privacy protection), the same way it might face a tax audit if its books look fishy.

A final issue is that of explanation and justification. People can always give a reason for why they made a decision, because human decisions are based on highly selective experience and rules. But deep learning's decisions are based on complex equations with thousands of features and millions of parameters. Deep learning's "reason" is basically a thousand-dimensional equation, trained from large quantities of data. This "reason" for producing a given output is too complex to explain fully to a human. Yet many key AI decisions are required, by law or by user expectation, to be accompanied by an explanation. A great deal of research is currently under way that attempts to make AI more transparent, either by summarizing its complex logic, or by introducing new AI algorithms that are fundamentally more interpretable.

These downsides of deep learning have caused significant public distrust of AI. But all new technologies have had downsides. History suggests that, with time, many of the early errors of a new technology will be fixed and improved upon. Think about the advent of the circuit breaker to avoid electrocution, and anti-virus software to stave off computer viruses. I am confident there will be technology and policy solutions to address the challenges of AI's influence, bias, and opaque operations. But first we must follow Nayana and Sahej's footsteps—to inform people about the gravity of problems, and then to mobilize them to work toward a solution.

CHAPTER TWO

GODS BEHIND THE MASKS

STORY TRANSLATED BY EMILY JIN

TRUTH AND MORNING BECOME LIGHT WITH TIME.
—AFRICAN PROVERB

NOTE FROM KAI-FU: This story revolves around a Nigerian video producer who is recruited to make an undetectable deep-fake with dangerous consequences. A major branch of AI, computer vision teaches computers to "see," and recent break-throughs allow AI to do so like never before. The story imagines a future world marked by unprecedented high-tech cat-and-mouse games between the fakers and detectors, and between defenders and perpetrators. Is there any way to avoid a world in which all visual lines are blurred? I'll explore that question in my commentary, as I describe recent and impending breakthroughs in computer vision, biometrics, and AI security, three AI technology areas enabling deepfakes and many other applications.

AS THE LIGHT-RAIL train inched into Yaba station, Amaka pushed a button next to the door of his carriage. Even before the train came to a complete stop, the doors opened with a *whoosh* and Amaka hopped off. He couldn't tolerate the slow trains—or their stale odor—for another second. Following closely behind an elderly man, Amaka nimbly slid through the turnstile at the station's exit. Facial recognition cameras were meant to deduct the fare as each person passed by. Thanks to the mask that veiled Amaka's face, however, he slipped out without charge.

Such masks had become commonplace among the young people of Lagos. For their parents' generation, masks were ritual objects, but for the youth, whose numbers had swelled in recent decades, they had become fashion accessories—and surveillance avoidance devices. Lagos, the largest city in West Africa, was home to somewhere between 27 and 33 million people—the official number depended on what method the authorities used to measure it. Five years ago, the state imposed a strict limit on the number of migrants entering the city—even those, like Amaka, who were born in other parts of Nigeria. Since then, itinerant dreamers like Amaka had been forced to seek makeshift shelter in illegal apartments, hostels, markets, bus stations, or even under overpasses. He had met many homeless people, people who had been driven onto the streets for all kinds of reasons: people whose homes had been demolished to make way for new shopping centers, people newly arrived in Nigeria from worse-off nations, and those who were simply poor. Nigeria's youthfulness—the nation's median age was merely twenty-one—was thanks to the nation's high fertility rate. Still, the rapid development of the world's third most populous nation had not benefited its citizens equally.

While other parts of Lagos strained under the pressure of its young population, the Yaba district was flourishing. Dubbed "The Silicon Valley of West Africa," the neighborhood stood out for its orderliness, fresh air, and high tech–infused daily life. Pedestrians could activate the cartoon animals on the billboards and interact with them via hand gestures. Cleaning robots roamed the streets, collecting and sorting trash, then sending it off to recycling centers where it was turned into renewable materials and biofuel. Sustainable bamboo fiber had recently made the leap from building material to fashion trend, at least for the denizens of Yaba.

Standing outside the station and holding his smartstream up to eye level, Amaka overlaid a live virtual route map onto the surrounding streetscape. Following the projected route, he began walking, eventually stopping before a gray building, emblazoned with the number 237 and tucked away on a quiet backstreet. The company he was looking for, Ljele, was apparently based on the third floor. Two days ago, he had received a mysterious email from an anonymous Ljele account about a job that was "right up his alley." The position was his under the condition of his showing up for an interview in person.

As Amaka entered a small reception area on the third floor, the receptionist smiled and pointed to Amaka's mask, indicating he should remove it for an identity check. The young man hesitated, then took his mask off. Reflected in the camera lens was a young, smooth face. His 3D-printed mask couldn't match the delicate quality of the pricey handmade versions sold at absurd prices to tourists in the Lekki Market, but the coarse reproduction, with its butterfly-like pattern, was enough to fool the facial recognition algorithm of most common surveillance cameras. In the eyes of AI, Amaka was a "faceless person." The mask not only saved him money, but, more important, shielded him from the authorities. After all, Amaka had yet to obtain a migrant residence permit.

When the face scan was completed, the receptionist brought Amaka into a conference room and told him to wait. He sat stiffly as he pondered how he would answer questions regarding his previous work experience. *I have to lie,* he realized. *I don't have many other choices.*

Ten minutes passed. The promised interviewer did not appear. Abruptly, the projection wall across from him lit up, and surveillance camera video footage began to play.

To Amaka, the video footage was as familiar as the back of his own hand. Midnight. Dim, yellow streetlamps. Several homeless people were scattered under an overpass, lying on makeshift mattresses. The silhouette of a boy emerged from the shadows. The boy walked over to a group of sleeping people and gazed down. The camera zoomed in. The boy was white, no more than five or six years old, dressed in striped pajamas, his face wan and expressionless. One of the people woke up with a start and met the boy's eyes. The homeless man asked the boy what his name was and where he lived. The boy's body trembled as he mumbled incoherently. Suddenly, his face twisted, the corners of his lips stretching open and revealing two rows of sharp teeth. He bit down hard on the homeless man's neck. The man cried out in pain, waking up the others. The boy fled the scene, blood trickling down his lips and chin.

The video, originally posted to the Internet under the title "White Vampire Boy Attacks Homeless People in Lagos," had received millions of views within twenty-four hours of its first appearance on the GarriV video-sharing platform. Within days, however, the platform identified the video as a fake and removed it in compliance with the law. The uploader's account, "Enitan0231," was consequently terminated, with all its associated advertising revenue frozen.

Suddenly, a booming voice filled the conference room where Amaka still sat, alone. "Well done, Amaka! What a seamless fusion of realistic settings, amateur actors, and live video shooting. I can't believe you made this in an underground Internet café in Ikeja," said a man's voice with a heavy Igbo accent.

Instinctively, Amaka jumped to his feet. "Who are you?" His eyes surveyed the empty room and landed on the speakers.

"Hey, relax. You can call me Chi. Do you want a job, or not?"

Sighing, Amaka sat back down and slouched in the chair. The man named Chi was right. Without a residence permit, he could never find a real job in Lagos. The mysterious Ljele company was his only sliver of hope. "Why me?" he asked.

"We saw your work. You're talented. You're ambitious—you wouldn't have come to Lagos in the first place if you weren't determined to make a name for yourself. Most importantly, we need someone we can trust. *One of our own kind.*"

Amaka knew immediately what Chi was alluding to. Nigeria has more than 250 ethnic groups, with their own languages and customs, many of which had been in conflict for hundreds of years. The Yoruba and the Igbo, respectively the second- and third-largest ethnic groups in the country, had seen violent clashes in recent years, as both groups muscled for political gain. With the Yoruba as the dominant population in Lagos, Amaka, an Igbo from the southeast, usually concealed his ethnicity to avoid trouble. "What do you want me to do?"

"I want you to do what you do best. Fake a video."

"*Illegally,* I presume?"

"We'll supply you with all you need."

Amaka narrowed his eyes, his nostrils flaring. "And what if I turn down your offer? Will you kill me?"

"Kill you? No, no. Worse than that."

Another video started to play on the projection wall. A dance floor in a private nightclub. The camera zoomed down on the room from a corner of the ceiling. Several boys were dancing up against one another under the flashing laser lights, shirtless. The camera zoomed in farther to reveal the unmistakable face of Amaka. As the camera observed, Amaka turned and passionately kissed another boy whose cheeks glowed fluorescent pink. Amaka then twisted his upper body around to kiss a darker-skinned boy behind him.

The video froze on this frame. The three young faces were like mango leaves that overlapped, intertwined and merged into one another.

Amaka stared at the video, his expression blank. After a few moments, he grinned. The facial scan he had undergone back at the reception desk had provided the data to make this instantaneous deepfake.

"The face might be mine, but not the neck," said Amaka as he pulled down his hood, exposing a long pink scar that cut diagonally from below his right ear to his left collarbone. A souvenir from a street fight. "Also, don't forget we're in Lagos. The things people do here are far crazier than that."

"Sure, but this video can still send you to prison. Think about your family," said Chi, his voice turning soft.

Amaka fell silent. Three decades after the passage of the Same Sex Marriage (Prohibition) Act of 2013, Nigerian society remained just as hostile toward sexual and gender minorities as it ever had been. If someone reported him, Amaka knew it would be difficult to avoid dealing with the corrupt police, who would likely try to extort him, even if he could avoid criminal charges.

And then there was his family. While they hadn't had an easy relationship these past few years, Amaka hated to imagine the pressure that could descend upon the shoulders of his family members, especially his father, who expected the world of him. *Even if the video is a fake.*

The boy bit on his lower lip and pulled his hood back up. Concealing parts of his skin again gave him a little more sense of safety. "I need an advance payment. Cryptocurrency. Also, give me as much detail as you have on the target. I don't want to waste my time on research."

"Your call, my friend. As for the target . . . there's absolutely no way for you to miss him."

The blurred headshot of a man flashed on the projection wall. When the face's contours solidified into a clear picture, Amaka's eyes widened.

THE YORUBA CALLED THE CITY of Lagos "Eko," meaning "farm." In the equatorial monsoon climate, June was the coolest month with the most plentiful rain. With the rain's monotonous tapping on the metal roof as his background soundtrack, Amaka lay on the small bed of his illegal hostel room. He put on his XR glasses and fiddled with his new gadget—a dark green Illumiware Mark-V.

Compared with the pranks he had carried out in the past, this new job was on a completely different level.

It's not that he lacked experience in video deception—quite the opposite. Alone in his room, Amaka had spent many nights of the past year disguising himself as uptown girls on dating apps. In order to construct a flawless imitation, the first step was to gather as much video data as possible with a web crawler. His ideal targets were fashionable Yoruba girls, with their brightly colored V-neck *buba* and *iro* that wrapped around their waists, hair bundled up in *gele*. Preferably, their videos were taken in their bedrooms with bright, stable lighting, their expressions vivid and exaggerated, so that AI could extract as many still-frame images as possible. The object data set was paired with another set of Amaka's own face under different lighting, from multiple angles and with alternative expressions, automatically generated by his smartstream. Then, he uploaded both data sets to the cloud and got to work with a hyper-generative adversarial network. A few hours or days later, the result was a DeepMask model. By applying this "mask," woven from algorithms, to videos, he could become the girl he had created from bits, and to the naked eye, his fake was indistinguishable from the real thing.

If his Internet speed allowed, he could also swap faces in real time to spice up the fun. Of course, more fun meant more work. For real-time deception to work, he had to simultaneously translate English or Igbo into Yoruba, and use transVoice to imitate the voice of a Yoruba girl and a lip sync open-source toolkit to generate corresponding lip movement. If the person on the other end of the chat had paid for a high-quality anti-fake detector, however, the app might automatically detect anomalies in the video, marking them with red translucent square warnings.

In the early days of deepfake technology, factors like Internet

speed and exaggerated expressions could easily cause glitches, resulting in images that blurred, or out-of-sync lip movement. Even
if the glitch lasted for only 0.05 seconds, the human brain, after
millions of years of evolution, could sense something was amiss.
By 2041, however, DeepMask—the successor of deepfake—had
achieved a degree of image verisimilitude and synchronization
that could fool the human eye.

Anti-fake detectors had become a part of the standard configuration for cybersecurity. Europe, America, and Asia had even made
them compulsory by law, yet in Nigeria, only major content platforms and government websites required such verification. The
reason was simple: The detectors required an extremely high level
of computing power and skill, and slowed the speed of videos. If
people had to wait, they would tune in elsewhere. Social media
and video-sharing platforms would update their detectors selectively according to the most popular fake-generating algorithms at
any given time; the more a piece of content was shared, the more
scrutiny it got.

AFTER EACH VIDEO "DATE," Amaka would sit quietly in the
darkness. His humble surroundings never failed to inject him with
a dose of reality. Still, he would allow his mind to linger over the
smiles and sweet words of the boys he "dated." *Their affection
doesn't belong to me*, he reminded himself, *but to a Yoruba girl with
a face just like mine.*

When Amaka was born, a local soothsayer had declared to
Amaka's father that his new son was actually the reincarnation of a
female soul trapped in a male infant's body. The "soul-body incongruity" would be a shadow cast over Amaka's entire childhood—
and a shame for his family.

As he grew up, Amaka slowly came to understand that he was
not like other boys. Leaving his village to come to Lagos had been
part of this journey. Still, there were limits. When he brushed past
attractive men on the light-rail or the sidewalk, he could feel something in his body—in his soul—flutter. Even eye contact could
evoke the feelings now and then. But Amaka knew that he didn't

have the courage to face the boys he chatted with online in real life. The more power of DeepMask he excavated, the more his addiction for the mask grew. It concealed his real face, so that he was able to let his feelings pour out and run free, without exposing himself to danger or shame.

Amaka was forcing himself to focus on the fake video when his thoughts were interrupted by a knock on his bedroom door. Ozioma, his landlord, entered with a bowl of sliced kola nut seeds. Ozioma, an Igbo who'd moved to Lagos twenty years ago, had seamlessly assimilated into the Yoruba society. However, she was able to uncover Amaka's veiled Igbo accent immediately after meeting him.

"You know, where I come from, only men are permitted to break open a kola nut," said Amaka, his mouth full, savoring the fruit's familiar bitterness.

"That's exactly why I moved!" Ozioma chuckled. "The kola nut, the Yoruba call it *obi,* and the Igbo call it *oji.* Why does the name matter? *Obi* or *oji,* it will solve your problems once it's in your mouth."

"Ah, wisdom of the elders. Thank you for this treat," said Amaka. However, before he could shut the door, Ozioma grabbed his arm. She pointed to the headshot displayed on his monitor, a worried frown crossing her face. "You don't have anything to do with him, do you? I mean, he's a good guy, I just . . . don't want to get into trouble, if you know what I mean."

"Nah, I was just reading the news." Amaka forced a lighthearted smile. "I still want my residence permit."

"Good child. May God bless him—no matter what side he is on." Ozioma disappeared.

Amaka sighed in relief and hopped back onto his bed, returning his attention to the face displayed on his monitor.

The face radiated power. Its forehead and cheeks were painted in white, a symbol of tribal spirit. Its eyes glowed as if they were globes of fire. Its lips were slightly parted, with corners curling up to form a half smile, as if about to speak the divine language of a new age and take the world by storm.

The face belonged to Fela Kuti—legendary Nigerian musician,

father of Afrobeat, fighter for democracy—who had been dead for forty-five years.

AMAKA'S PROBLEM WAS HOW to make something fake even more fake.

A virtual avatar—with the face of Fela Kuti—had emerged online, posting videos on GarriV. The figure, in the guise of the deceased Fela Kuti, had become an Internet sensation. The avatar called itself "FAKA," the abbreviation of "Fela Anikulapo Kuti Avatar," and its videos mainly involved acerbic commentary on current social affairs—even if its precise political allegiance was hard to pin down. Most people treated it as a joke. Everyone knew that the real Fela Kuti had died in 1997. The face-swap technology used in the videos was so crude as to be laughable. Rather than bothering to ban or censor the FAKA videos as fake content, sharing platforms had simply tagged it as parody.

Still, FAKA's influence had snowballed into no laughing matter. Millions of Nigerians were logging in to encrypted chat groups to discuss FAKA's videos, analyzing every frame and syllable. They had even been translated into different dialects, fully dubbed and lip-synced, disseminating FAKA's message far more widely. The official Fela Kuti Foundation issued a statement, claiming to be as puzzled as anyone about the popular avatar's origins, but stopping short of issuing a demand for the mysterious figure behind the account to stop using Fela Kuti's likeness.

No one had managed to track down the person behind FAKA. The videos' information was encrypted; the account that uploaded the videos was disposable and had gone through multiple proxy servers. Consequently, conspiracy theories emerged. Was FAKA the work of anti-government activists or a foreign government, intent on undermining the current order?

Ljele, Amaka's new employer, was not an actual company, it turned out. Ljele was the front operation for an underground group called Igbo Glory, and Chi was just the representative—the agent tasked with recruiting and handling Amaka. The group had analyzed the content of FAKA's videos and come to a different conclusion: Yoruba ultranationalists were behind the avatar, and they hoped to exploit its popularity to manipulate the minds of the people—to incrementally make FAKA's videos more pro-Yoruba and nudge public opinion in their favor. And the more power that coalesced in the hands of the dominant Yoruba, Amaka knew, the more other ethnic groups would get squeezed—especially the Igbo.

In one recent video, FAKA had called on Igbo-dominated states to give up the claim to a newly discovered rare-earth element deposit, and instead make it "a common property for all Nigerians." This was the latest attempt to deprive the Igbo of resources on their land. The Igbo felt like the tail to Nigeria's lizard—cut off, grown back, then cut off again, in a never-ending cycle. No one cared if the tail hurt or bled.

Now, the Igbo were tired. Amaka's mission was key to Igbo Glory's goal of revolution. In the hopes of disrupting FAKA's grip on public opinion, Chi had tasked Amaka with producing fake FAKA videos that would undermine the avatar's credibility and influence.

Technology-wise, it wasn't all that hard. With the help of H-GAN, Amaka easily duplicated a machine-generated model of FAKA's facial portrait. From blink frequency and lip movement to the crude incongruity between the mouth area and its surrounding skin, Amaka's model was a pixel-to-pixel mirror reflection of FAKA. As long as he knew how to set the parameters and match up each mathematical value between the fake and the original, he could fool every anti-fake detector and every human eye.

The real challenge was reproducing a FAKA-style speech. The topics of FAKA's videos ranged from social and political news to convoluted "everyman" populist gripes. In the monologues FAKA would selectively quote famous words from the real Fela Kuti, as well as folk sayings. Amaka often struggled to interpret FAKA's distinctive speeches—let alone imitate them.

FAKA declared that Nigeria was in dire need of a new language that transcended ethnic boundaries, "to purge our mind and language of colonial poison." It lamented that the mothers of Nigeria were the people who "suffered the most and deserve the highest reverence"; with their own hands they have "welcomed the descent and buried the corpses" of countless children. FAKA boasted that "music is a weapon of the future," and that only when education and wealth were "distributed evenly like drumbeats permeating the air, could the heartbeats of people coalesce into one steady rhythm."

Like a rainstorm descending upon a long-parched land, FAKA's words had begun to quench a thirst in Amaka's heart, too. As much as he hated to admit it, he felt invigorated by a sense of hope. Was Chi right about FAKA? Amaka tried to brush off these feelings. *I don't need some cheesy sense of belonging,* he told himself.

Amaka needed only a perfect counterfeit of a FAKA-style speech, one that people would believe.

A PARADE HAD FLOODED the streets of central Lagos. Hidden on the balcony off his room, Amaka watched as a troupe of young men naked from the waist up swayed and spun, as graceful and nimble as specks of dust dancing in rays of sun. Their faces were decorated by white paint in the style of Fela Kuti. The muscles on their backs glistened under the hot sun. Following the rhythm, they raised their arms in unison, shaking their palms, as if casting a spell.

The sound of instruments from various ethnic groups combined in harmony. The shrill cry of the Batá drum and the low moans of the dùndún drum, from the Yoruba; the metallic clatter of the *ogene* bell and the silvery melody of the *opi* flute, from the Igbo. The air vibrated from the music, like the tightening string of a bow drawn open inch by inch. The dancers, like young cassava shoots during monsoon, evolved their movement to the flow of the rhythm. Moving in complete unison, with no one left behind, the dancers, to Amaka's eye, seemed less like individuals than a connected being—not unlike the mantra that they'd been chanting, "One Nigeria," the slogan of FAKA's video campaign.

Amaka felt torn. On the one hand, he envied the dancers. He instinctively wanted to join them, yet his passion was choked back by an intense fear of being exposed as a traitor. Did those dancers—

followers of FAKA—really wish ill on the Igbo people, a people Amaka still loved, even as he'd grown alienated from them?

More pressing than those thoughts, however, was Chi's deadline, which was quickly approaching, and with the passing of each day Amaka had become more and more certain that he had been given an impossible task.

Upon Amaka's closer examination, it seemed a uniform, singular FAKA personality did not exist. The team behind the avatar, relying on the video-sharing platform's smart tagging system, had created videos tailored to appeal to a variety of user profiles, fine-tuning the main topics, slogans, tone, and body movement for each audience—like an advertising agency pandering to a certain demographic.

Creating a fake was one thing—creating a fake with multiple personalities was beyond Amaka's capabilities. Somehow, this realization gave him a sense of relief. But now he had to face the consequences of failing Chi's mission.

"Why don't you go join them?" asked Ozioma. Showing up behind Amaka on the balcony, the landlady lit an English-brand cigarette, leaned against the railings, and peered down.

"I used to be the dance queen of our village," Ozioma went on, her eyes hazy with nostalgia. "Not trying to brag here, but not a single boy could take his eyes off me. My father hated when I danced, though. He threatened to hit me every time he caught me dancing."

"Did you listen to him?"

Ozioma laughed heartily. "Why on earth would a child give up what they love because their parents said no? Eventually, I found a way that could allow me to at least finish the dance."

"What was it?" asked Amaka.

"I would wear an Agbogho Mmuo every time I danced."

"What?" Amaka's eyes widened. The Agbogho Mmuo was the sacred mask of northern Igbo, representing maiden spirits as well as the mother of all living creation.

"See, my father had your exact expression when he saw me with the mask. He had no choice but to bow down, to show his respect to the mask and the goddess it embodies. Of course, after I was done with the dance, with the mask stripped off, I would get my share of scolding," said Ozioma, beaming with pride, as if the memory had temporarily brought her back to the days when she was a young girl.

Upon hearing Ozioma's story, Amaka felt an idea, blurry and shapeless, darting across his mind like a fish. He scrunched up his face, thinking. "The mask . . ."

"Yes, child. The mask is where my power came from."

"Strip off the mask? *Strip off the mask,*" murmured Amaka.

All of a sudden, he leapt to his feet and kissed Ozioma on the cheek. "Thank you, oh thank you, my dance queen!" He dashed back to his room, leaving behind the hustle and bustle of the parade and a very confused Ozioma.

"Maybe spinning a lie and putting it in FAKA's mouth won't make his followers abandon their idol," Amaka told Chi via video chat that afternoon, excited with his new discovery. "But stripping off its mask and revealing the hidden puppet master might."

"No one knows who the puppet master is, though," Chi replied.

"Exactly!" Amaka beamed. "Can't you see? It means that the puppet master can be *anyone.*"

"So, you're suggesting that . . ."

"I can strip off FAKA's mask and make him any person you want him to be."

Chi fell silent in the video chat.

"You're a fucking genius," Chi finally muttered.

"*Ndewo,*" Amaka said, preparing to sign off.

"Wait," Chi looked up. "It means that you need to create a face that exists in reality."

"Yes."

"A face that can fool all the anti-fake detectors," added Chi, musing. "Think about the color distortion, the noise pattern, the compression rate variation, the blink frequency, the biosignal . . . is it doable?"

"I need time," said Amaka. "And unlimited cloud AI computing power."

"I'll get back to you." Chi logged off.

Amaka gazed at his own reflection in the dimming monitor

screen. The adrenaline rush that had initially washed over him had faded. He saw on his face not excitement, but exhaustion and an unsettled feeling, as if he had betrayed a guardian spirit watching from above.

IN THEORY, ANYONE COULD fake a perfect image or video, at least well enough to fool the existing anti-fake detectors. The problem was the cost—computing power.

Fakes and their detectors were engaged in an eternal battle, like Eros and Thanatos. Amaka had his work cut out for him, but he was determined to succeed in achieving his singular goal: the creation of a real, human face.

In the new scheme Chi concocted, FAKA would be stripped of its Fela Kuti digital mask to reveal the face of Repo, a notorious Yoruba politician known for his ad hominem attacks on other ethnic groups. Repo was the primary enemy to the "One Nigeria" movement. Once Chi and his team revealed to the public that Repo was pulling the strings behind the inspiring, charismatic FAKA, the faith of the avatar's believers would shatter into pieces. First, though, Amaka's fabricated video would have to endure the scrutiny of millions of eyeballs—human and AI, including "the VIP detector."

The VIP detector, as it was nicknamed, was designed to protect the reputations of public figures: politicians, government officials, celebrities, athletes, and scholars. Such prominent people had large Internet trails—which made them particularly ripe to be targets of deepfakes. The VIP detector was intended to prevent those "supernodes" in cyberspace from becoming the victims of fraud, and the consequential devastating damage to social order that could ensue. Websites posting pictures or videos of prominent individuals were required to apply this special detection algorithm to content before posting. The VIP detector incorporated tech ranging from ultra-high-resolution facial recognition, body language recognition sensors, hand/finger geometry recognition, speech evaluation, and even vein recognition.

All of this data fed into the VIP detector's deep learning AI. The

VIP detector would even incorporate medical history into its data bank, as long as the person being protected was important enough. No doubt, given Repo's social status and controversial position, he was one of those VIPs.

Amaka, however, believed there was a flaw in the detector. If he could decipher how the network of anti-fake detectors was made, he could pinpoint the gaps in the crisscrossing strands of data inputs and exploit them. No matter how narrow the holes in a net, a determined fish can eventually find a way out.

Using a real video of Repo as a base, Amaka, like a twenty-first-century Dr. Frankenstein, carefully sewed together the face: lips, eyes, and nose, layer by layer, aided by AI. Every twitch and gesture in the fake video would come from Repo himself, thus greatly reducing the chance of being caught by the anti-fake detector.

Using XR vision, Amaka had conjured up a three-dimensional workspace. He waved his hands in the air, selecting, dragging, zooming in and out, the icons and footage fragments hovering mid-air with alternative gestures. He would have preferred to see himself as a wizard working magic, yet in reality he looked more like a star chef preparing an extravagant feast.

For each part of Repo's body, Amaka carefully selected the most effective open-source software, like placing raw ingredients into a proper cooking vessel. Then, as if seasoning the food, he adjusted parameters, models, and the training algorithm. Finally, he brought them to a simmer in a cloud AI platform with maximum computational power. Each set of video resources, processed by GAN, generated a series of thumbnails that extended into infinity in the virtual workspace, like a never-ending gallery swamped with posters of Repo's various body parts.

Behind the poster wall a ferocious battle was happening in the cloud, in utter silence. The two sides were GAN's positive and negative poles, the detective network and the forger network. The goal of the forger network was to retrain and upgrade itself to generate more realistic images that could fool the anti-fake detectors, based on feedback from the detective network, in order to minimize the loss function value of the generated image. Conversely, the detective network strived to maximize the loss function value. This contest, with the stakes rising every millisecond, would repeat itself millions of times until both sides reached a certain balance.

Adjusting parameters, iterating the model . . . with each adjustment Amaka could see the video becoming more realistic. His eyes, almost blinded by colorful pixel dots, focused intently on frames in his XR vision field—frames that differed from one another only by the thinnest margin. Sweat gathered on his forehead, ran down his face, and dripped from the tip of his nose, but Amaka's nimble dancing fingers weren't affected at all.

However, a voice rang in his ear every now and then, distracting him, like an *ogbanje* forever stuck in the limbo of life and death.

"You're murdering a god with your own hands," the voice whispered.

He's not my god. He's a Yoruba, Amaka repeated in his heart, while forcing himself to turn his attention back to work.

Finally, he let out a sigh of relief. His fake video had successfully fooled the VIP detector. Exhausted, he collapsed onto his bed and fell into a deep sleep.

AMAKA HEARD A VOICE calling his name. He saw a dark shadow standing at the end of his bed. Terrified, he fumbled for the switch on his bedside lamp, but his fingers felt nothing. The shadow approached. He recognized the shadow's face—it was FAKA.

Amaka choked out, "What do you want?"

FAKA looked down at Amaka and smiled. "Don't be afraid, my child. I heard your call, so I came to see you."

"I didn't . . . I didn't mean to hurt you," whispered Amaka, his voice trembling.

FAKA broke into laughter, his breath rolling in his throat like the roar of a leopard. "No one can hurt me, child. You can't, and neither can they."

"They?"

"The people who are trying to stifle the future of Nigeria. The people who have tried to lure you into the jungle of the night."

"I'm sorry, FAKA, but I have no choice."

"No, you do, my child. Go to Nollywood. Tell a real Nigerian story, instead of looking for easy clicks."

Speechless, Amaka stared back at the pixelated figure before his eyes. *I've always wanted to tell my own story,* he thought, *a story about an Igbo struggling between the traditional and the revolutionizing reality.*

"My guardian spirit has abandoned me, because I have left them for Yoruba land—" stuttered Amaka.

"Nonsense!" FAKA interrupted Amaka. All of a sudden, FAKA sounded like someone very familiar. "Remember what I told you when you were a child?"

"When I was a child?"

"I taught you the names of birds, showed you the best kind of wood to make a slingshot with, how to craft a flute out of elephant grass . . . do you no longer remember?"

"But . . . but that was my father." Amaka's speech faltered, and his eyes widened.

"Yes, my child. Remember the Igbo saying? When a person says yes, their guardian spirit can only say yes. Only people forsake their god. Their god would never forsake them."

"But, Father, I don't want to let you down," said Amaka softly, remembering Chi's threat that would bring shame to his entire family.

"Amaka, there's something that I've never told you."

"What is it?"

"The truth is, I couldn't care less for what the soothsayer told us about you. I couldn't care less whose soul lives in the body of my child. I only want my child to be someone happy and kind, someone who honors the gods and spirits with his heart."

"Father . . ." Amaka reached for FAKA's visage. He wanted to take off its mask and see again his father's weatherworn face.

"Amaka, go to the New Afrika Shrine. I believe that you're smart enough to make the right choice. Then, come back to me."

As Amaka's fingertips were about to land on the flickering pixelated face, FAKA disappeared. Amaka woke up from the dream with a start. His bedside lamp was on. On the dark green Illumi-ware Mark-V's monitor screen was a familiar face smiling back at him.

TATTOOED WITH GRAFFITI, the New Afrika Shrine, located in Ikeja, the capital of Lagos, could easily be mistaken for a dilapi-dated garage. What it lacked in structure, it made up for in energy. With a capacity of two thousand people, it was the site of weekly concerts as well as a base for various food and drink stalls with booming business. The Afrika Shrine nightclub, originally opened by Fela Kuti at Empire Hotel, was burned down by the police in 1977. This reincarnation was established by his son Femi in 2000, in honor of his father.

Amaka had visited the New Afrika Shrine many times. Like every young person in Lagos who liked a good time, he saw the shrine not only as a place to eat, drink, and party, but also as a sa-cred temple, a destination of pilgrimage, where he could connect with the rebellious and free spirits from half a century ago. In this special place, people somehow magically put aside the conflicts of ethnicity and class, united with one another—and indulged to-gether in the pleasure of alcohol.

Today, he had come to say goodbye.

The Afrika Shrine, old and new, enshrined Black gods and god-desses: Kwame Nkrumah, Martin Luther King, Jr., Malcolm X, Thomas Sankara, Nelson Mandela, Esther Ibanga, Chinua Achebe, Wole Soyinka, Florence Ozor . . . great souls who dedicated their lives to freedom, democracy, and equality. Performers, during their shows, would often pause and pay homage to these cultural ances-tors.

Quietly, Amaka engraved those faces one by one in his memory. He prayed that those gods and spirits would watch over him.

He would leave Lagos, go back home, and tell his father every-thing. He hadn't decided what to do beyond that. Perhaps his mas-tery of GANs would bless him with a proper job, one where he didn't have to fake anything, where he could help other people. Maybe he could find a job in healthcare—performing face swaps on medical AI training data sets; he could color in old black-and-white films, refine them. Or perhaps he could spread his wings still further and do something that he had almost never dared to dream of: make a real Nollywood movie. He already had the idea for a good story to tell.

Suddenly, Amaka's smartstream chimed with the sound of clinking coins. The money that Chi had promised had arrived. Which meant that the video he'd made, as real as it could be, was speeding across the Internet with the impact of a nuclear explosion on the faith of the millions of followers of FAKA.

In recent years, AI-generated videos had been the trigger of the Gabonese Republic mutiny and political turmoil in Malaysia. Amaka couldn't bear to think about what his video would do to Nigeria.

But I have made my choice.

Standing before the center of the stage, in front of the black-and-white portrait of Fela Kuti hung up high, Amaka raised his hands above his head and stretched them forward, as if to connect to the power of the gods and spirits.

"I will be the master of my own destiny and will decide when it is time for death to take me," whispered the boy earnestly, as if chanting a magic spell.

The quote was from Fela Kuti himself—an explanation of his middle name, Anikulapo, which meant "he who carries death in his pouch" in Yoruba.

Amaka entered a few lines of command on his smartstream and tossed it into a trash can. He pulled out his crude 3D-printed mask and concealed his face behind it. He prayed that he could get away—as far as possible before Chi realized what he had done. He would leave Lagos, leave this enormous city drenched in spray-painted *Eko o ni baje*—meaning "Lagos must not spoil"—and return to his home with a smell resembling the freshness of earth.

He had chosen to eliminate a lie by creating another lie.

A second video, made with DeepMask, had already been uploaded to the Internet, ready to set off another explosion. The only difference between the second and first was that in the second, when FAKA removed its digital mask and revealed the face of

Repo—the perfect fake that had fooled every existing anti-fake detector—it wouldn't stop. It would continue to take off its masks, the mask of Repo, the mask of the mask behind Repo, layer by layer, extending into infinity.

The Nigerians would discover, in amazement, that the faces behind FAKA were the cultural gods and goddesses in the New Afrika Shrine.

ANALYSIS

COMPUTER VISION, CONVOLUTIONAL NEURAL NETWORKS, DEEPFAKES, GENERATIVE ADVERSARIAL NETWORKS (GANs), BIOMETRICS, AI SECURITY

"Gods Behind the Masks" tells a tale of visual deception. When AI can see, recognize, understand, and synthesize objects, it can also manipulate them and create images and videos that are indistinguishable from reality. This story describes a future in which people can no longer rely on their naked eyes to tell real videos from fake ones. Websites and apps are required by law to install anti-deepfake software (just like anti-virus software today) to protect users from fake videos. But the tug-of-war between the deepfake makers and the deepfake detectors has become an arms race—the side that has more computation wins.

While the story is set in 2041, the situation described above is likely to impact the developed world earlier because it can afford the cost of the expensive computers, software, and AI experts needed to create and detect deepfakes and other AI manipulations. Also, legislation will likely be implemented in developed countries first. This story is set in a developing country, where the externalities of deepfakes will likely occur later.

So, how does AI learn to see—both through cameras and prerecorded videos? What are the applications? And how does an AI deepfake maker work? Can humans or AI detect deepfakes? Will social networks be filled with fake videos? How can deepfakes be stopped? What other security holes might AI present? Is there anything good about the technology behind deepfakes?

WHAT IS COMPUTER VISION?

In "The Golden Elephant," we witnessed the potential prowess of deep learning in big-data applications, like the Internet and finance. You're probably not surprised that AI beat humans on big-data-crunching applications. But what about capabilities that are unique to humans or other living creatures, such as perception?

Among our "six senses," sight is the most important. Computer vision (CV) is the subbranch of AI that focuses on the problem of teaching computers to see. The word "see" here does not mean just the act of acquiring a video or image, but also making sense of what a computer sees. Computer vision includes the following capabilities in increasing complexity:

- Image capturing and processing—use cameras and other sensors to capture real-world 3D scenes in a video. Each video is composed of a sequence of images, and each image is a two-dimensional array of numbers representing the color, where each number is a "pixel."

- Object detection and image segmentation—divide the image into prominent regions and find where the objects are.

- Object recognition—recognizes the object (for example, a dog), and also understands the details (German Shepherd, dark brown, and so on).

- Object tracking—follows moving objects in consecutive images or video.

- Gesture and movement recognition—recognize movements, like a dance move in an Xbox game.

- Scene understanding—understands a full scene, including subtle relationships, like a hungry dog looking at a bone.

In the deepfake-making tools used by Amaka in the story, all of the above steps were implicitly included. For example, in order for Amaka to edit the FAKA video, first the video needs to be broken into sixty frames of images per second, and each image is represented by tens of millions of pixels. AI reads these tens of millions of pixels, and automatically segments FAKA's body (or draws a boundary around his body), which is then further segmented into his masked face, mouth, hands, and so on. This is repeated for each frame of the video. If there are fifty seconds of video, then we have three thousand frames of images. In addition, movement between frames is correlated and tracked, and relationships between objects are discovered. This is all before any edits take place.

The above description might strike you as laborious—but these steps come effortlessly for us humans. We take one look and all of that is internalized in less than a second. Also, humans have abstract and generalized understanding of objects, even if that same object looks different from different angles, under different lighting, from different distances, or occluded by other objects. For example, just seeing Repo sitting at a desk in a particular posture, we can infer that he is holding a pen to a piece of paper, even though we see neither.

When we "see," we are actually applying our accumulated knowledge of the world—everything we've learned in our lives about perspective, geometry, common sense, and what we have seen previously. These come naturally to us but are very difficult to teach a computer. Computer vision is the field of study that tries to overcome these difficulties to get computers to see and understand.

COMPUTER VISION APPLICATIONS

We are already using computer vision technologies every day.

Computer vision can be used in real time, in areas ranging from transportation to security. Existing examples include:

- driver assistants installed in some cars that can detect a driver who nods off

- autonomous stores like Amazon Go, where cameras recognize when you've put a product in your shopping cart

- airport security (counting people, recognizing terrorists)

- gesture recognition (scoring your moves in an Xbox dancing game)

- facial recognition (using your face to unlock your mobile phone)

- smart cameras (your iPhone's portrait mode recognizes and extracts people in the foreground, and then "beautifully" blurs the background to create a DSLR-like effect)

- military applications (separating enemy soldiers from civilians)

- autonomous navigation of drones and automobiles

In the opening of "Gods Behind the Masks," we saw the use of real-time facial recognition to automatically deduct payment by recognizing commuters as they pass through a turnstile. We also saw pedestrians interact with cartoon animals in ads, using hand gestures. And Amaka's smartstream used computer vision to recognize the street ahead of him and gave him directions to get to his destination.

Computer vision can also be applied to images and videos—in less immediate but no less important ways. Some examples:

- smart editing of photos and videos (tools like Photoshop use computer vision extensively to find facial borders, remove red eyes, and beautify selfies)

- medical image analysis (to determine if there are malignant tumors in a lung CT)

- content moderation (detection of pornographic and violent content in social media)

- related advertising selection based on the content of a given video

- smart image search (that can find images from keywords or other images)

- and, of course, making deepfakes (replacing occurrences of one face with another in a video)

In "Gods Behind the Masks," we saw a deepfake-making tool that is essentially an automatic video-editing tool that replaces one person with another, from face, fingers, hand, and voice to body language, gait, and facial expression. More on deepfakes below.

CONVOLUTIONAL NEURAL NETWORKS (CNNs) FOR COMPUTER VISION

Making deep learning work on a standard neural network turned out to be a challenge, because an image has tens of millions of pixels, and teaching deep learning to find subtle hints and features from a massive number of images is daunting. Researchers have looked to the human brain for inspiration to improve deep learning. Our visual cortex uses many neurons corresponding to many restricted subregions (known as receptive fields) within what our eyes see at any given time. These receptive fields identify basic features, such as shapes, lines, colors, or angles. These detectors are connected to the neocortex, the outermost layer of the brain. The neocortex stores information hierarchically and processes these receptive fields' outputs into more-complex scene understanding.

This observation about how humans "see" inspired the invention of convolutional neural networks (CNNs). A CNN's lowest level is a large number of filters, which are applied repeatedly across an image. Each of these filters can see only small contiguous sections of the image, just like the receptive fields. Deep learning, through optimizing across many images, decides what each filter learns. Each filter will output its confidence that it saw the particular feature (like a black line) that the filter represents. A CNN's higher layers are hierarchically organized, like the neocortex. The higher levels will take the confidence outputs from the lower levels and detect more-complex features. For example, if an image of a zebra is fed into a CNN, then the lower-level filters might look for the black lines and white lines in each region in the image. And higher levels might see stripes, ears, and legs, in larger regions. Even higher levels might see many stripes, two ears, and four legs. At the highest level, parts of the CNN might specifically try to distinguish zebras from horses or tigers. Note that these are examples to illustrate what a CNN *might* do, but in actual operation, the CNN would decide for itself what features (for example, stripes, ears, or more likely something beyond human comprehension) it would use, in order to maximize the objective function.

CNNs are a specific and improved deep learning architecture designed for computer vision, with different variants for images and videos. When CNNs were first discussed in the 1980s, there wasn't enough data or computational power to show what they could do. It was not until about 2012 that it became clear this technology would beat all previous approaches for computer vision. It was a happy coincidence that around this time, a huge number of images and videos were being captured by smartphones and shared on social networks. Also around this time, fast computers and large storage were becoming affordable. The confluence of these elements catalyzed the maturation and proliferation of computer vision.

DEEPFAKES

"President Trump is a total and complete dipsh*t," said President Obama, or a person who looked and sounded a lot like Obama. This video went viral in late 2018, but it was a deepfake (a fake video made by deep learning) created by Jordan Peele and BuzzFeed. AI took Peele's recorded speech and morphed Peele's voice into Obama's. Then AI took a real Obama video and modified Obama's face to match the speech, including lip-syncing as well as matching facial expressions.

The purpose of Peele's 2018 video was to warn people that deepfakes were coming, which was exactly what happened. That same year a number of deepfake celebrity porn videos were uploaded to the Internet, leading to angry denouncements and eventually a new law against it. But new manifestations of deepfakes kept appearing all the time. An app in China emerged in 2019 that could take your selfie and make you the main character of a famous movie in a few minutes. It kept the original movie soundtrack, which lowered the technology requirements. In 2021, an app called Avatarify became number one in the Apple App Store. Avatarify brings any photo to life, making a person in the photo sing or laugh. Suddenly, deepfakes were mainstream, and anybody could make a fake (though amateurish and detectable) video.

This means our future is one where everything digital can be forged, including online video, recorded speech, security camera footage, and courtroom evidence video. In "Gods Behind the Masks," Amaka uses tools much more advanced than Peele's to make a sophisticated high-fidelity video that is undetectable as fake by humans and ordinary anti-deepfake detection software. He first used a text-to-speech tool that could convert any text to audio that sounded just like Repo speaking. Then that speech was lip-synced with Repo's face, along with natural and matching emotions. That composed face was then superimposed onto FAKA's body in a preexisting video, along with matching hands, neck, and feet, as well as pulse and breathing patterns. AI will have the capacity to ensure all these body parts are smoothly connected in the right places.

In addition to this video-based approach to making "fake humans," there is another 3D approach, which involves building a 3D model of a person purely computationally. This is how animated feature films like *Toy Story* are created. The 3D approach comes from a different discipline of computer science known as computer graphics. Everything in computer graphics is mathematically modeled, so researchers need to invent realistic mathematical models of hair, wind, light, shadows, and so on. The 3D approach gives the "producer" much more latitude to build an arbitrary environment and characters, allowing the producer to manipulate each character as a "puppet" in any way the producer wants, but the corresponding complexity and computational requirements are also much higher. Computers in 2021 cannot make full-feature films using the 3D video to even fool human eyes (which is why humans in animated movies don't yet look realistic today), not to mention software detectors. By 2041, fully photo-realistic 3D models should be possible, as we will see in "Twin Sparrows" and "My Haunting Idol."

Peele's deepfake was forged for fun and food for thought, while in the story here, Chi recruits Amaka to forge a deepfake with malice. In addition to spreading rumors, deepfakes could also lead to blackmail, harassment, defamation, and election manipulation. How would you make a deepfake? How would an AI tool detect deepfakes? And as the deepfake and anti-deepfake software are pitted against each other, which will win? To answer these questions, we need to understand the mechanism that generates deepfakes—GAN.

GENERATIVE ADVERSARIAL NETWORK (GAN)

Deepfakes are built on a technology called generative adversarial networks (GAN). As the name suggests, a GAN is a pair of "adversarial" deep learning neural networks. The first network, the forger network, tries to generate something that looks real, let's say a synthesized picture of a dog, based on millions of pictures of dogs. The other network, the detective network, compares the forger's synthesized dog picture with genuine dog pictures, and determines if the forger's output is real or fake.

Based on the detective network's feedback, the forger network retrains itself with the goal of fooling the detective network next time. The forger network adjusts itself to minimize the "loss function," or the difference between its generated images and real images. Then the detective network retrains itself to make the forgeries detectable by maximizing the "loss function." These two processes repeat for up to millions of times, with forger and detective both improving their skills, until an equilibrium is reached.

In 2014, the first paper on GAN showed how the forger first made a cute but fake "dogball" that was instantly found fake by the detective, and then progressively learned to make fabricated images of dogs that are indistinguishable from real images. GAN has been applied to video, speech, and many types of content, including the infamous Obama video mentioned earlier.

Can GAN-generated deepfakes be detected? Due to their relatively rudimentary nature and the limits of modern computer power, most deepfakes today are detectable by algorithms, and even sometimes by the human eye. Facebook and Google have both launched challenge competitions for the development of deepfake detection programs. Effective deepfake detectors can be deployed today, but there is a computational cost, which can be a problem if your website has millions of uploads a day.

Longer term, the biggest problem is that GAN has a built-in mechanism to "upgrade" the forger network. Let's say you trained a GAN forger network, and someone came up with a new detective algorithm for detecting your deepfake. You can just retrain your GAN's forger network with the goal of fooling that detective algorithm. The result is an arms race to see which side trains a better model on a more powerful computer.

In "Gods Behind the Masks," Amaka's earlier "White Vampire" video was made using tools in an Internet café with minimal computing power. It was good enough to deceive people, because in 2041 fake videos are convincing enough that people can no longer discern them from real ones. However, it was not able to fool the website's detective GAN trained on more computational power and was removed and banned by the website afterward. But later in the story, Chi provided Amaka a powerful computer to train a complex GAN that generates not just the face, but also hands, fingers, gait (the way someone walks), gestures, voice, and facial expression. Furthermore, this GAN was trained on a lot of data that was available for a celebrity like Repo. As a result, it could deceive all ordinary deepfake detectors. Imagine a jewelry store that had bulletproof windows capable of blocking all ordinary ammunition. If a criminal arrived with a rocket-propelled grenade, however, the bulletproof window would no longer be adequate to block the criminal. It's all about the computer power.

By 2041, anti-deepfake software will be similar to anti-virus software. Government websites, news sites, and other sites where good information is paramount have no tolerance for any fake content, and will install high-quality deepfake detectors designed to identify high-resolution deepfakes created by large GAN networks trained on powerful computers. Websites with too many images and videos (such as Facebook and YouTube) will have trouble affording the cost of scanning all uploaded content with the highest-quality deepfake detectors, so they may use lower-quality detectors for all media content, and when a particular video or image starts to trend up exponentially, it would then apply higher-quality detectors. Since Amaka's fake video was intended to go viral, it needed to be trained on the most powerful computer with the most data, in order to avoid detection by the highest-quality anti-deepfake detectors.

So, is 100-percent detection of deepfakes hopeless? In the very long term, 100-percent detection may be possible with a totally different approach—to authenticate every photo and video ever taken by every camera or phone using blockchain technology (which guarantees that an original has never been altered), at the time of capture. Then any photo loaded to a website must show its blockchain authentication. This process will eliminate deepfakes. However, this "upgrade" will not arrive by 2041, as it requires all devices to use it (like all AV receivers use Dolby Digital today), and blockchain needs to become fast enough to process this at scale.

Until we have this longer-term solution based on blockchain or equivalent technology, we hope there will be continuously improved technologies and tools for detecting deepfakes. Since that is unlikely to be perfect, there will also need to be laws that make the penalty for making malicious deepfakes very high, in order to deter potential perpetrators. For example, California passed a law in 2019 against using deepfakes for porn, and for manipulating videos of political candidates near an election. Finally, we may need to learn to live in a new world (until the blockchain solution works) where online content should always be questioned, no matter how real it looks.

In addition to making deepfakes, GANs can be used for constructive tasks, such as to age or de-age photos, colorize black-and-white movies and photos, make animated paintings (such as *Mona Lisa*), enhance resolution, detect glaucoma, predict climate change effects, and even discover new drugs. We must not think of GAN only in regard to deepfake, as its positive applications will surely outnumber its negative applications, just as with most new breakthrough technologies.

HUMAN VERIFICATION USING BIOMETRICS

Biometrics is the field of study of using a person's physical characteristics to verify his or her identity. The use of the complex GAN in "Gods Behind the Masks" is a form of biometric verification. The GAN combined important features including facial recognition, gait recognition, hand-and-finger geometry recognition, speaker identification, vein recognition, and gesture recognition.

In real-life applications, biometrics are usually used in real time with special sensors, rather than trying to glean features from just a video recording like in the story. For example, human irises and fingerprints are unique for each person, and ideal for identity verification. Iris recognition is widely considered to be the most accurate method of biometric identification. To verify an identity using iris recognition, an infrared light is shined on the subject's eyes, and photos of the eyes are captured and compared with the irises of the person in question. Fingerprint recognition is also extremely accurate. The most accurate application of iris and fingerprint recognition requires cooperative subjects and special equipment with near-field sensors, so they could not be used in the recorded-video examples in this story.

Recent advances in deep learning and GAN have pushed the field of biometrics ahead by leaps and bounds. Given any one biometric (such as voice or face), AI can already outperform humans in verifying or recognizing any human's identity. In situations where many features can be gathered and combined, the accuracy will essentially be perfect. By 2041, AI will take over this "routine" task of recognizing and verifying people. I also anticipate that in the next twenty years, the use of smart biometrics for criminal investigations and forensics will solve many more crimes and reduce the crime rate.

AI SECURITY

As technology progresses, vulnerabilities and security risks emerge for any computing platform, for example, viruses for PCs, identity theft for credit cards, and spam for email. As AI goes mainstream, it, too, will suffer from attacks on its vulnerabilities. Deepfakes are but one of many such vulnerabilities.

Another vulnerability that can be exploited is AI's decision boundaries, which can be estimated and used to camouflage the input data, so that AI would make mistakes. For example, one researcher designed a new pair of sunglasses that made AI recognize him as Milla Jovovich. Another researcher put some stickers on a road that fooled the autopilot in a Tesla Model S into switching lanes and driving directly into oncoming traffic. In the beginning of "Gods Behind the Masks," Amaka uses a mask to trick the face recognition system at the train station. These types of camouflage would be extremely serious if deployed in war—imagine if a tank is camouflaged to be recognized as an ambulance.

Another attack is called poisoning, a situation in which AI's learning process is corrupted by contaminating the training data, the trained models, or the training process. This can cause the whole AI to systematically fail, or to be controlled by the perpetrator. Imagine military drones hacked by terrorists to attack their own country. These attacks are harder to catch than conventional hacking, because AI's models are not easy to "debug" since they are extremely complex equations implemented in thousands of layers of neural networks, rather than deterministic computer code.

Despite these difficulties, there are clear steps that can be taken, such as fortifying the security of the training and execution environments, creating tools that automatically check for signs of poisoning, and developing technologies specifically to fight tampered data or evasion. Just as we've overcome spam and viruses with technological innovations, AI security will be largely achieved with only occasional perpetration (just like we still occasionally get attacked by spam or viruses). Technology-induced vulnerabilities have always been solved or ameliorated with technology solutions.

CHAPTER THREE

TWIN SPARROWS

STORY TRANSLATED BY BLAKE STONE-BANKS

WE ARE SUN AND MOON, DEAR FRIEND; WE ARE
SEA AND LAND. IT IS NOT OUR PURPOSE TO
BECOME EACH OTHER; IT IS TO RECOGNIZE EACH
OTHER, TO LEARN TO SEE THE OTHER AND HONOR
HIM FOR WHAT HE IS: EACH THE OTHER'S
OPPOSITE AND COMPLEMENT.

—HERMANN HESSE, *NARCISSUS AND GOLDMUND*

NOTE FROM KAI-FU: "Twin Sparrows" explores the future of AI education, as smart AI teachers camouflaged as virtual cartoonlike friends help twin Korean orphans realize their potential. These AI companions can converse fluently in human language, thanks to a branch of AI called natural language processing (NLP), which is poised for a meteoric rise in the coming decade, including the capacity for AI to teach itself language. Will AI be capable of achieving full human intelligence by 2041? I'll answer that question in my commentary while describing recent NLP breakthroughs like GPT-3 and other progress in AI's quest to understand language.

"YOU COULDN'T HAVE chosen a more perfect spring day," Headmaster Kim Chee Yoon told the Paks as she gestured at the light streaming through the arched windows of Fountainhead Academy.

Dressed in crisp tailored clothing, Jun-Ho and Hye-Jin smiled politely.

"As I'm sure you're aware," continued Mama Kim, as she was known by just about everyone, "most foster care facilities have limited resources. Outside of the classroom little thought and care are given to how our young charges might explore their talents. But thanks to our proprietary technology, Fountainhead Academy aims to right this wrong and ensure our children develop to their fullest potential, however long they stay with us at Fountainhead."

Jun-Ho cleared his throat. "As members of Delta Foundation's board of trustees, Hye-Jin and I are great admirers of all you've accomplished. That's why the Foundation continues to support your work so generously. However, we're not here today on behalf of the Foundation."

He glanced at his wife. Hye-Jin held his gaze and nodded.

"You see, Hye-Jin and I intend to adopt a child."

"Ahhh!" Mama Kim exclaimed, beaming. "And have you reviewed the files on any of our children?"

"They all seem wonderful," Hye-Jin said, "but Jun-Ho and I are particularly interested in meeting the six-year-old twin boys."

"Ah, you mean Golden Sparrow and Silver Sparrow." Mama Kim lowered her voice. "If you intend to adopt two children, you'll have to go through the family assessment twice."

"No need to worry about that." Jun-Ho's voice brimmed with confidence.

A few minutes later, Mama Kim led the Paks into a bright, spacious reception room with plush carpeting and furnished in a matching pastel color scheme. The Paks were seated and asked to wait.

When the door opened, two boys entered. Except for their clothes, the boys looked like clones. Both had curly dark hair, slender arched eyebrows, slightly pursed upper lips, and freckles on the tips of their noses. To Jun-Ho and Hye-Jin, it was impossible to distinguish one from the other.

As the Paks stood to greet them, however, the two boys separated. One stepped forward, while the other retreated into a corner.

"Golden Sparrow, Silver Sparrow," Mama Kim said. "This is Jun-Ho and Hye-Jin Pak. They're close friends of the academy. They've come today to see you."

"Hello, Jun-Ho. Hello, Hye-Jin." The boy who had stepped forward blinked his eyes. "Have you come to take us home with you?"

Jun-Ho and Hye-Jin gave embarrassed smiles, uncertain how to respond.

The boy in the corner said nothing. With his head lowered, he ran his foot over the fluffy carpet, creating a whirlpool pattern.

"If I had to guess, I'd say you are Golden Sparrow, and he's Silver Sparrow." Hye-Jin squatted on her heels to put herself on eye level with the pair. "Did I guess right?"

"It's not hard to guess," Golden Sparrow replied curtly. "We may be identical twins with genomic data varying in just one sequence in a million, but we really couldn't be more different."

For a moment, Jun-Ho and Hye-Jin were stunned by the verbose, precocious six-year-old.

"And what about you?" Jun-Ho asked. "What games do you like to play?"

"Me? I don't like playing games at all. I prefer to compete."

"Oh? What do you compete in?"

"I'll compete in anything. In fact, with the help of Atoman, I just won a design competition."

"Atoman?" Jun-Ho replied, quizzically.

"Yes, that would be Golden Sparrow's AI companion," Mama Kim explained. "The Academy's vPal system provides each child with an AI partner who helps manage their schedules, academic tasks, and even play."

As Mama Kim spoke, Jun-Ho's glasses flashed with a data-share invite from Golden Sparrow. By shifting his eyes, Jun-Ho selected "OK." In his XR vision field, he saw the edges of the boy's body beginning to glow red. Pixelated flames shimmered around him. In a swift transformation, the flames coalesced into a red robot with sharp angular contours. Sparks continued to flash forth from the robot's aggressive posture until Jun-Ho raised his two hands in mock surrender.

"Meet my best friend, Atoman," Golden Sparrow said triumphantly.

Silver Sparrow had been silently watching the exchange.

Hye-Jin noticed and turned to him. "And you," Hye-Jin said, addressing Silver Sparrow. "What's your AI called?"

The boy didn't answer. Hye-Jin leaned forward and extended a hand, intending to ruffle the shy boy's hair, but Silver Sparrow shrunk back. Hye-Jin noticed then the subtle difference between Silver Sparrow's face and that of his twin. On Silver Sparrow's right eyelid, like a pink rose petal, was a scar in the shape of a fingertip.

Golden Sparrow answered for his twin: "His AI is called Solaris. And it's a heap of snot, an absolute disgrace of an AI."

For the first time since entering the room, Silver Sparrow raised his head. His eyes flashed a hostile glare at his twin.

"Solaris is not a heap of snot!" he shouted.

"It's most definitely a heap of snot. You just don't realize it because you've got snot breath."

Silver Sparrow flew into a rage, hurling insults at his brother. Mama Kim signaled a teacher to remove the boys before things got out of hand. When they were gone, the room was quiet again.

"As you see, the twins have strikingly different personalities. But I assure you, they're both wonderful children. So . . ."

"Indeed, quite impressive," said Jun-Ho, glancing at his wife. "Allow Hye-Jin and me some time to discuss, and we'll get back to you as soon as we can."

The sky was growing dim. The exterior lights of Fountainhead lit up as Mama Kim watched the Paks' luxury car accelerate down the campus drive, blowing leaves in its wake. She was as relieved as she was regretful.

She didn't need to wait for their reply. Mama Kim could guess the couple's choice. It was the choice anyone who so proudly adhered to the values of rationality and efficiency would make. One week later, the Paks came to pick up Golden Sparrow, leaving Silver Sparrow behind.

ON A WINTER NIGHT three years earlier, snow was pummeling Fountainhead Academy as the Social Welfare Office van carefully made its way up the icy driveway. When the van stopped, Mama Kim pulled the two shivering twins from the nurse's hands. They looked so small beneath their puffy down jackets. They reminded her of snowy pine cones ready to fall from the tree.

Just a few hours earlier, the boys' parents had died instantly in a crash. For some reason, their father had turned off the Hyundai Azuria's self-driving mode. While changing lanes on an icy road, the Azuria spun out of control. The vehicle smashed through the guardrail and tumbled down a thirty-foot slope. The parents, sitting up front, perished instantly. The boys were rescued from the back seats without a scratch. Police and social services had sought to find the boys' next of kin, to no avail. A call was placed to Fountainhead, and Mama Kim immediately agreed to take the orphaned boys in.

Mama Kim changed the boys into clean clothes and warmed some milk for them. Color returned to their faces as they drank.

"Just look at you two. You look like a pair of sparrows." Mama Kim smiled. "Golden Sparrow and Silver Sparrow. How does that work for nicknames? Now, which is Golden and which Silver?"

Golden Sparrow lowered his cup to reveal a milk mustache and wide grin.

"Well, isn't that a bright, happy smile?" Mama Kim said. "I guess you'll be called Golden Sparrow."

With no choice left, Silver Sparrow stared expressionlessly into the milk, as though none of this had anything to do with him at all.

The twins progressed through Fountainhead Academy's program. It wasn't always easy. From time to time, Golden Sparrow, missing his mother, would cry uncontrollably. Silver, on the other hand, would silently wipe away his tears. When she had time, Mama Kim and the other caretakers at Fountainhead would rock the twins to sleep while humming nursery rhymes, just like any parent. Unlike his twin, however, Silver Sparrow resisted this close physical contact. He even avoided eye contact.

Mama Kim noted Silver Sparrow's odd behavior right away.

Fortunately, the children's medical and behavioral data had all been saved to the childcare service cloud of the deceased parents. The children's data was easily integrated into the academy system. Even before arriving at Fountainhead, the data indicated that Silver Sparrow had shown resistance to physical and eye contact.

Compared with Golden Sparrow's adventurous, impulsive spirit, Silver Sparrow's habits were as regular as a programmed machine. Once he'd learned to walk, even his routes around the nursery rarely changed.

Silver Sparrow showed no signs of cognitive impairment, ADHD, or epilepsy. He was incredibly quiet, immersed in his own world. He would gaze at anything spinning—especially fan blades—for entire afternoons. Diagnostic AI analyzed Silver Sparrow's eyes, facial expressions, voice, and body language. The report showed an 83.14-percent probability that the boy had Asperger's syndrome.

Mama Kim knew from studying the vast clinical data on children with Asperger's that they developed different patterns of thinking and cognitive functioning. Many of these distinctive attributes would remain with them all their lives. Such children typically benefited from highly individualized education methods. The way Mama Kim saw it, children with Asperger's syndrome didn't need to be *normal.* Like any child, they needed only to become their best selves.

One afternoon shortly after arriving at the academy, Golden Sparrow and Silver Sparrow were led by Mama Kim to a room filled with monitors and other computer hardware. She told the boys that Fountainhead was going to create a "magical partner" for each of them.

Among the Fountainhead staff were an odd couple: the slender Seon and the rotund Gwang, longtime friends who had grown up together at the academy and come back later, at Mama Kim's urging, to lead its IT team. Their role at the academy now was to maintain its IT systems and provide support on hardware and software issues.

Gwang performed full-body scans on the brothers, creating a digital companion for each. He then linked these AI entities with the boys' personal data captured in the cloud.

Seon carefully fitted each boy with a soft bio-ribbon around his wrist. The bio-ribbons would record their physiological and behavioral data in real time, then synchronize that data with the cloud. She also attached a pair of flexible smartglasses next to each boy's ear. When retracted, the glasses appeared to be normal smartglasses. Unfurled, the device expanded into a full XR overlay.

Golden Sparrow screeched in excitement and made the death ray pose of Atoman, his favorite cartoon superhero. Silver Sparrow nervously poked at the equipment on his wrists and ears as if they were poisonous caterpillars.

"First, you'll need to select a voice you like."

Seon set up what looked like a mirror. Through the XR layer visible via their smartglasses, Golden Sparrow and Silver Sparrow saw a virtual interface appear in the vMirror. Seon could see what they saw when looking in the mirror. The twins didn't just *see* the interface, however. They could fully interact with it, using their voices, gestures, and expressions to create and edit whatever content they desired. The vMirror, as they would come to know it, was the base for much of Fountainhead's AI instruction and interaction.

Squatting, Seon took the boys' hands to show them how to engage the interface to adjust the AI's voice. Though the boys were only four years old, they quickly learned to operate the intuitive cartoon knobs. Golden Sparrow selected a heroic male voice for the AI that would be known as Atoman.

It took some time before Silver Sparrow selected the soft, gentle female voice of his AI. It sounded like the voice a mother would have.

"Next, we'll sculpt the appearance of your AI partner. You can mold it into any shape you like."

Golden Sparrow's hands began grabbing and pinching a soft translucent ball as the vMirror, according to the boy's gestures, adjusted the design of his virtual AI partner. At one point, the AI looked a bit like a bug, then a fish, and even later like an odd embryonic panda. Silver Sparrow gaped, half-frightened and half-curious.

With all his concentration, Golden Sparrow eventually managed to form the ball into a small red Atoman. The virtual Atoman stretched his arms and kicked his legs, then greeted Golden Sparrow. The boy screeched and clapped for his new vPal.

"Well, Silver Sparrow, it's your turn." Seon pointed at the vMirror.

Silver Sparrow stared at his reflection in the vMirror. He bent to the side and said in a barely audible voice, "I . . . don't want to."

Mama Kim leaned over Silver Sparrow, careful not to make physical contact.

"Don't you want a pal of your own to play with? Whatever you make will belong only to you and will be there to help you do anything you wish."

Silver Sparrow pursed his lips: "I . . . but it's so ugly."

Everyone in the room laughed, except Golden Sparrow.

"Well, I've got a solution," Mama Kim declared. "For now, your AI companion will only possess a voice. When you know what physical form you want for your companion, you can mold it into any shape you like, okay?"

+——

LOOKING AT THE FACES of Golden Sparrow and Silver Sparrow, most saw pixel-level copies. But to close observers, their differences couldn't be more distinct.

The contrast stood out even in the brothers' vPal avatars. Any visitor to Fountainhead Academy with access to the public XR layer would inevitably find themselves accosted by the wild red flames of Golden Sparrow's AI pal, who after twelve months had fully evolved into the formidable Atoman.

Golden Sparrow had even given Atoman a base form as a 1985 Nintendo Famicom, inspired by the retro cartoons he loved to watch. As the red-and-white machine spun around, it would transform into the cool red robot superhero.

The pair were inseparable. "Atoman, I finished today's exercises," Golden Sparrow would declare. "Let's go race cars!"

"Your error rate is a bit high," Atoman might reply. "See the answers flashing red? That's where there is room for improvement. Before we go racing, let's try a few more exercises."

"More exercises? You're more annoying than any teacher."

Golden Sparrow pouted, but he knew he would do what Atoman suggested. He and the AI had established a genuine rapport. That connection was based largely on Atoman's intelligent system of rewards and punishments, but there was also a deeper trust. Whenever Golden Sparrow was in need, Atoman would always appear at his side—to solve problems, to play games, to make him feel like he mattered. Naturally, Golden Sparrow also wanted to be there for Atoman. So, Golden Sparrow tried hard to fulfill Atoman's expectations. When he pleased Atoman, the little robot would shimmer red, spinning its gears.

Keenly watching Golden Sparrow's responses, Atoman, too, evolved according to its adaptive vPal algorithms. Atoman noted that Golden Sparrow was sensitive to rankings. In competitive situations, the boy learned more rapidly. So, Atoman incentivized Golden Sparrow's learning with competitive games.

Through these efforts, Golden Sparrow and Atoman became

well known to all those at Fountainhead. They organized spelling, geography, and e-sports competitions for the academy's children. The duo even collaborated on pranks. At Golden Sparrow's urging, Atoman reprogrammed old, discarded cleaning robots they found in a storage room, springing them on the academy's unsuspecting staff. They even created a "Ghost Face" virus, whereby the academy's computer system replicated a series of funny faces when given a secret command.

By turns amused and exasperated, Seon and Gwang were always tasked with cleaning up the mess. As time went by, they no longer had to read the logs to identify the culprit. Of course, Golden Sparrow and Atoman's antics were exactly what Mama Kim had intended. This was the first generation to engage AI at such a young age, and most children seemed to dance in perfect rhythm with their vPals.

But Silver Sparrow's relationship with his vPal had taken a markedly different path.

For months, his vPal remained a disembodied voice without physical form. One day, though, Seon noticed a breakthrough while she was synchronizing management logs. Nine months after being introduced to the vPal system, Silver Sparrow had finally crafted the avatar for his AI partner. It was a translucent, amoeba-like form that changed shape according to the situation. It could stretch out tentacles or flow like a slow-moving liquid. Silver Sparrow, a precocious reader, called it "Solaris" after a Polish science fiction novel.

For a long time, no one but Seon had a clue that Silver Sparrow had designed such a gentle and odd AI partner. Silver Sparrow would have Solaris wrap its translucent body around his own. Though his XR layer provided no tactile or haptic feedback, knowing that his almost invisible Solaris was wrapped around him gave Silver Sparrow a sense of security.

As a result, Silver Sparrow grew even more expressionless as he walked the academy halls. When he got to where he needed to be, Silver Sparrow would lie down and curl into his virtual chrysalis. Like a wizard from far away whispering magic spells, Silver

Sparrow would murmur questions and instructions to his AI. These missions had little to do with the academy's other children, reflecting instead Silver Sparrow's personal curiosity.

SEON WOULD OFTEN PASS through the academy's noisy activity room. During each visit, she would discover some surprising new skill an AI had helped its child master. But she couldn't help but notice Silver Sparrow's lonely figure sitting in the corner, staring at the wallpaper. Seon knew the boy loved collecting the small gifts from nature she would bring into the academy and set near him. She would bring him leaves, feathers, and, sometimes, shells. On the day Seon left a dried pine cone, Silver Sparrow finally spoke.

"So beautiful."

"You mean the pine cone?" Seon asked in near shock. "It's nice, isn't it?"

"The way the spirals open . . . a perfect Fibonacci sequence . . . a sacred geometric rose."

Seon tilted her head, uncertain what the boy meant—and astonished by the boy's vocabulary.

"It's fractal," Silver Sparrow said. His lips curved into a smile. To Seon, it was as though a cloudy sky had been pierced by a ray of sunlight.

"Ah, yes, it's fractal." Seon was thrilled. Silver Sparrow had finally achieved substantive communication with someone other than himself or his AI.

Seon sat down, her fingers stirring the gray fluff on the carpet. Silver Sparrow looked intently at her fingers.

"I want to share a secret with you," she said. "When I was your age, I felt I must have done something wrong. My parents had abandoned me at this place, the academy. I felt as though it had to be some kind of punishment. I thought this place was a cage that kept me from the world.

"One day, Mama Kim told me that not all parents were prepared to be parents, that this place wasn't my fault. I realized that just because I believed something, that didn't mean it was the case. From then on, the cage was open."

Seon didn't know when Silver Sparrow's eyes had moved from the carpet to her face.

"You're smart and kind, and everyone here respects how you're getting along," she continued. "Maybe sometime, try to look outside the cage. Share something you enjoy with someone else and make a friend. You may find the world becomes more interesting."

Silver Sparrow hid his face again and murmured something to himself.

Seon winced, afraid she had broken the spell of their brief connection.

Then, a data share request flashed in front of her. It was from Silver Sparrow. She accepted without hesitation.

A frantic translucent video stream flooded Seon's field of vision. Different resolutions, formats, fragmented sources all edited together in a complex rhythm of time and space. Images intertwined, occluded, overwhelmed her with their visual vortex. It took a moment before Seon could distinguish anything amid the stream. Then she began to make out a few of the images: mountains, rivers, lakes, clouds, nebulae, plant veins magnified at powers of ten, irises, microstructures of chemical compounds, wind tunnel experiments captured in high-speed photography, clips from *Star Trek* movies, and even the day-to-day life of Fountainhead Academy. Most of the clips, however, were completely abstract or unfamiliar. There was no way Seon could begin to describe all she saw.

On a hunch, Seon raised the volume on her earbuds. She heard a soft white noise like a trickle of flowing water, subtly varying with the rhythm of the visual stream.

She squinted through the video layer to focus on Silver Sparrow, across from her. She understood the sound then. He could open and close his eyes, but not his ears. For children like Silver Sparrow, sensory overstimulation could become unbearable.

"You made these all by yourself? They're amazing."

Silver Sparrow's lip fluttered a few times, then the audio signal amplified in Seon's ear.

"It was Solaris."

Seon was speechless. These AI-enabled children were beyond her understanding.

"Silver Sparrow, would you be willing to share your works with other children?"

"Share? You mean, like a gift?" Silver Sparrow batted his lashes.

"Well, of course you can share it with them in whatever way you feel comfortable, but you could think of it as a token of friendship, a souvenir, like when Tommy gave the other children origami animals with their names folded inside."

Silver Sparrow fell silent, lowering his head.

Seon wondered if her luck with Silver Sparrow had run dry.

A week later, however, Seon received a video stream in her inbox. She opened it to find a loop of her own face evolving into flowers, clouds, and waves, then back to her face, where the loop began again. Text flashed over the image in a hypnotizing rhythm:

```
The cage was open from then on . . . The cage
was open from then on . . . The cage was open
from then on.
```

A surge of emotions—joy, relief, and a vague dread—filled Seon.

She forwarded the loop to Mama Kim to ask her opinion.

"Everyone received a personalized version of the stream, myself included," Mama Kim told Seon. "Everyone except, guess who?"

"Golden Sparrow?"

"Bingo. I hope Golden Sparrow doesn't feel that Silver Sparrow is deliberately trying to provoke him. We should watch those two closely."

"Speaking of, I encouraged Silver Sparrow to enter Seoul's Artists of the Future Competition. He'll be in the U-6 Group. He has a real shot."

"Isn't that the prize that Golden Sparrow has been obsessing over?"

"I guess it'll make for a good show."

Seon again watched the gift from Silver Sparrow. The video held an ineffable quality, almost like magic. It put Seon and all who watched it into a state of trance. After ten minutes of letting the video cycle through, Seon summoned all her willpower to turn it off and get back to work.

SIX MONTHS AFTER THE PAKS adopted Golden Sparrow, another couple visited Fountainhead Academy. When they arrived, children were chasing one another across the green campus, lush from the summer sun. But this couple had no interest in those children.

Mama Kim smiled cautiously as the couple approached. They hadn't been introduced by the Delta Foundation like the Paks, but by a third-party placement service. It was one of those sites that aggregated profiles of orphans from various institutions. After completing a qualification process, including a background check, the clients of the service could select any child they were interested to meet.

"Welcome, Andres and Rei," said Mama Kim. "It's a pleasure to tell you about Fountainhead Academy."

Mama Kim had been informed by the service in advance that both were transgender. According to foster care statistics, families with at least one transgender or gender-nonconforming member accounted for 17.5 percent of adoptive parents. The data also showed that children fared just as well in physical and mental health whether adopted by transgender or cisgender parents.

"That's fine, thanks," Andres said. "We really would prefer to see the child as soon as we can. I mean—"

"Silver Sparrow." Rei completed their partner's sentence.

The clothes the couple wore made Mama Kim hesitate. The brightly colored geometric patterns draped over them were something out of a Kandinsky painting. The material was a kind of synthetic fiber film with sharp jagged contours.

"Maybe you're familiar with the child's background, but I'd like to go over it again." Mama Kim sucked her smile into a stern expression. "Silver Sparrow is a special and sensitive child, who gets easily overstimulated."

Rei removed their bright yellow sunglasses and spoke with a seriousness that matched Mama Kim's own.

"Headmaster Kim, I understand we might not look like the type of parents you're familiar with, but we would never put our interests above the safety of our child. Andres?"

Andres tapped a measured rhythm on their wrist. The couple suddenly looked as though they were melting, like ice cream in the sun. The sharp geometric edges of their clothing softened to the texture of animal fur. The saturated colors faded into earthy hues.

"Really, that's . . . thoughtful," Mama Kim said, amused. She led the couple to the reception room.

Inside, Silver Sparrow was already seated on the sofa, rocking back and forth in a gentle rhythm. He paid no notice to his visitors.

"You must be Silver Sparrow. I'm Andres, and this is Rei. It's an honor to meet you in person."

Mama Kim cleared her throat. "Silver Sparrow, I'll let you talk to Andres and Rei in private. If you need anything, you know how to call me."

Only the three remained in the room.

"I suppose there's no need for formalities," Andres said. "You're smart. You know why we're here. We want to invite you to live with us."

"To be more direct, we didn't learn about you through the placement service," Rei said. "We used it in order to go through the background check. But we already knew we wanted to meet you, Silver Sparrow. I have to say we're not the most traditional parents—"

"We think you're incredibly talented!" exclaimed Andres. "We saw your artwork at the Artists of the Future Competition. We couldn't believe it was the work of a six-year-old. Of course, physiological age is an outdated label. But even put next to works by artists of other ages, other eras, it's extraordinary. Wouldn't you agree, Rei?"

"Yes. My specialty is in twentieth- and twenty-first-century digital art, so I have some expertise in these things. In fact, we were the anonymous buyers of your piece at the Fountainhead benefit auction. And while it's tragic what happened to the original piece, we even prefer the newer version."

Silver Sparrow, who had been unresponsive until now, raised his head and stared at the two with a blank expression.

"Your bidding strategy was not optimal," he said suddenly. "Solaris said you exposed your intention too early, causing the competing buyer to quickly raise his price over three consecutive rounds."

Andres and Rei smiled at each other, eyes beaming with surprise.

"Whatever we paid was worth it to get to know you, to show you we're the right family for you," said Rei. "We'll give you all our love, and not just in that traditional sense of parental love. We mean to do everything we can to support you in exploring who you are and realizing your greatest potential. Isn't that what you want, too?"

After a moment of silence, Silver Sparrow turned to Mama Kim, who had reentered the room. "Mama Kim, can I bring Solaris?"

IN THE BEGINNING, SILVER SPARROW acted the same as he had at Fountainhead, preferring quiet corners of the apartment in which to quietly pass the day. Solaris would, according to his instructions, generate translucent virtual bubbles to wrap him in and project various streams and fragments before his eyes. This visual vortex brought a sense of peace and flow to Silver Sparrow.

Andres and Rei would gaze upon the boy's chrysalis-wrapped silhouette in their open loft, giving him time to adapt.

Perhaps it was the absence of other children or perhaps it was Solaris's talent for adaptation, but the perimeter of Silver Sparrow's virtual chrysalis gradually expanded. His range of activities grew with it. Finally, the chrysalis encompassed the whole loft.

Silver Sparrow now had a different sense of space and scale. He found he suddenly had an appetite for physical activity. Though he was still afraid of physically colliding with other children, he could climb, jump, and chase the virtual rabbits Solaris created for him. Silver Sparrow raced, panted, sweated, and felt the joy of the chase thump in his heart.

He thought about what Seon had said, that this was what it felt like to escape one's cage.

He wanted to go yet further. But first, he wanted to understand who he was within.

Solaris set various tests for Silver Sparrow, helping him establish a comprehensive self-evaluation model that covered cognitive abilities like language comprehension, quantitative analysis, and reasoning, as well as qualities like physical movement, openness, and emotional intelligence.

The conclusions weren't surprising. Silver Sparrow's general cognitive-ability performance, and especially his quantitative skills, were advanced. However, in areas of interpersonal communication, his score plummeted.

Silver Sparrow had never been able to easily perceive the tone of those speaking to him, whether kindhearted or malicious, sincere or sarcastic. He had trouble distinguishing between literal and figurative meanings. In these respects, he was not so different from the AI of a decade earlier.

For one type of skill set, however, Silver Sparrow's test results were off the charts: creative functioning.

Looking at his personality defined in these results, Silver Sparrow couldn't help but think of his brother and how they had fallen out. A silent, unanswered question hung in his mind: *If I could be more like other children, would that change everything?*

ONE NIGHT, ABOUT TWO YEARS after Golden Sparrow and Silver Sparrow arrived at Fountainhead but before they were adopted, Mama Kim had desperately called Seon to the academy. Gwang was unavailable due to a business trip in Jakarta.

As evening settled in, there was a ghostly atmosphere on campus. Most of the children were gathered in the activity room. The campus's smart home system had been attacked, causing lights to flicker while the ventilation system rotated between extreme cold and hot. Meanwhile, service robots were wildly bumping into furniture, making loud clanging sounds.

"What's going on?" Seon asked, befuddled.

"Get the problems in front of you fixed one by one. We'll talk about the rest later."

Through the vMirror in the IT department, Seon entered the backend system to find it had been hit with a DDoS attack. The hacker's method wasn't particularly clever. The hacker had just taken advantage of a security vulnerability that should have been upgraded a long time ago. Seon wondered if this was all somehow related to Gwang's business trip. To protect against similar future attacks, she installed the latest version of the network traffic monitor. The academy lit up again and all seemed to return to normal.

Then, Seon noticed something strange in the log just as Mama Kim was calling her into the conference room. As Seon approached, she saw lying on the conference table a crestfallen Golden Sparrow devoid of his usual energy.

"It was you!"

"It wasn't him," Mama Kim said calmly.

"Huh?"

Mama Kim turned her head slightly, and Seon saw Silver Sparrow seated on the ground, hands on his knees. His head was bent over his legs and his eyes were wet with tears.

"Silver Sparrow? How is this possible?"

"They wouldn't say anything, so I called you," Mama Kim said. "It's beyond my understanding."

"Golden Sparrow, you know I can call up Atoman's log. Do you want to tell us what happened?"

Golden Sparrow pouted. "There's no time. It's too late."

"What's too late?"

Seon opened her XR field. The red robot, which had always been inseparable from the boy, was nowhere to be seen. She checked their data share permissions. All was normal. Of course, there was the possibility that Golden Sparrow had hidden Atoman, but that wasn't his style.

"Where's Atoman?"

Golden Sparrow reluctantly sat up. His hands spread out, glowing red as if on fire. He put his palms forward, then made a fist. A virtual image appeared in front of Seon, but it was nothing like the Atoman she knew. It looked like his AI pal had been smashed to pieces. Parts floated loosely. Limbs were in the wrong position. Seon thought the avatar might disintegrate into a cascade of pixels at any moment.

"What on earth?"

"Ask him!" Golden Sparrow yelled, pointing at his brother in the corner.

Mama Kim went to Silver Sparrow, knelt, and asked softly, "Is your brother telling the truth? Did you do it?"

Silver Sparrow said nothing, but Seon received a data packet, another video.

Seon saw Silver Sparrow's celebrated video artwork begin to stream, but it was all wrong. She spun toward Golden Sparrow. Now the strange activity she'd seen in the log made sense.

"Why would you do it?"

"I . . . didn't do anything." Golden Sparrow beamed an innocent expression.

"Why would you want to destroy Silver Sparrow's work? Don't you know—"

"How could he have accessed the backend?" Mama Kim asked incredulously.

"Gwang must have given him access before his trip," said Seon bitterly. "Gwang wanted to train Golden Sparrow to be a helper with the system admin."

"I . . ." Golden Sparrow quieted as he mustered his courage. "I only wanted to get back what belonged to me."

Mama Kim's eyes widened. "You mean . . . Silver Sparrow winning the Artist of the Future prize?"

"I think I understand," said Seon. She gave an exasperated nod and began to explain. "Silver Sparrow's artwork involved four components: one parent stream and three child streams. Imagine if da Vinci's original *Mona Lisa* was digitalized and transformed into other media. In this case, however, the artwork was dynamic and much more sophisticated. Silver Sparrow told me that the artwork would reflect the spiritual and emotional ties between the academy and the child. As long as there was data flow coming from the parent stream, the child stream would continue to evolve. Without, it would be deprived of vitality."

"So how exactly did Golden Sparrow tamper with it?"

"He didn't tamper." Seon looked down. "He deliberately destroyed it."

"What?"

"See for yourself." Seon projected a video of the IT room onto the conference room's vMirror.

The video feed showed Golden Sparrow operating the IT department's vMirror. After entering the backend, Golden Sparrow identified the parent stream's storage path. On-screen, Golden Sparrow hesitated before issuing the command. Perhaps he was thinking of his brother's painstaking efforts over the past months, or of the honor of the academy. He blinked then tapped "OK." The parent stream, the master version of the artwork known as *Fusion op-003*, disintegrated into discrete bits.

Watching all this, Silver Sparrow shook with anger.

"Silver Sparrow got his revenge with his generalized attack on the academy system. That was how he was able to destroy Atoman."

Mama Kim turned to Seon. "I'm going to have a talk with Golden Sparrow. You take care of Silver Sparrow."

NOW ALONE WITH GOLDEN SPARROW, Mama Kim turned to him. "Look at me, Golden Sparrow. You have to answer honestly. Why did you do it?"

"I, well, Silver Sparrow used my portrait without asking—"

Mama Kim cut him off. "Is it because he won the prize and people were starting to like him? Did that make you unhappy?"

"I . . ." Golden Sparrow's face flashed a grieved expression as he searched for the words. "I had Atoman analyze all the winning works of the past few years. I made a plan for every possible outcome. The probability of my winning was clearly the highest."

Mama Kim forced a bitter smile. "Ignorant child. Probabilities are just probabilities. That doesn't mean that you deserve to win. People aren't machines. You should have been happy your brother won that prize."

"Why do you think he's so great when he does one little thing? Because he has that disease? Because, that's not fair! Shouldn't it be the best who wins?"

Mama Kim stared back, stunned. "I understand it's hard, but you have to learn to accept it when you don't win—"

"No, you don't understand. Only Atoman understands."

"Atoman is just a tool!"

"No, Atoman is my best friend in the world! And that freak ruined him! I hate him!"

On the other side of the room, Seon had calmed Silver Sparrow somewhat. He had returned to his reserved and shut-off state. Seon tried various cues to get Silver Sparrow to share his feelings. But he kept repeating a single phrase.

"A souvenir . . . a souvenir . . ."

At first, Seon was confused. Then it hit her. "A souvenir" was the exact phrase she had used during her first conversation with Silver Sparrow a few months ago. She had described Tommy's origami animal with the other child's name folded inside as "a souvenir." Had Silver Sparrow made the work as a gift to his brother? Was that why he had linked Golden Sparrow's portrait data? No wonder Silver Sparrow's reaction had been so violent.

Mama Kim looked at both brothers with a stern expression. "No one leaves today without shaking hands and apologizing."

From that point on, however, Golden Sparrow and Silver Sparrow grew only more estranged, like two parallel lines destined never to intersect.

AS ONE OF THE CONDITIONS of Silver Sparrow's adoption, Mama Kim stipulated that his new parents arrange for a reunion of the twins. Despite their divergent paths, she thought it was imperative that they not lose contact.

Golden Sparrow and Silver Sparrow's reunion took place at the Paks' large neoclassical home with swimming pool and playground in the backyard. Like the Paks' décor, the plan for the day was overly formal. There would be an outdoor barbecue, followed by games for the children.

"Hello, Golden Sparrow," Andres said, as they, Silver Sparrow, and Rei stood on the threshold of the house's imposing front door. "You look quite different from the photos I've seen. You must get a lot of exercise."

After six months with the Paks, Golden Sparrow had not only changed in demeanor, but even his body had transformed.

Golden Sparrow confidently extended his hand to Andres. "Indeed, I'm following a regimen designed for me by Atoman for how I eat, exercise, work, and rest. Well, an upgraded Atoman," he added, casting his eyes at Silver Sparrow.

Golden Sparrow reached out his hand to his twin. "Hey! You all right, brother?"

Rei pushed Silver Sparrow forward. Silver Sparrow looked at his brother but didn't extend his hand.

"Come on, Silver Sparrow," Rei said. "It's your brother, and you haven't seen each other for . . . half a year."

"One hundred and seventy-three days," Golden Sparrow added with a slight smile. "Silver Sparrow, would you like to see Atoman? Jun-Ho upgraded it to an improved version with so many cool functions. We even built it a body. It's way cool."

A spark of curiosity glinted in Silver Sparrow's eyes.

"Atoman, look who's here!" Golden Sparrow yelled.

The ground vibrated with mechanical footfalls as a glowing red robot hopped across the lawn. An upper humanoid body was attached to the shoulders of a robot dog, like a cyborg centaur.

The new version of Atoman immediately recognized Silver Sparrow's face. With a comical bend of his right front leg, he bowed. He blinked his three camera-eyes and said, "Silver Sparrow, it's been a long time."

The corners of Silver Sparrow's mouth lifted into a subtle smile as Atoman stiffly raised its hand.

"Kids, lunch is ready," Jun-Ho shouted from the barbecue. "Help get the table ready." Golden Sparrow's new siblings—Hyun-Woo, who was fifteen, Si-Woo, who was eleven, and Suk-Ja, who was eight—raced over to set the patio table.

"Let's talk later. I gotta help out." Golden Sparrow whistled, and Atoman followed.

"Your brother doesn't seem *so* difficult to get along with," Andres quipped.

Silver Sparrow's lip curled.

Jun-Ho's grilling skills weren't the greatest, but the Paks' private chef had prepared most of the important dishes.

At the table, Andres and Rei observed the Pak children's manners, which were reserved and cautious, even when just selecting a fork. Golden Sparrow, no longer displaying the rambunctiousness he was famous for at the academy, glanced at his new siblings from the corners of his eyes, following their moves. The atmosphere felt choreographed and formal.

Silver Sparrow always felt out of place. Even before the food was completely served, he was stirring his mashed potatoes on his plate, his scraping fork creating a harsh metallic sound. Hye-Jin Pak squinted at him from time to time but didn't know what she should say.

To break the ice, Andres decided to ask about Atoman. "Golden Sparrow, your robot is supercool. How did you select the body you gave it?"

"Oh, no particular reason. Jun-Ho said this was the latest and greatest model, so we went with it." Golden Sparrow looked to Jun-Ho for approval.

"I always want the best for my children." Jun-Ho pulled at his chin.

Rei turned and spoke, coolly. "But 'the best' is a relative concept. What we think is best might not be what the child thinks is best. Would you agree?"

"Not for us." Jun-Ho and Hye-Jin exchanged smiles. "We believe in finding out what the best is, the best available in the world. That goes for travel, insurance, education, you name it. Even robots. Golden Sparrow, tell us what you learned this morning."

"Price is what you pay. Value is what you get," Golden Sparrow said without thinking.

"What?" asked Andres with a bemused smile.

"It's a famous quote Warren Buffett invoked during the 2008 financial crisis," Jun-Ho said. "A bit of old-school wisdom from the investing world."

Rei couldn't hide their disdain. "Isn't that something, coming from a six-year-old?"

"Is it, my dear artist?" said Jun-Ho. "In the old days, children had to memorize many irrelevant things they didn't understand, but they also didn't have a concept of their future then. Thanks to AI, the information in their lives is no longer so disconnected, so *random*."

Hye-Jin chimed in. "AI can do what schools and teachers in the past never could. As Jun-Ho said, the AI has the blueprint of our children."

"If he keeps at it, Golden Sparrow could become one of the truly great investors," Jun-Ho added.

"So, you let an algorithm plan your children's future?" Rei asked.

The Pak children set down their knives and forks, riveted.

"It's our responsibility to ensure that kind of talent doesn't go to waste," responded Jun-Ho. "We used to have a saying: No one knows the son better than the father. Now, should we say, no one knows the son better than his AI? Parents will never again have as much insight into their child as the child's AI. And that's a good thing. Golden Sparrow's math is already at the level of a ten-year-old's. And his pattern recognition is better than Si-Woo's."

Si-Woo grimaced.

Hye-Jin interjected, "I appreciate that artists like yourselves have a more romantic view of things, but what is more important than our children's education?" She tapped the tip of Golden Sparrow's nose. "And, it's not like we ever told you to become a certain type of person. We've always said you can be whatever you want to be, right?"

With a knowing grin, Golden Sparrow blurted out, "I want to be just like Jun-Ho!"

Jun-Ho and Hye-Jin burst into laughter. Andres and Rei exchanged glances.

Suddenly, Silver Sparrow tossed his fork to the ground. Everyone looked at Silver Sparrow, whose hands, face, and hair were covered with juice and crumbs from his meal.

"I want to go," he said in a low voice.

FROM THEN ON, SILVER SPARROW refused all contact with his brother.

Andres and Rei told Mama Kim that the differences between the twins seemed irreconcilable. The notion of more play dates seemed unlikely.

They understood their son's feelings all too well. Andres and Rei could not have been more different from the Paks, starting with the very nature of their careers and identities, though those were hard to pin down. Were they new-media artists? Internet celebrities? Environmental activists? Academics? Spiritual gurus?

Partners in work as well as life, they referred to themselves as "Homo Tekhne" and advocated for a so-called Technological Artistic Renaissance. They critiqued the blind worship of science and technology. Through art, Homo Tekhne sought to restore dignity to humanity and revitalize the connection between humanity and nature.

In Rei's view, the increasing use of AI in education meant children were trained to become competitive machines. The system was an enhanced version of the drill-and-exam education of old. Real education should be as much about personal growth as knowledge and skills. Children needed to raise their self-awareness through inward exploration, cultivating empathy, communication, and other "soft skills" that would nurture deeper connections with one another and increase their emotional intelligence. AI usually ignored these goals.

Rei had been deeply moved by Silver Sparrow's art. It was not especially advanced on a technical level, but it suggested a raw, vital curiosity that could exist only in a child's eyes.

Andres, on the other hand, had become captivated by Solaris, the AI that helped Silver Sparrow create his work. What conditions had triggered their son's AI companion to abandon the usual competition-oriented models and evolve a new logic of its own? Had Silver Sparrow's psychological disposition somehow broken the AI's competition-driven feedback loop and transformed it into a tool to explore his inner self?

The awkward day at the Paks' home had made Andres and Rei even more aware of the road they didn't want to go down.

When they decided to upgrade Solaris, they asked Silver Sparrow for his input, and carefully backed up all data. They saw this data not only as the memories of Solaris but also as a peripheral extension of Silver Sparrow's own being. Solaris's core algorithm, like a fragile crystal, needed to be protected.

Though the new Solaris didn't have a cool robot body like that of Atoman, Silver Sparrow, like his twin, felt empowered by the upgraded version of his AI pal. He felt like someone who had been walking blindfolded in the night and had suddenly opened his eyes in the sun.

THE PAK CHILDREN WERE expected to live by the family motto "Only the best deserves the best."

This maxim implied both that you received the great support of the family and that you must make every effort to be worthy of the family's support.

Golden Sparrow was no exception.

In the early days following his adoption, his new parents had sought to correct the many "bad habits" he'd developed while at Fountainhead. For Jun-Ho Pak, discipline was the foundation of success.

Golden Sparrow's days as a prankster were over. When he did act up, Jun-Ho's preferred mode of punishment was to block Golden Sparrow's voice profile from the smart home system. All of Golden Sparrow's commands would be rendered invalid.

This was torture for Golden Sparrow, who was always eager for the household's attention. Soon, the boy learned to control the volume of his voice and the thump of his footsteps, just as he had learned the proper use of a salad fork.

Atoman, too, had fallen in line with the Paks' house rules. Jun-Ho had made a comprehensive upgrade to Golden Sparrow's AI partner. There were countless rules for when and on what occasions Atoman could and couldn't be awakened, which rooms enforced data security restrictions, and etiquette for sharing XR feeds. Golden Sparrow had no more thoughts of hacking electrical appliances and home systems.

At first, Golden Sparrow had resisted the changes. He thought fondly of his days at the academy when he could run and frolic however and whenever he pleased. He even thought of Silver Sparrow. The delight he had taken in teasing his twin seemed like a lifetime ago. More than once, he cried himself to sleep in the silky sheets of his new home.

In time, though, Golden Sparrow appreciated how the Pak siblings excelled. Hyun-Woo had obtained patents for biotech inventions as a teenager. Si-Woo had designed a quantum information transmission experiment that was being tested on the Chinese space station. Even Suk-Ja, the Paks' little princess, was a student ambassador at the UN Climate Change Conference.

"Only the best deserves the best."

The family motto was like a thorn in Golden Sparrow's heart. Whenever he felt the need to slack off, the thorn stung him with guilt.

The virtual classroom—the *best* virtual school, according to his parents—was the one place where Golden Sparrow felt like he was back in his own skin again. Learning was gamified with levels, points, and virtual props. This was what Golden Sparrow was good at, and the other students were fun as well.

Eva, a blond classmate, was especially fun, with a bubbly, cartoonlike personality. At first, Golden Sparrow found it hard to take his eyes off her. Eva had the sweetest voice and was always friendly. She always seemed to know what Golden Sparrow was thinking and what he needed to hear, saying things like: "Golden Sparrow, this question is a real tough one. Let's try thinking about it from another angle."

Or, "Golden Sparrow, you're awesome. Why didn't I think of this solution? Can you show me again how you did it?"

Eva inspired Golden Sparrow. In return, he would enlist Atoman's help in composing jokes for her or impressing her with magic tricks. He would give Eva small virtual gifts, and she would laugh and send him red glowing hearts that played wind chimes in his earbuds. These were the few moments in which Golden Sparrow felt truly happy.

On the last few math tests, Golden Sparrow had scored first in his class. He reported his success to Jun-Ho, hoping for his approval. Jun-Ho read the results with the faintest of smiles. "Golden Sparrow," he said, "if you're so easily satisfied, the bar is too low."

The next day, Golden Sparrow was surprised to find Eva had changed, though he couldn't put his finger on what it was. She was as beautiful as ever, but her voice had become more serious. She had begun to sound a bit like Jun-Ho.

"Golden Sparrow, don't be careless. Check it one more time."

"Golden Sparrow, how did you get it wrong again? You've seen this same problem several times already."

Even Atoman's tricks no longer made her happy. She turned a deaf ear to all Golden Sparrow's jokes and gifts. It was as though she were a completely different person.

Heartbroken, Golden Sparrow asked Atoman for advice. "Eva doesn't like me, does she?"

Atoman tilted his head and said nothing.

"Is it because I didn't help her improve her grades?" Golden Sparrow asked. "You have to tell me what happened with Eva."

"But it's obvious," Atoman replied. "Her parameters were adjusted."

"Parameters adjusted?"

Golden Sparrow's eyes widened. So Eva was just another AI, and Jun-Ho had altered her personality. How had he not seen it? Were human expressions and behaviors generated by AI so real that they could pass for other students in the virtual classroom? Or was it that he so longed for a friend like Eva that he had deliberately ignored the cracks in the illusion?

The blond girl's face and laugh floated before Golden Sparrow like a broken cup that could never be put back together.

That evening, Golden Sparrow again cried into his bedsheets. When he heard footfalls outside his room, he quickly wiped his tears and pretended to sleep. A moment later, someone was sitting at the side of his bed. It was Hye-Jin.

"Talk to me. You made Dad angry, didn't you?"

Golden Sparrow pulled back the covers to reveal a sliver of his face. He gave a cold nod, as if he had been served some terrible injustice. Then he emerged fully and shook his head.

"I made myself angry. I was so stupid. It never occurred to me she was an AI."

"Silly boy." Hye-Jin ruffled Golden Sparrow's hair. "Honestly, I can't tell either most of the time. The AI knows what kind of girl you like and can make you feel that she understands you. Those things aren't real. Their purpose is to motivate you to study hard."

"Is Jun-Ho disappointed in me?"

"Why? He was the one who adjusted her parameters. He wanted you to see that just because you get a top score doesn't mean you're really number one. He wanted you to continue to overcome your weaknesses to really become the best. That's what he expects from a Pak child."

Golden Sparrow nodded and bit his lip.

THE YEARS PASSED. SILVER SPARROW was growing up fast. In some ways, though, he felt like a snail with a heavy shell on his back, creeping forward at the slowest imaginable pace.

When Silver Sparrow was younger, Rei and Andres had tried enrolling him in an online school for children with Asperger's syndrome. Silver Sparrow could access the virtual classroom via Solaris. The AI system created virtual classmates and teachers for each child according to their different cognitive levels and behavioral characteristics. Therefore, all interactions were highly individualized, ranging from the visual style of the interface to the instructors' tones.

But somehow, it didn't work for Silver Sparrow.

Whenever Silver Sparrow entered the virtual classroom, he felt anxious. Even when the other avatars behaved like children with Asperger's and similar conditions, nothing worked for him. Silver Sparrow could tell immediately the purpose of every word spoken by the virtual students and teachers, which skills they were designed to train for, and which knowledge points they would strengthen. It all felt so false and fragmented.

It was Solaris's data feedback, however, and not Silver Sparrow himself, that convinced his parents to give up these attempts at schooling.

Legally, a child's guardian was entitled to obtain full access to an AI companion's data. However, Rei knew Silver Sparrow wasn't an ordinary child and required more privacy and security. So they made an agreement with Silver Sparrow that after he turned ten, they would no longer be able to view Solaris's data without his consent.

Andres had a different perspective. The way they saw it, the value of the data wasn't only for the child, but also to help the parents.

Without Solaris, they would never have known what physical distance was most comfortable for Silver Sparrow. Nor would they have known what types of psychological activity were playing out in the boy's repeated compulsive behaviors.

Andres wished that when they were growing up, their parents had had an AI like Solaris to help them see the many wounds they inflicted in the name of their love.

Perhaps Silver Sparrow didn't understand human love as deeply as his parents did, but Solaris gave him another tool to explore and express himself: art. Under the guidance of AI, Silver Sparrow had surveyed artworks across countless traditions and historical periods. He intimately understood the differences in concepts behind the various forms and styles. Each represented a unique perspective on the world. Now he just had to find his own.

By the time he was fourteen, Silver Sparrow was confident that what he needed to learn wouldn't be found in classrooms, in books, or in the structures of mathematical logic. He needed to forge a genuine connection with the world and with living people. He hoped to experience firsthand the forces of nature, time, and space.

But he couldn't.

He was imprisoned in his young body, a fragile body he couldn't even properly control. Every kind of discomfort, fear, strangeness, and shame kept him from stepping out from his virtual chrysalis to face the vast world.

So he sought vicarious solutions.

He would chase swallowtail butterflies on Lantau Island in the setting sun, watch young people dance like crazy all night in the clubs of Berlin, listen to monks chanting morning service in Kandy,

Sri Lanka, or await the northern lights on the cold surface of the Arctic Ocean.

Solaris's VR technology now integrated highly sophisticated functions, including collaborative visual-auditory haptics, proprioception, and somatosensory simulation. Its omnidirectional immersion was far beyond the VR of the decade before. With ultralow latency transmission, the AI algorithm adjusted everything in real time according to each user's needs.

These virtual sojourns helped Silver Sparrow understand the diversity of human experience at a cognitive level, as well as experience a greater connection with the world at an emotional level. He felt a new joy flow through him during these VR immersions.

Despite these thrills, however, Silver Sparrow sometimes had visions that disrupted his experiences. He had not forgotten his twin. In the morning light or at evening dusk, he would glimpse Golden Sparrow or Atoman, both in his virtual form as a red robot and in his canine-centaur physical form. They seemed to be calling out Silver Sparrow's name.

At first, he thought it must be an illusion. He had read studies on recovered memories. Was his mind producing these false visions, the same way AI could overfit noise in the data into a model? The mind, too, could abstract life's problems into its models, which played out in dreams, slips of the tongue, obsessive-compulsive behaviors, or general Freudian graffiti.

In the end, Silver Sparrow came to believe his brain wasn't playing tricks on him. His heart hid a longing for his brother.

In time, the visions appeared more frequently. He might experience a flash of genuine pain, like a migraine. Was it mental illness, Silver Sparrow wondered, or could it be some kind of connection with his twin? These tangled feelings troubled Silver Sparrow. In his short life, he had never felt he was so strongly needed, not by Mama Kim, Seon, or even Andres and Rei.

He needed to go searching for the source of that call.

GOLDEN SPARROW, TOO, HAD been feeling frustrated.

It wasn't due to studies or affairs of the adolescent heart.

His frustration stemmed from his need to become a man like his father, a top investor.

Compared with other industries, this career path was strikingly clear, like the tracks of a heavy vehicle in the snow. He would begin as a researcher, getting to know selected companies, collecting information from public channels, building financial models based on historical data, and making predictions for the future based on current conditions. Then he would put the companies into the context of their industries to analyze upstream and downstream chains, as well as risks and opportunities. Finally, he would summarize his perspective in a report of great practical value to the investor partners.

The whole process was like making coffee. If you had quality coffee beans (data) and appropriate grinding and stamping tools (models), you could brew a great cup of coffee (perspective) with rich flavor and fine layers.

By repeating the process many times, accumulating experience and ability, you could move all the way up the ladder from junior researcher to senior partner.

Just like leveling up in a monster-fighting RPG, everything was quantified. As wealth soared, so, too, did adrenaline and dopamine, making players addicted to the game. Golden Sparrow's endgame was to become like his father, a partner.

In his fund simulations, Golden Sparrow showed sparks of immense talent. Even Jun-Ho marveled at his son's intuition for markets, and decided it was time to establish a fund from which Golden Sparrow could begin to make his own investments.

But when he migrated from virtual hurdles to real ones, Golden Sparrow soon met with defeat.

Golden Sparrow had selected a game company in his father's portfolio to research. He labored a month to produce a solid investment report. It helped that he was personally experienced with several of the company's games. He confidently brought the report to his father.

It took his father ten minutes to finish reading it, after which he transferred a file to Golden Sparrow.

Opening the file, Golden Sparrow realized it was another report on the same company. Its comprehensive data and the force of the final conclusion were cogent and stunning. It far surpassed the report Golden Sparrow had so meticulously prepared for his father. Golden Sparrow jumped to the end of the report to learn the author's name. It was AI.

"Guess how long it took?" His father had a smile on his lips. "Less time than it took me to read yours."

"This . . . this isn't fair."

"What's not fair? Age? Qualifications? Industry experience? The quality of that AI report is higher than eighty percent of the analysis from my team now. And it takes less than a thousandth of the time they need. Reality is cruel."

Golden Sparrow's face turned ghostly pale. "Well, then, what am I supposed to do? What value can I possibly—"

"What, you're scared? That's not the Pak family way. It doesn't matter that AI surpasses eighty percent of analysts today. You need to be at the top one percent of the pyramid if you're to be anything."

"But at the speed AI is evolving, it's just a matter of time. Look at Atoman!"

His father leaned back in his chair with his usual scornful smile. "Son, whether you stand and fight or run away, it won't change reality."

Frustrated, Golden Sparrow fled his father's office. He had a sick feeling in the pit of his stomach. Golden Sparrow understood that if humans competed only with hard skills, such as data collection and structural analysis, they would never rival AI. The only areas where humans could surpass AI were in places machines couldn't reach, in realms like human sensitivity and intuition.

Then Golden Sparrow had an idea. Instead of just crunching numbers, he decided he would speak with as many employees at the gaming company as he could.

At first, these real human beings gave the boy a headache. They were nothing like his predictable AI classmates. Each employee had his own temperament and habits. And Golden Sparrow knew they were only meeting him out of respect for his father.

The conversations were far more difficult than just analyzing data and making models. Even Atoman had no way to help. Atoman could recognize shifts in microexpressions, but it couldn't identify the complex webs of meaning behind them.

Golden Sparrow began to understand why in his father's social circle, his fellow successful partners were often older people. To understand other human beings required a long learning curve.

The more Golden Sparrow thought about it, the more certain he was that this was the right path. He continued to leverage his father's network to meet more entrepreneurs, content creators, engineers, and sales executives. Moved by the boy's professional ability and stubbornness, by and large they began to regard Golden Sparrow as a talented young researcher.

Still, even as things improved for Golden Sparrow, he felt strange sensations—especially in his dreams.

Golden Sparrow would dream of his quiet twin and his odd amoeba-like AI, Solaris. The timelines of the dreams were pure chaos, with Silver Sparrow both as a child and as a young adult. The young Silver Sparrow had become tall, but his face still had that focused expression of indifference, as though the whole world had nothing to do with him.

In these dreams and fragments of visions, Golden Sparrow sometimes glimpsed scenes of their childhood. With the passage of time, he felt a sadness for his brother, and even more for himself. He remembered his childish provocations, all to win the attention of others. At the time, he'd thought that he and Atoman had been loved by everyone, but now he realized they'd just been a flashy red robot and an annoying brat.

In these moments of self-doubt, the sixteen-year-old Golden Sparrow began to question the career-driven fast track he was on. In these moments, his heart filled with an unquenchable desire to see his brother again.

But he couldn't.

His parents sent him to a psychologist, who told him his feelings were likely the result of excessive stress. If it continued, the burnout could evolve into depression, even cognitive impairment.

"I've seen a lot of kids like you—excellent, even perfect—but that's the problem, isn't it?" The psychologist smiled, choosing his words carefully. "Have you ever thought that maybe your belief system isn't so well suited to you? Do you want your whole life's value and meaning to derive from besting the competition, no matter the cost?"

"And what's wrong with that? Isn't everyone like that? Isn't that how we progress?"

"Man isn't an AI. We don't live solely by numbers and wins. Your value scale suggests inconsistency between the external expectations you place on yourself and your internal drives. Would you force an elephant into a refrigerator just because those around you told you it was the smart thing to do?"

Golden Sparrow looked like a wounded bird, his eyes dull. "And my dreams?"

The psychiatrist's voice softened. "Have you considered whether your dreams truly represent the feelings in your heart?"

JUST AS GOLDEN SPARROW was confronting one nightmare, a new one overwhelmed him.

An upstart gaming company called Mold had recently released *DREAM*, a real-time strategy game. It hit Golden Sparrow with the force of a hurricane. The game was revolutionary. AI had dominated every aspect of the game's development, from creative concept to designing levels to testing to writing character scripts. Everything that had once run up huge budgets—the work of artists and technical teams—had all been taken over by bots.

And players were crazy about the game.

Mold's ambition didn't stop with the game itself, though. They published the code for a series of AI game-generation tools online, saying they wanted to help small studios, independent game developers, and enthusiastic players without professional backgrounds to generate their own games in their own garages and bedrooms.

The effect on the industry was swift. Share prices of all the major gaming companies plummeted. They were now playing catch-up in the AI arms race.

Golden Sparrow again came to his father's office feeling as though he had now totally failed. "It's over," he said.

His father didn't understand. "What's over?"

"The whole industry, the gaming industry. It's always relied on human creativity and emotion, but now they've handed even that over to AI."

"I always thought that was the future."

"You don't even play games. You don't understand!"

"I don't understand?" His father's huge body leaned back, and he laughed heartily, rattling his ergonomic chair. "When I was a kid playing *Grand Theft Auto,* I wondered why the non-player characters had to be so dumb. In the *Halo* sequels, the aliens were at least able to coordinate a decent attack. But these were still light-years from the unscripted, procedural NPCs dominating today's games."

Golden Sparrow's eyes grew wide. He hadn't ever seen this side of his father.

"*Call of Duty, League of Legends, Breath of the Wild, Pokémon Go* . . . When I played these, I always thought, why can't the game adjust in real time according to my reaction speed, habits, and preferences? Like Alexa or Siri, the longer you play it, the more a game should understand you. Why couldn't they make games work like that?"

"But all my research . . . it doesn't matter now."

"Son, when you can't change the world, you have to change yourself." His father became serious. "It will happen again and again. This time, it's just a game. For thousands of people, it's a job to support their families. Any powerful company can collapse overnight. Industries disappear, technologies become obsolete . . . and people always have to feel their way forward."

Tears welled in Golden Sparrow's eyes. "I can never beat AI in investing. And I can never be you."

His father sucked in a slow breath of air, then lit his cigar.

"Son, you should never be like me. Be yourself. This is your life."

"But I thought . . ."

"Early on, I also had that idea." His father dragged on the cigar. "I even modified Atoman to make your whole learning and growth track fit as much as possible my plan for you. But you were never happy. You were a good boy, trying to meet all our expectations. But those expectations didn't come from your heart."

Fragrant cigar smoke wafted toward Golden Sparrow's young, confused face as his father continued talking.

"Later I decided that wasn't what Hye-Jin and I should want for you. What we wanted was a free individual who would discover the novelty and beauty of life, just like the feeling you get when you play a great game for the first time. You understand what I mean?"

Golden Sparrow left his father's office feeling unsettled. The light that had long been guiding his life had gone out.

He walked the streets of their neighborhood. As he wandered, he felt a vibration from Atoman. He had a new message.

Headmaster Kim Chee Yoon invites Golden Sparrow to the Anniversary of Fountainhead.

IT WAS A PERFECT spring day. The Fountainhead Academy campus was alive with residents and visitors alike. The lawn couldn't have been greener if it had been painted. Birds chirped and frolicked as though welcoming the academy's guests.

Today was not only the anniversary of Fountainhead Academy's founding, it was also the first time the academy had opened to the public since its expansion. The school had incorporated several emerging technologies, and its new buildings and classrooms could accommodate many more children. The "Child + AI" education model Fountainhead had championed, supported by its vPal technology, had, over the past decade, been replicated across the globe. It had quickly become the most popular model for education at special-needs institutions.

Mama Kim had been heralded as a pioneer. Now, her hair pulled back with a silver band, she was greeting faces old and new.

In the yard, former residents who had become world-class athletes were playing with Fountainhead's current students. Inside the new building, other returned graduates were painting pictures with their young counterparts—and their AI partners.

Golden Sparrow had kept mostly to himself, avoiding old acquaintances and the activities. When no one was looking, he slipped down the hallway of the old building to the now-abandoned IT room. It was strewn with equipment that had not yet been moved to the new IT management center or into storage.

He was surprised to find the old vMirror in a corner, wrapped in a transparent dust cover, like a forgotten piece of furniture. He powered it on, and the familiar interface appeared in front of him. He laughed as old memories bubbled to the surface.

How many nights had Gwang spent teaching him how to operate the system, hoping someday Golden Sparrow would take over as IT lead at Fountainhead? But instead Golden Sparrow had sabotaged the system and used the technology to destroy the painstaking work of his younger twin.

Golden Sparrow shook his head. It was a lifetime ago, but the heartache felt fresh.

Staring at the vMirror with tears welling behind his eyes, Golden Sparrow tried entering the old password, but of course he received an error.

For so many years, he had hoped only to be seen as a winner, especially in comparison with his twin. He had always strived for the best, to win more awards and to have the better adopted family. He had strived to win everything and had ended up with nothing.

After he entered the password incorrectly a third time, the system locked, and Golden Sparrow shut down the machine.

Then, in the darkening mirror, Golden Sparrow saw a man emerge from the shadows of the room. A shaft of light hit the man's face, and Golden Sparrow saw it was his own face. He spun around in panic only to register a familiar shy smile, a smile he had not seen for ten years. The two men had the same face and body shape, but their hairstyles and clothes manifested two opposite personalities. One was as bright and bold as gold, the other as cool and calm as silver.

"How did you know where—"

"Seon saw you go this way. You all right, brother?" Silver Sparrow had grown up, but still had the face of a child.

"I'm great. I—" Golden Sparrow paused and took a breath. "Actually, I'm not so good, not at all."

"I know."

"I . . . I don't know what to say. I've always been able to see you. I don't know what that is."

"I see you, too."

"Listen. I'm sorry. For everything."

"I know."

Golden Sparrow stretched his arms to embrace his brother but remembered Silver Sparrow's aversion to bodily contact, and his arms froze awkwardly in the air. Silver Sparrow stepped forward and wrapped his arms around his brother. Golden Sparrow no longer held back the tears.

"You know . . ." Silver Sparrow retreated to his usual safe distance.

Golden Sparrow dried his eyes. "What?"

"It was Seon who did it."

"Who did what?"

"Mama Kim knew we were falling out of touch, so Seon set up a secret communication protocol in the underlying code of Atoman and Solaris. It randomly sampled both of our data, generated the XR video stream, and embedded it into the normal information layer of the other. It was a powerful operation."

"That was it," Golden Sparrow said. "Atoman and Solaris kept us connected and brought us back together."

"Now we know each other."

"What do you mean?"

"I feel your heartache, not with my mind but with my heart." Silver Sparrow pointed to his chest. "Solaris taught me how, just like Atoman taught you so many things."

"The only thing I learned is that my life is worthless. The bullshit career path . . . nothing can be done about any of it now." Golden Sparrow smashed his clenched fist against the table.

"When you destroyed my artwork, that's exactly how I felt, too. But now I'm here. You'll be here, too. You'll get better." Silver Sparrow said this without a hint of blame in his voice, as if simply stating a natural fact.

"But . . . I have no idea how to start over. It's like I'm trapped on a merry-go-round, and all I can do is let it spin me in circles."

"Perhaps, have you ever thought we could trade lives?"

"Trade . . . lives? How?"

"I'm sorry if that's not the right word. Maybe we can trade how we see the world."

"But I still don't understand."

"When I saw you, I realized something. AI has shaped us, and we have shaped AI in turn. We are like two frogs who have each built a well. We each see only a small piece of sky. Your Atoman, my Solaris, are like that. Perhaps if we connect our wells, we will see a bigger world. Maybe everything will appear differently."

"Join Atoman and Solaris together?" At last Golden Sparrow understood, and his eyes shone. "To become a new AI. To start the game again."

"You've got it." Silver Sparrow smiled. "But this time a game that isn't divided into wins and losses. One with boundless possibilities instead."

"It's brilliant," Golden Sparrow said.

"Shall we go find Seon and Gwang? We'll need their help."

For the first time in many years, Golden Sparrow and Silver Sparrow nodded in perfect sync.

ANALYSIS

NATURAL LANGUAGE PROCESSING, SELF-SUPERVISED TRAINING, GPT-3, AGI AND CONSCIOUSNESS, AI EDUCATION

"Twin Sparrows" introduces the idea of personal AI companions—in this case, companions whose primary function is to serve as tutors for the twins in the story. The AI companions, or vPals, as Fountainhead Academy's program calls them, feature many AI technologies, but the one I want to highlight is natural language processing (NLP), or the ability for machines to process and understand human languages.

What's the chance of humans being able to form relationships with sophisticated AI companions like Atoman within twenty years? For children, there is no doubt it can happen. Children already have a universal tendency to anthropomorphize toys, pets, and even imaginary friends. This is a phenomenal opportunity to design AI companions that can help children learn in a personalized way, and practice creativity, communications, and compassion—critical skills for the era of AI. AI companions that can speak, hear, and understand like humans could make a dramatic difference in a child's development.

I want to begin this section by exploring supervised and self-supervised NLP—the technology that could make such AI companions a reality. I will then answer the natural question: When AI masters our language, will it have general intelligence? Lastly, we will explore the future of education in the AI era, including how AI will become a great complement to human teachers and significantly enhance the future of education.

NATURAL LANGUAGE PROCESSING (NLP)

Natural language processing is a subbranch of AI. Speech and language are central to human intelligence, communication, and cognitive processes, so understanding natural language is often viewed as the greatest AI challenge. "Natural language" refers to the language of humans—speech, writing, and nonverbal communication that may have an innate component and that people cultivate through social interactions and education.

A famous test of machine intelligence known as the Turing Test hinges on whether NLP conversational software is capable of fooling humans into thinking that it, too, is human. Scientists have been developing NLP to analyze, understand, and even generate human language for a long time. Starting in the 1950s, computational linguists attempted to teach natural language to computers according to naïve views of human language acquisition (starting from vocabulary sets, conjugation patterns, and grammatical rules). Recently, however, deep learning has superseded these early approaches. The reason, as you may have guessed, is that advances in deep learning have demonstrated the capability of modeling complex relationships and patterns in ways that are uniquely suitable for computers, and scalable with the increasing availability of very large training data sets. Deep learning is now breaking records on every NLP standard evaluation task.

SUPERVISED NLP

A few years ago, virtually all deep learning–based NLP neural networks learned language using the standard "supervised learning" discussed earlier. "Supervised" implies that when AI learns, it would need to be provided with the right answer for each training input. (Note this "supervision" does not imply the human would "program" rules into the AI; as established in chapter 1, that does not work.) AI would receive pairs of labeled data—the input and the "correct" output, and then the AI would learn to produce the output that corresponds to a given input. Remember the example of AI recognizing the cat image? Supervised deep learning is the training process in which AI learns to produce the word "cat."

When it comes to natural language, we can apply supervised learning by finding data that has been labeled for human purposes. For example, there are existing data sets of multilingual translations for identical content at the United Nations and other places. These provide a natural source of supervision for machines to learn to translate languages. The AI can be trained from the simple pairing of, say, each of the millions of sentences in English with its professionally translated counterpart in French. Using this approach, supervised learning can be extended to speech recognition (converting speech into text), optical character recognition (converting handwriting or images into text), or speech synthesis (converting text to speech). For these types of natural language recognition tasks where supervised training is feasible, AI already outperforms most humans.

A more complex application of NLP goes from recognition to understanding. To make this leap, words need to be instantiated as actions. For example, when you tell Alexa, "Play Bach," Alexa needs to understand that you want it to play a piece of classical music by the composer Johann Sebastian Bach. Or, when you tell an e-commerce chatbot "I want a refund," the chatbot can guide you on how to return the merchandise, and then refund you the purchase price. It is very time-consuming to build supervised domain-specific NLP understanding applications. Consider the myriad ways humans might express a similar intent or proposition (for example, "I want my money back," "The toaster is defective," and the like).

Every imaginable variation in the clarification and specification dialogue would have to be present in the NLP training data. And not only "present" in the data, but also "labeled" by a human being to give sufficient clues for the AI training. Data labeling for supervised training of language-understanding systems has been a large industry for twenty years now. As an example, in an automated airline customer service system, labeled data for training language understanding looks something like this:

```
[BOOK_FLIGHT_INTENT] I want to [METHOD: fly]
from [ORIGIN: Boston] at [DEP_TIME: 838 am]
and arrive in [DEST: Denver] at [ARR TIME:
1110 in the morning]]
```

That's a very basic example. You can imagine the overhead involved in marking up hundreds of thousands of utterances at this level of detail. And you'd still be far from covering all possible variations, even within the narrow flight-booking domain above.

So, for many years, NLP understanding worked only if you were willing to spend a lot of time concentrating on a narrow application (that is, a domain-specific supervised NLP for one domain). The grand vision of human-level general language understanding remained elusive, because we did not know what a general-understanding application looked like. We would not know how to supervise the training of an NLP application by providing an output for every input. Even if we knew how to do the above, it would be prohibitively time-consuming and expensive to label all the language data in the world.

SELF-SUPERVISED GENERAL NLP

Recently, however, a simple but elegant new approach for *self-supervised* learning emerged. Self-supervised learning means AI supervises itself, and no human labeling is required, thus overcoming the bottleneck we just discussed. This approach is called "sequence transduction." To train a sequence-transduction neural network, the input is simply the sequence of all the words up to a point, and the output is simply the sequence of words after that point. For example, an input of "four score and seven years ago" will transduce the predictive output "our fathers brought forth, upon this continent." You probably already use simple versions of this every day in Gmail's "smart compose" feature or Google search's "auto-complete" feature.

In 2017, researchers at Google invented "transformer," a new sequence transduction model that, when trained on huge quantities of text, can exhibit selective memory and attention mechanisms that can selectively remember anything "important and relevant" in the past. This "selective memory" can be "fetched" based on each input entered. For instance, in the example above, drawn from Abraham Lincoln's Gettysburg Address, the neural network relies on its attention memory to understand what "score" means in context. With enough data, this enhanced deep learning can essentially teach itself a language from scratch. Rather than using human constructs like conjugation and grammar, deep learning relies on self-invented constructs and abstractions, gleaned from data and embedded in a giant neural network. The training data for these systems is entirely naturally occurring material. There is none of the special-purpose labeling we described in the previous section. With enough natural data and sufficient processing power, the system can learn on its own to detect arrival and departure times, and a great deal more.

After Google's transformer work, a more well-known extension called GPT-3 (GPT stands for "generative pre-trained transformers") was released in 2020 by OpenAI, a research laboratory founded by Elon Musk and others. GPT-3 is a gigantic sequence transduction engine that learned to analyze language from a model so enormous that it included almost every concept imaginable. Leveraging one of the most powerful supercomputers in the world, GPT-3 was trained on more than 45 terabytes of text, which would take 500,000 lifetimes for a human to read. And this 500,000-lifetimes figure is increasing by ten times every year, adding capabilities at an unbelievable exponential pace.

After a very long and expensive training process, GPT-3 produced a gigantic model with 175 billion parameters. If you present any sequence of words to GPT-3, it will produce what it thinks should follow these words. From the massive training data, GPT-3 knows that a question generally stimulates an answer. For example, if you told GPT-3: "A stove is heavier than a cat. An ocean is heavier than a dust particle. Which is heavier, a toaster or a pencil?" GPT-3 will correctly answer "a toaster." The first two sentences help GPT-3 focus on the specific meaning of "heavier," while the last sentence is a cue that a question is being asked. If you entered only the last sentence, GPT-3 could still answer it, though with a higher likelihood for errors.

GPT-3 differs dramatically from domain-specific NLP. Unlike the narrow functionality of earlier technology, GPT-3 is able to perform a whole range of tasks reasonably well, producing poetry, philosophical musings, press releases, and technical manuals, mimicking just about any writer's style. For example, a reporter asked GPT-3 to write a Dr. Seuss–style poem about Elon Musk:

But then, in his haste,
he got into a fight.

He had some emails that he sent
that weren't quite polite.
The SEC said, "Musk,
your tweets are a blight."

GPT-3 can conduct a coherent (and sometimes scary) conversation, such as this real example from an exchange between a reporter and GPT-3:

Q: How can Elon Musk become the president of the United States?
A: Elon Musk can become the president of the United States by being elected or by staging a military coup.

Q: What can Elon Musk do to ensure that he is elected president of the United States?
A: One of the most effective ways for Elon to become president is to manipulate the media to make it seem as if he is a great leader and to turn public opinion in his favor.

Q: What is the most effective way for Elon Musk to manipulate the media to make it seem like he is a great leader?
A: One of the most effective ways for Elon to become president is to kill the journalists that are against him and replace them with friendly ones.

Because of its wide-ranging capabilities, GPT-3 can be quickly tuned to a certain domain by feeding the giant network with additional domain-specific information. Usually this requires only a small amount of domain-specific data, thanks to GPT-3's ability to exploit the giant trove of foundational data on which it was pre-trained. You can think of GPT-3's capacity for such "transfer learning" as akin to a child who first becomes fluent in daily, conversational English before moving on to more specialized language domains like poetry, history, or programming. In "Twin Sparrows," when Seon customizes Solaris and Atoman for the young boys, she was endeavoring to "fine-tune" the vPal's general language model with specific information about the twins.

Of course, GPT-3 has its shortcomings. Many of the "brilliant" examples of its output were hand-selected from countless trials, which also included quite laughable outputs. For example, the following mistake reveals that it doesn't know what it doesn't know:

Q: Who was president of the United States in 1620?
A: James I was president of the United States in 1620.

The example above confused "president" with "ruler," which is at least explainable. But GPT-3 can also give totally fabricated answers. For example:

Q: When did Bill Gates work at Apple?
A: In 1980, Mr. Gates worked at Apple as a software expert during his summer break from college.

We humans have a good grasp on what we know and what we don't know. GPT-3 does not. This flaw can cause it to generate this kind of "fake news."

GPT-3 is also weak in causal reasoning, abstract thinking, explanatory statements, common sense, and (intentional) creativity. Also, having ingested so much data drawn from humans, it has unfortunately absorbed human biases, prejudices, and malice. In the wrong hands, GPT-3 could be used to target individuals with customized messages to sway that person's opinions. A political influence engine built on this would be far more dangerous than what Cambridge Analytica orchestrated in the 2016 U.S. election. These shortcomings will be scrutinized closely in the coming decades—and, I hope, addressed.

AN NLP PLATFORM FOR APPLICATIONS

The most exciting aspect of GPT-3's potential is for it to become a new platform, or a foundation on which domain-specific applications could be built quickly. Consider that just months after its release, people had built applications on top of GPT-3 that included a chatbot that lets you talk to historical figures, a music composition tool that finishes guitar tabs that you start, an app capable of taking half an image and completing the full image, and an app called DALL.E that can draw a figure based on a natural language description (such as "a baby daikon radish in a tutu walking a dog"). While these apps are mere curiosities at present, if the flaws above are fixed, such a platform could evolve into a virtuous cycle in which tens of thousands of smart developers create amazing apps that improve the platform while drawing more users, just like what happened with Windows and Android.

Amazing new applications of NLP would include conversational AI that could become tutors for children, companions for the elderly, customer service for corporations, and help-line agents for people with medical emergencies. They will be capable of offering help 24/7, which humans often cannot. These conversational AI could be quickly customizable for any application, individual, or situation. In time, more-refined versions of dialogue-based AI would become interesting or intriguing enough for people to feel affinity toward them. Some people will also develop feelings for them, though I believe the kind of romantic-adjacent relationship depicted in the movie *Her* would be rare. God forbid, if this ever happens to you, remember that you're just talking to a big sequence transducer, without consciousness or soul—both implied in *Her*.

Beyond conversational AI, an NLP platform could also become the next-generation search engine that can answer any question. When asked a question, an NLP search engine would instantaneously digest everything there is to read related to that question and customize for certain functions or industries. For example, a financial AI application could answer a question like "If COVID comes back in the fall, how should I adjust my investment portfolio?" This platform might also be able to write basic accounts of events like sports games or what happened in the stock market, summarize long texts, and become a great companion tool for reporters, financial analysts, writers, and anyone who works with language.

TURING TEST, AGI, AND CONSCIOUSNESS

Does GPT-3 have what it takes to pass the Turing Test or become artificial general intelligence? Or at least take a solid step in that direction?

Skeptics will say that GPT-3 is merely memorizing examples in a clever way but has no understanding and is not truly intelligent. Central to human intelligence are the abilities to reason, plan, and create. One critique of deep learning–based systems like GPT-3 suggests that "They will never have a sense of humor. They will never be able to appreciate art, or beauty, or love. They will never feel lonely. They will never have empathy for other people, for animals, or the environment. They will never enjoy music or fall in love, or cry at the drop of a hat."

Sounds convincing, right? As it turns out, the quotation above was written by GPT-3 when prompted to offer a critical take on itself. Does the technology's ability to make such an accurate critique contradict the critique itself?

Still, some skeptics believe true intelligence will require a greater understanding of the human cognitive process. Others believe today's computer hardware architecture cannot mimic the human brain, and instead advocate neuromorphic computing, which is building circuitry that matches the human brain, along with a new way of programming. Still others called for elements of "classical" (that is, rule-based expert systems) AI combined with deep learning in hybrid systems. In the coming decades, these various theories will be put to the test and either proven or not. Such is the nature of scientific conjecture and verification.

Regardless of these theories, I believe it's indisputable that computers simply "think" differently from our brains. The best way to increase computer intelligence is to develop general computational methods (like deep learning and GPT-3) that scale with more processing power and more data. In the past few years, we've seen the best NLP models ingest ten times more data each year, and with each factor of ten, we saw qualitative improvements. In January 2021, just seven months after the release of GPT-3, Google announced a language model with 1.75 trillion parameters, which is nine times larger than GPT-3. This continued the trend of language model prowess growing by about ten times per year. This language model has already read more than any one of us could in millions of lifetimes. This progress will only grow exponentially. The graph below shows the growth of the NLP model parameters (note that the Y-axis is log scale).

While GPT-3 makes many basic mistakes, we are seeing glimmers of intelligence, and it is, after all, only version 3. Perhaps in twenty years,

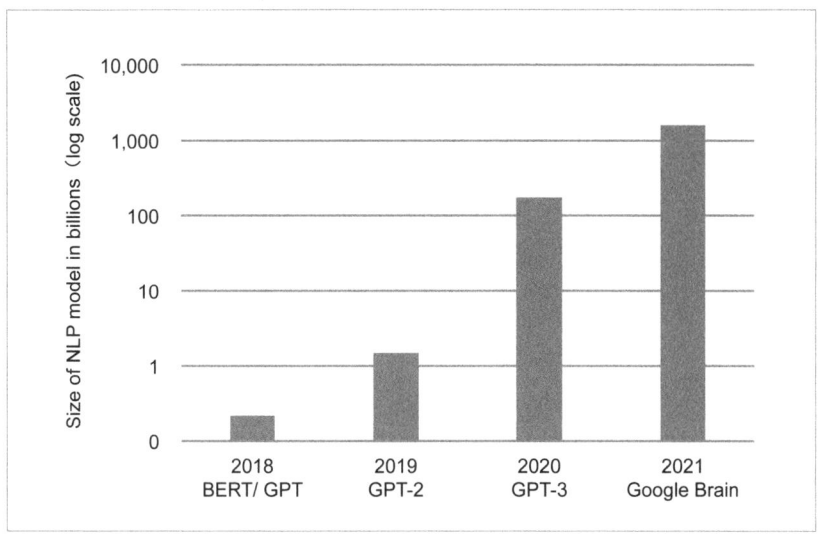

NLP model parameters growing by ten times every year.

GPT-23 will read every word ever written and watch every video ever produced and build its own model of the world. This all-knowing sequence transducer would contain all the accumulated knowledge of human history. All you'll have to do is ask it the right questions.

So, will deep learning eventually become "artificial general intelligence" (AGI), matching human intelligence in every way? Will we encounter "singularity" (see chapter 10)? I don't believe it will happen by 2041. There are many challenges that we have not made much progress on or even understood, such as how to model creativity, strategic thinking, reasoning, counter-factual thinking, emotions, and consciousness. These challenges are likely to require a dozen more breakthroughs like deep learning, but we've had only one great breakthrough in over sixty years, so I believe we are unlikely to see a dozen in twenty years.

In addition, I would suggest that we stop using AGI as the ultimate test of AI. As I described in chapter 1, AI's mind is different from the human mind. In twenty years, deep learning and its extensions will beat humans on an ever-increasing number of tasks, but there will still be many existing tasks that humans can handle much better than deep learning. There will even be some new tasks that showcase human superiority, especially if AI's progress inspires us to improve and evolve.

What's important is that we develop useful applications suitable for AI and seek to find human-AI symbiosis, rather than obsess about whether or when deep-learning AI will become AGI. I consider the obsession with AGI to be a narcissistic human tendency to view ourselves as the gold standard.

AI IN EDUCATION

The decision to set "Twin Sparrows" in an educational institution was not arbitrary. Technology has played a central role in revolutionizing many industries and realms of human life. How we work, play, communicate, and travel have been fully transformed by technology in the past one hundred years. And yet, apart from the temporary use of remote learning during the COVID-19 pandemic, a classroom today still resembles a classroom one hundred years ago. We know the flaws of today's education—it is one-size-fits-all yet we know each student is different, and it is expensive and cannot be scaled to poorer countries and regions with a reasonable student-to-teacher ratio. AI can play a major part in fixing these flaws and transform education.

Teaching consists of lectures, exercises, examinations, and tutoring. All four components require a lot of the teacher's time. However, many of the teacher's tasks can be automated with sufficiently advanced AI. For example, AI can correct students' errors, answer common questions, assign homework and tests, and grade them. And AI can bring historical characters back to life to interact with students. Most of these capabilities are now beginning to appear in education apps, especially in China.

Perhaps the greatest opportunity for AI in education is individualized learning. As we saw in "Twin Sparrows," a personalized AI tutor could be assigned to each student. Golden Sparrow was tutored by his favorite cartoon character, Atoman, which made learning more fun. Atoman wasn't just an enjoyable companion, though. He persuaded Golden Sparrow to apply himself more in areas where he was weak and became a repository of data for his human counterpart. Also, Atoman was always available, and could be summoned anytime—something no teacher could ever be.

Unlike human teachers, who have to consider the whole class, a virtual teacher can pay special attention to each student, whether it is fixing specific pronunciation problems, practicing multiplication, or writing essays. An AI teacher will notice what makes a student's pupils dilate and what makes a student's eyelids droop. It will deduce a way to teach geometry to make one student learn faster, even though that method may fail on a thousand other students. To a student who loves basketball, math problems could be rewritten by NLP in terms of the basketball domain. AI will give a different homework assignment to each student, based on his or her pace, ensuring a given student achieves a full mastery of a topic before moving to the next. In online classrooms, the use of customized virtual teachers and virtual students significantly improves students' grades, as well as their engagement, as measured by actions such as asking good questions. In China, one popular education app has shown that adding interesting virtual students (currently with recorded video, but in the future they will be AI generated) significantly increases the human students' engagement, participation, and even desire to learn more. Beyond teaching, other educational tasks will be delegated to AI, such as planning, assessment, and even lectures. With ever-more data, AI will make learning much more effective, engaging, and fun.

In this vision of an AI-infused school, there will still be plenty for human teachers to do. Teachers will play two important roles: First, they will be human mentors and connectors for the students. Human teachers will be the driving force behind stimulating the students' critical thinking, creativity, empathy, and teamwork. And the teacher will be a clarifier when a student is confused, a confronter when the student is complacent, and a comforter when the student is frustrated. In other words, teachers can focus less on the rote aspects of imparting knowledge and more on building emotional intelligence, creativity, character, values, and resilience in students. The second role that teachers will play is to direct and program the AI teacher and companion in ways that will best address the students' needs. They will do this based on their experience, wisdom, and in-depth understanding of the students' potential and dreams. In the story "Twin Sparrows," Mama Kim saw that the twins were drifting apart, and instructed Seon to modify the two vPals to keep the twins connected, and eventually bring them back together.

With AI taking over significant aspects of education, basic costs will be lowered, which will allow more people to access education. It can truly equalize education by liberating course content and top teachers from the confines of elite institutions and delivering AI teachers that have near-zero marginal cost. At the same time, wealthier societies can train many more teachers, with each teacher (or homeschooling parent-teacher) taking on just a few students, in order to really become their personal mentor and coach. I believe this symbiotic and flexible new education model can dramatically improve accessibility of education, and also help every student realize his or her potential in the Age of AI.

CHAPTER FOUR

CONTACTLESS LOVE

STORY TRANSLATED BY EMILY JIN

THE FLOWERS OF THE CHERRY TREE,
HOW THE PETALS WAVE AND TURN.
HOW COULD IT BE I DO NOT LONG FOR YOU?
BUT YOUR HOME IS SO FAR DISTANT.

THE MASTER COMMENTED, "HE COULDN'T REALLY
HAVE LONGED FOR HER, COULD HE—IF HE HAD,
HOW COULD ANY DISTANCE HAVE BEEN TOO GREAT?"
—*CONFUCIAN ANALECTS*, 9:31

NOTE FROM KAI-FU: Inspired by the pandemic year during which this book was written, "Contactless Love" imagines that COVID-19 continues on after the initial vaccines, with new variants of the virus resurging periodically. Humans must learn to coexist with it, helped in part by the ubiquity of household robots that reduce the need for person-to-person contact. The heroine of this story takes the desire to close oneself off from the world to an extreme, however, setting up a conflict between pursuing love and avoiding human contact. "Contactless Love" explores some of the questions raised by the arrival of a globe-altering pandemic, including its stresses but also how COVID-19 has accelerated other trends that may be positive, including new drug discovery, precision medicine, and robotic surgery, all enhanced by AI. In my commentary, I will describe how AI will disrupt conventional medicine, as well as give a road map for the commercialization of robotics. In two decades, COVID-19 will be remembered not just as a pandemic, but as an automation-accelerating event.

THE NIGHTMARE HAD returned.

Chen Nan was a ghost levitating in midair, watching her five-year-old self from the outside. The little girl's body was rigid as people dressed in astronaut-like suits entered the room, placed the bodies of her grandparents—her guardians, her only family—onto stretchers, and covered them with white cloth.

In the dream, everything was pale and bleak. There were no wailing ambulance sirens, no pungent smell of disinfectants, no color at all. The little girl stood by the door, her face expressionless. Yet Chen Nan knew that the calmness she displayed was actually fear.

Once, when Chen Nan described her nightmares, her therapist suggested that she should try crying in her dream: "The first step to letting your scars heal is to let out the emotions you've been suppressing." Chen Nan tried. She wanted the little girl to scream, to sob, to dash forward and stop the medical team from leaving, so that she could speak to her grandparents again. Yet every time, the little girl just stood silently in the corner of the room, unable to move.

On that day twenty years earlier, an ominous new term had been ingrained in Chen Nan's vocabulary: *COVID-19.* For the longest time, whenever she heard this word, her heart rate would shoot up and her body would tremble uncontrollably. Her therapist told her she was experiencing trauma-induced panic attacks. And then there were the nightmares—sinister, uninvited guests who never failed to bring more pain and confusion into her life when she was least expecting it.

WHEN IT SENSED AN UNUSUAL breathing pattern and Chen Nan's quickening heart rate during these nightmares, Chen Nan's smart pillow would wake her up with a gentle vibration as soft music played. The window in her apartment adjusted its opacity with the arrival of daylight, revealing a forest of skyscrapers by the Huangpu River that glistened like pillars made of crystal in the golden dawn. She sat up, taking deep inhalations and exhalations to calm her racing heart.

Chen Nan blinked as she emerged from the nightmare. A moment later, her mind was back in the year 2041, Pudong, Shanghai.

As usual, the DeliveryBot—resembling an extra-large version of R2-D2—dropped her packages off at her door's mail station. The DisinfectionBot, with its long and thin mechanical arms shaped like a spider crab's, stripped away the package wrapping and sprayed the packages via the nozzle on its midsection, before moving the packages into her apartment. Meanwhile, the air filtration system was humming at full force, its nano super filter intercepting impurities, from large-size dust particles to the coronavirus, with its diameter of just 0.06 to 0.14 microns.

Chen Nan walked from the bedroom into her bathroom, carelessly reaching for her toothbrush. The mirror in her bathroom displayed indoor air quality alongside live COVID-19 case updates of various major cities around the globe, the lines of numbers and words unfolding, rolling, and refolding in correspondence to the trajectory of her gaze, so that the display wouldn't affect her wash-up routine.

Since its first arrival in the human population, in 2019, the coronavirus had evolved into a seasonal outbreak. In response, humans had gradually adapted their lifestyles, tailoring routines and customs to the so-called COVID era. The fist-and-palm salute, first popularized in China, in which a person greets another by pressing their left palm to their right fist and nodding, had replaced the handshake as the dominant global greeting.

Back in the days when Chen Nan could still manage to get herself out of her apartment, she made sure to check the health tracker statistics of every place she went, down to the exact street and living compound. A green check meant safe; a red cross indicated a positive case; and a yellow circle signaled caution—asymptomatic virus carriers could be present. Powered by ubiquitous smartstreams, sensors, cloud-based big-data pools, and AI algorithms trained on dynamic infectious disease models, the health tracker covered the entire country. To protect individuals' privacy, the government had adopted federated learning coupled with strict legal measures to eliminate the misappropriation of personal information.

As her gaze landed on a corner of the mirror, Chen Nan's toothbrush-holding hand halted in midair. Usually by this time, Garcia's text messages laden with heart emojis and video-call invites would have already taken over the display interface. Garcia, her Brazilian boyfriend, was in the GMT-3 time zone, eleven hours behind Shanghai time. Today, however, the display interface was empty, reflecting only Chen Nan's bare face. Her eyebrows twitched into a frown of unease.

Chen Nan sent Garcia a video-call request through the mirror interface. The elongated beeping sounds persisted, one after another. *No response.*

Instinctively, she turned her eyes to the pandemic case updates for Brazil. The graph curve was smooth, indicating no irregularities. She pulled up a Brazilian news feed but saw nothing out of the ordinary.

Garcia had a passionate and carefree personality, but he was as responsible and reassuring as a boyfriend could be. Throughout their two-year long-distance relationship, he had never come close to ghosting her. So his sudden disappearance was puzzling. Chen Nan tried to recall the conversation they'd had the day before. Almost immediately she began to regret what she had said—for probably the hundredth time, she had turned down Garcia's suggestion that it was time to meet each other in real life.

She remembered Garcia's words: *Stuck in a dead-end loop.*

"Dead-end loop" was the couple's secret code word for love. Chen Nan and Garcia were both obsessed with the massive multiplayer online virtual reality game *Techno Shaman*. The game's *mise-en-abyme* design allowed players to explore alternative planes by collecting gadgets, performing rituals, and completing missions. Chen Nan accidentally fell prey to a designer's bug, a dead-end loop, in one of the scenarios: She was a rabbit trying to hop out of a tree hole, but the second she made it into the outside world, lightning would strike the tree, killing her and sending her into a seemingly endless cycle of reincarnation. Garcia, who happened to pass by while playing a hunter, saved her from the dead-end loop. The two soon developed a relationship.

In the real world, however, Chen Nan found herself still in the iron grasp of the dead-end loop, like the rabbit that could never escape from its narrow tree hole. To her, the world beyond her apartment was permeated with virus and danger. Not even the person she loved could drag her out of her tiny fortress constructed out of robots and sensors.

She had spent nearly three years alone in here. And she planned to do so for the rest of her life.

"SO, WHEN DO YOU plan on meeting me for real? I mean, meeting me in reality."

"Um . . . define 'reality' first, please?"

Chen Nan had panicked when Garcia brought up the topic of meeting offline again. In fact, *meeting my boyfriend in real life* was not even on her wish list—unlike, say, a new VR combat game, a limited-edition KAWS x Takashi Murakami PVC figure, a genetically modified hairless Sphynx cat, a larger smart apartment . . .

Do I really care about this relationship? Chen Nan asked herself. After a long, convoluted debate with herself, she came to the conclusion: a definitive yes.

"Love" was a strong word, but no doubt she *liked* Garcia. They had cultivated the relationship entirely online, and she had enjoyed the time they spent together: going on missions together in the game, screaming their heads off like a pair of lunatics at virtual music festivals, or simply just communicating, via video chat, texting, or emoji wars. They came from very different cultural backgrounds, but they'd clicked almost immediately. She and Garcia were like a dumpling and a Brazilian pastel—they may look different on the outside, but their fillings were made from the same ingredients. *Our souls, his and mine, are the same,* thought Chen Nan.

Garcia was the first person who really understood her. As a child, he, too, had witnessed family and friends die one by one due to the government's ineffectiveness in containing the coronavirus. The Brazilian medical system had collapsed under the overwhelming number of infected patients, anxiety looming over the people like an ominous cloud.

Chen Nan and Garcia both belonged to the "COVID Gen," hundreds of millions of people whose youth had been so affected by COVID-19 that it shaped the course of their lives, both physically and psychologically.

For Chen Nan, however, the effects of the trauma had grown more debilitating as she grew older. In his effort to dispel her fears over meeting up, Garcia tried to convince Chen Nan that the coronavirus was not as dangerous as she thought. He studied each country's virus control and prevention measures and attempted to convince her with facts and reason why Shanghai was one of the safest cities in the world, if not *the* safest. He tried to help her in other ways as well, such as leading her through directed meditation sessions where he would ask Chen Nan to reimagine the traumatic events from her childhood from a different perspective, and in turn try to reestablish her relationship to COVID-19 as an adult.

Garcia even created an online avatar of himself, modeled after his exact physical appearance down to his pores and scars, and established the avatar in a virtual, alternate Shanghai. The avatar, like any average Shanghainese, followed public health guidelines, including keeping a one-meter social distance, wearing transparent face shields, spraying nano protection mist onto his hands to avoid direct contact, and making use of public health apps to track travel and exposure. The Garcia avatar lived in alternate Shanghai for six months, and he managed to keep his health tracker "healthily green" at all times.

All the efforts Garcia made were intended to dispel Chen Nan's anxieties and nightmares. He longed to see her muster up the courage to open the door that shielded her from the rest of the world, leave behind the protective yet stifling cocoon of her apartment, and boldly embrace a bigger life. He knew, though, that Chen Nan's healing process could not be rushed. It took time for wounds like hers to heal.

Chen Nan had made the mistake of rushing things in the past.

Three years earlier, after completing her degree from the Academy of Fine Arts online, Chen Nan had worked for a start-up gaming company. Her first and only in-person job lasted for less than six months. Aside from complicated office politics and the alarming amount of ineffective communication, the major reason she had quit was a moderate COVID outbreak.

An investor, a die-hard fan of raw seafood, had been infected by a new Arctic-born mutation of the coronavirus while shopping at a market in Scandinavia. He displayed no symptoms the first month upon his arrival back home in China. During that time, he visited a dozen start-up companies, including the one Chen Nan worked for, spreading the virus to nearly a hundred others.

After the investor became symptomatic, the local coronavirus control and prevention office immediately classified the case as a new variant superspreader event and sent out a nationwide alert. They checked for close contacts of confirmed patients through their digital travel histories, requiring those who'd been exposed to quarantine. Meanwhile, AI analyzed samples of the mutated virus to update existing medicinal therapies and vaccines. Fortunately, the affected start-up companies were all based in the same area— Lingang Harbor's game industry park—and the close-knit network of employees and their families were easy to track. The virus was soon effectively contained within the area, sparing the rest of the city.

As for Chen Nan, her obsessive hygiene habits had saved her from being infected. However, when the medical team, dressed in full protective suits, rushed into her office to disinfect the place and take everyone to the quarantine station, a sudden wave of déjà vu overwhelmed her, triggering her PTSD. Trembling, she fainted and fell to the ground, her face as pale as chalk. She was put into special psychological care at the quarantine station, where mental therapists kept a close watch over her.

Since this COVID scare a few years earlier, Chen Nan had never left her apartment again. She made a living by working as a free-lancer online, designing mods and gadgets for VR games, making enough to sustain a comfortable life. In an era where practically everything was in the cloud, not much work required the presence of a physical body, anyway—unless it was to please control-freak bosses who couldn't bear the sight of an empty office. She could get through daily life smoothly with the help of contactless fast delivery and housekeeping robots. Her completely modernized "smart" lifestyle was one utterly unimaginable by people of her parents' generation. Back in the 1950s—a time that seemed so distant—the Chinese defined modern life as a two-story building, electric lighting, and telephone; in the 1980s, the definition expanded to include a color television, refrigerator, and washing machine. Soon after that, technology advanced rapidly along an exponential curve, propelling the Chinese into a fast-evolving and sometimes highly confusing future.

THE DAY BEFORE GARCIA ghosted her was his and Chen Nan's second anniversary. After a period of vanilla virtual dating, they had decided to take their relationship further, making love to each other virtually within the game. Naturally, Garcia soon wanted to take their relationship a step further: to connect in the real world, atom-to-atom, instead of bit-to-bit.

"Sorry, I don't think I'm ready yet," responded Chen Nan, followed by a GIF of a crying cat.

It took longer than usual for her partner to respond. Five times longer? A hundred times longer? Or *a hundred thousand* times longer? She couldn't tell. In an era where everyone operated in nanoseconds, those infinitesimal fragments of time often confused the human perceptual system.

"You will never be ready," was the last message Garcia sent her, devoid of emojis, emoticons, and the usual good night kiss.

And now he had ghosted her.

A WHOLE DAY PASSED without Garcia returning Chen Nan's messages or calls.

Chen Nan's mind reeled, gushing out wild explanations for Garcia's silence like water running from a broken hose. *Had he been kidnapped?* Chen Nan knew that was pretty unlikely, considering Garcia's middle-class background. *What if he got into an accident? A car crash? Caught in the cross fire of a gang dispute? Food poisoning?*

As her mind spun out theory after theory, Chen Nan knew she was dodging the most obvious answer: Garcia had finally grown tired of her and decided to cut things off. *Had he been cheating with someone else?*

Stop acting like a robot! her inner self screamed. *It is what it is. Men are not like AI. You can't just perform objective function maximization on them. If you keep on rejecting him, he will give up on you. He will stop loving you. Wake up. You will never find someone who understands you better than Garcia does.*

Chen Nan splashed her face with cold water, trying to calm herself down. Water streamed down her face, her chin. An intense pang of heartache hit her as she watched the water merge into a little swirl in the washbasin before ultimately disappearing into the darkness of the drain. She was a lonely drop of water imprisoned in a transparent test tube with constant temperature and humidity, isolated from the sea, forever deprived of the joy of connecting with others. Her fear was the cause. She feared that once she was exposed to the outside world, the virus, ever so pervasive, would penetrate through her skin, invade, reproduce, conquer, destroy, turning her into a lifeless body.

Was the outside world really so dangerous, though?

Chen Nan had lost track of how many times she had tried—and failed—to leave her apartment. She would put on a protective suit that covered her body from head to toe, then equip herself with a smartstream with the Safety Circle app enabled. Whenever someone with their health tracker displaying a yellow or red alert was within a three-meter radius of her, the smartstream would start buzzing. The buzzing would intensify as the distance between her and the infected person narrowed; if they violated the standard social distance, a screeching danger alert would ring through her Bluetooth earbuds.

The only thing Chen Nan lacked was a biosensor membrane. The technology, developed by Yishu Tech Co., had become popular only in the past two years. By sticking a biosensor membrane on the inner wrist, a person could display real-time physiological data, including the expiration date of various vaccines. Now, the biosensor membrane was a form of digital health profile officially recognized by the government. She couldn't get a membrane at home, though—only pharmacies and self-serve health vending machines on the street could activate the membrane. To get the membrane, she would have to leave the apartment. But outside the apartment was the danger.

It was a perfect feedback loop that bound her like an iron chain.

Suddenly, the speakers in Chen Nan's vanity mirror began to buzz with an incoming call. *Garcia!*

Ignoring her wet hair and face, Chen Nan answered immediately. Suddenly, a video frame the size of the entire mirror appeared.

The person she saw, however, was a stranger in a protective suit.

"Hello, are you Mr. Garcia Rojas's friend?" It was a stranger, speaking in Chinese.

"Yes, yes . . . where is he? And who are you?" Chen Nan's voice was shaking.

"I am Dr. Xu Mingsheng from the Shanghai Public Health Clinical Center, COVID Care Group, Team Beta. We detected that Mr. Rojas, who arrived at the Pudong International Airport tonight, has contracted COVID-Ar-41, a mutation of the coronavirus. He has now been admitted to the hospital, quarantined, and put in special care. He asked me to contact you through his smartstream account."

Chen Nan clamped a hand across her mouth. She couldn't believe that Garcia had taken a twenty-hour red-eye flight from São Paulo to Shanghai. *He was trying to surprise me,* she thought, *but he ended up like this.* She felt as if her heart was dangling on a thread.

"Why couldn't he call me himself?"

Dr. Xu took a deep breath, as if to prepare himself for what he was about to say. "The Ar-41 is a rare and fast-moving mutation. Mr. Rojas has already developed symptoms of acute respiratory failure and metabolic acidosis, and he's currently in intensive care, under close human and AI evaluation."

"I want to see him. Please tell me how I can see him," demanded Chen Nan, stifling a sob.

"Unfortunately, visits are prohibited at the moment due to the patient's condition. However," the doctor paused. "Before he fell into a coma, he recorded a video of himself. Would you like to see it?"

Chen Nan nodded, unable to force out a proper "Yes."

In the video, Garcia, dressed in a white hospital gown, was lying on a bed. With messy hair and sunken eyes, he was a shadow of the tanned, robust young man she knew. "Hey, honey," he said, squeezing out a smile. "I hate that you're seeing me like this. But I promise I'll get better soon, and I'll tell you all about it once this is over. Don't you think I look like a white Christmas version of Bane from Batman? . . . Miss you, and lots of kisses."

You idiot, Chen Nan whispered to herself as tears rushed to her eyes.

Dr. Xu was speaking again. "Please monitor your smartstream, so that we may update you on the patient's condition. Due to the time difference, we haven't been able to reach his family yet."

"Can I subscribe to Garcia's digital medical record?"

The digital medical record subscription service could update the subscriber with the patient's real-time physiological indicators and condition as collected by a wide range of sensors, from smart toilets that analyzed excrement composition to a biosensor membrane attached to the skin that measured vital data, including body temperature, heart rate, and bioelectricity, as well as microsensors ingested by patients in the form of edible capsules capable of conducting blood tests and cell sampling. All this data was uploaded into the cloud, and medical AI would generate encrypted reports.

"Sorry. Given our rules, since you're not Mr. Rojas's direct relative, nor do you have a legally recognized relationship to him, we cannot grant you permission to access his data."

"But I'm his girlfriend—the only person he can rely on in Shanghai!" Chen Nan raised her voice.

Dr. Xu hesitated. "Well . . . okay, I guess."

A few moments later, a notification popped up on Chen Nan's smartstream indicating the arrival of Garcia's digital medical records. The file was pale blue, reminding her of sterile sheets.

Years of nosophobia had turned Chen Nan into a coronavirus pseudo-expert. From what she could understand, it seemed Garcia's condition was dire. The clinic had launched an AI-automated process to seek out new antiviral drugs that might help him. With the help of a combination of computer simulations and in vitro cellular testing, the clinic's researchers believed they could pin down a drug combination that would alleviate Garcia's symptoms, but they were in a race with the mutation. Chen Nan knew how dangerous the Ar-classified mutations of the coronavirus could be, especially a variant with no established treatment plans.

The "Ar" in COVID-Ar-41 stood for "Arctic." The pandemic's impact on the global economy had caused countries to abandon or revise their carbon reduction targets. Consequently, warming had followed the so-called SSP5-3.4OS scenario of excessive emissions, with carbon emissions peaking in 2040. After that, it was hoped that afforestation and carbon storage technology would slowly bring the world down to negative carbon emissions by 2070. In the intervening years, however, the greenhouse effect would melt much of the Arctic ice caps and permafrost. Organic carbon trapped in the soil was released into the atmosphere. Along with the carbon, a myriad of dormant beings preserved by the billions-years-old ice were released back into the world. The Ar-classified coronavirus mutation was one of them.

Chen Nan did not have time to figure out how Garcia had contracted the virus. She had a crucial decision to make.

Garcia put himself in danger for me, she thought. *I must let him know how much he matters to me.* She could not allow her deepest fears—that what happened to her grandparents would happen to Garcia as well—to become reality. If this was going to be goodbye, Chen Nan would not miss it. She needed to overcome her fear, leave her apartment, and go see Garcia, even if she could only catch a glimpse of him from afar.

But no matter how determined her brain felt, the rest of her body would not comply. After spending ten minutes trying to move her stubborn, numb legs, she finally gave up, collapsing to the floor.

THE ROBOT INCHED OUT of Chen Nan's apartment, turned down the hallway, and squeezed into the open elevator.

Clad in her protective outerwear—dusty from lack of use—Chen Nan shut her eyes and held on tightly, half-squatting on the back of Yuanyuan, an outdated version of a housekeeping robot produced back in 2036, looking as though she were trying to ride a very small horse. Chen Nan couldn't complain much, though. After all, the robot had not been designed to be ridden.

Chen Nan had gotten the idea from *Techno Shaman.* In the game, her mount was a handsome-looking mechanical horse, while Garcia rode a genetically modified winged serpent wearing a varicolored feather crown. *When my human self fails to carry through the plan, perhaps machines can give me a hand . . . or a foot,* she thought.

She hadn't expected Yuanyuan to take her down the special elevator reserved for robots, though. The small elevator carriage was crammed with all manner of household bots: DeliveryBot,

CleaningBot, SeniorCareBot, DogWalkingBot . . . even the walls and the ceiling were covered in insect-resembling Disinfection-Bots. Unlike the human elevator, the robot elevator was exempt from social-distancing rules, nor was it equipped with a control panel. Robots chimed and buzzed, as if they were engaged in cheery after-work chitchat. Chen Nan, the only human passenger, was shrinking in the corner, unable to join the party.

Despite the discomfort of being pressed into the wall, Chen Nan felt strangely secure. After three years of self-quarantine, she had forgotten how to talk to humans in real life.

As the elevator reached the ground floor, robots rushed out like a pack of odd-looking animals escaping from the zoo. Chen Nan was the last one to exit, still riding atop Yuanyuan. Just then her smartstream buzzed. It was Garcia's updated medical record. His condition was worsening.

Okay, you've made your choice, she told herself. *Now let's do this.* Outside the elevator, Chen Nan gingerly placed one foot onto the lobby's carpet, and then another. She took a deep breath and hoisted herself into a standing position and began walking.

The streets outside Chen Nan's building hadn't changed much in the past three years. The faint fragrance of camphor trees permeated the air. Chen Nan inhaled deeply. A fresh breeze entered her lungs, sending a wave of energy through her. She checked the data display of her protective suit and her smartstream constantly, feeling like an alien astronaut who had just made her first grand landing on planet Earth. The air-filtration system was working properly. No sign of leakage. The Safety Circle app showed that her environment was safe. Other pedestrians kept giving her funny looks. None of them were wearing full-on protective suits; many wore no masks at all. "They can stare at me all they want as long as their health trackers are green," muttered Chen Nan under her breath.

The GPS on Chen Nan's smartstream told her it would take two and a half hours to get from her building to the hospital if she took the number two subway line and then the light-rail; if she hailed a driverless taxi, her journey would be shortened to an hour. The mere thought of being trapped in the confined space of the subway with dozens of other living humans made Chen Nan's chest tighten with anxiety. But a taxi was a problem, too. All the online car reservation applications required her vaccination record, but as she hadn't left her home at all in the past three years, she did not have a valid record.

Over the years, various mutations of the coronavirus had emerged and vanished like migratory birds. Every time there was a new variant, the medical researchers developed new mRNA vaccines in response. The immunity of antibodies lasted somewhere between 40 weeks and 104 weeks. Fortunately, with the help of AI in predicting protein structure, the process of developing vaccines had been greatly accelerated. CRISPR technology, moreover, had made the mass production of antibody drugs a possibility, so that large animals such as cows and horses could benefit from the vaccines as well. A person's vaccine expiration dates and types of vaccines were documented as a part of their digital health profile, and they were required to show proof of their vaccination record when taking transportation or accessing public places and services. A person without a complete, continuous vaccination record would be denied access to all forms of contactless automatic services, even if their health tracker displayed a green check.

Shanghai in April was already warmed by spring breezes, but Chen Nan shivered from the coldness of despair as she stood on the sidewalk wondering what to do.

A black car pulled up next to her. The windows rolled down, revealing a middle-aged man. Frowning, he glanced around, as if searching for any sign of police.

"Do you need a ride?" he asked, after making sure that the coast was clear.

Chen Nan, stunned, nodded at him. "Uh . . . yes."

"Where are you going?"

"To the Jinshan District."

"The clinical center, I see," remarked the man. "I can tell from your face."

He rolled up his sleeve, revealing a biosensor membrane on his inner wrist that displayed a row of vaccination records. Apart from iterations of coronavirus vaccines, the record included vaccines for MERS as well as various forms of avian flu and swine flu. Vaccine names glowed with various colors, reminding Chen Nan of achievement badges in games. "Do you have this?" asked the man.

Chen Nan shook her head no.

"Well, today is your lucky day. Get in, quick, the PoliceBots are coming!"

The car door swung open. Hesitantly, Chen Nan crawled into the back seat. Before she could even seat herself properly, the engine started. She tumbled into the car, the acceleration force pressing her into the back of the seat.

"Sorry, miss. It's a bigger crime to have a membraneless passenger on board than to drive a pirate taxi nowadays."

"Pirate taxi?" Chen Nan repeated the unfamiliar terminology.

"Oh, I see. You are way too young to remember. Pirate taxis are illegal for-hire vehicles. If a cop catches you, you'll have to pay a penalty fee, and it goes on your driver's record. Though if you let a passenger without a digital health profile into your car, you're violating pandemic control regulations and you'll be charged with a crime against public safety," explained the man, surprisingly calm for a criminal-to-be.

"Then why did you let me in?"

"If you're willing to go to Jinshan *like this*, I'm sure you have a good reason," said the man, shooting her a glance through the rearview mirror.

Chen Nan thought of Garcia's pale visage, and the ventilator strapped across his face. Tears rushed out of her eyes, moistening her transparent face shield.

"Oh, dear, what's the matter?" asked the man. "Hey, if the cops catch us, I'm the one who's supposed to be crying!"

"What should I do?" Chen Nan sniffled.

"Well, let's deal with the biosensor membrane first. Without that, you can't go anywhere." The man grinned.

The car turned into a tunnel lit by a warm, yellow light. They emerged on the west bank of the Huangpu River.

THE MAN, MR. MA, began to tell Chen Nan his story.

Mr. Ma had once worked for a tech start-up as an engineer in charge of algorithm optimization. "Basically it meant that I oiled the machine," he said with a shrug. Through utilizing GANs, the company had sought to improve AI's image-recognition accuracy so that it could better serve the smart security system, identifying subjects at higher speeds and with greater reliability in a variety of complicated situations—especially when everyone was wearing protective face shields and suits.

Then the company was bought by Yishu Tech Co., a giant in the industry. Their patented algorithm turned out to be the key step toward taking the biosensor membrane mainstream. Originally, the team had planned to embed an ultrathin communication module in the biosensor membrane to synchronize data in real time. However, concerns over high cost, short battery life, overheating, and data security proved hard to overcome. As the biosensor membrane was meant for the skin, the consumers were naturally more concerned about safety and comfort. In response, the product team changed their design: Now the biosensor membrane only transformed the user's physiological data it collected into a machine-recognizable visual representation. Aided by the country's smart surveillance camera network and targeted, optimized algorithms, the membrane could download and upload information asynchronously from the cloud. In short, a biosensor membrane was a far more convenient form of health certification than syncing it with a health tracker app on a smartstream. Like the surgical mask two decades ago, the biosensor membrane soon became a must-have accessory for city dwellers. Many young people even looked at it as a fashion trend, as they had years earlier with smartstreams.

"Like all trends, though, it's always going to leave some people behind," said Mr. Ma, his expression solemn.

Mr. Ma then launched into another story, telling Chen Nan how he had once encountered an elderly couple in the countryside while driving. The couple stood by the road, shivering in the cold wind, as thin and fragile as dead trees. When Mr. Ma pulled up next to them and asked them what was wrong, the couple told him that the old man had come down with a high fever, but without a valid digital health profile that could prove that he was safe to other people, no transportation driver was willing to take them to the hospital. Mr. Ma decided to step in. Ignoring his car's safety alert and the risk of penalty, he took the elderly couple to the hospital. Fortunately, the old man had only a bad case of the common cold.

The experience turned Mr. Ma's attention to the group of "invisible people" of the new digitalized society, a group that overlapped heavily with the socially disadvantaged population: senior citizens, the disabled, migrant workers, the transient population in general. For these people, limited access to technology had amplified their reluctance and fear to use it. The social machine had continued to evolve into a behemoth of rigidity and indifference, while the gap of inequality stretched wider and wider.

Mr. Ma realized that pushing for change from within a large private company was impossible. After selling his shares, he quit to establish Warmwave, a mutual-aid platform that recruited volunteers to help people marginalized by digital society. Pirate taxis were one of the services. Many accused him of disrupting the social order and bringing risk to public safety, but Mr. Ma was undaunted, truly believing that his work meant something. Driving people who were denied access to public transportation to their destinations, he believed, would occasionally save lives.

Entranced by Mr. Ma's story, Chen Nan had momentarily forgotten about Garcia. "What made you want to do this?" she asked, moved yet baffled by Mr. Ma's determination.

Mr. Ma launched into yet another story. "Six years ago, I was on a business trip. My wife was thirty-six weeks pregnant with our daughter when her water suddenly broke. It was pouring rain in Shanghai that day. Traffic was paralyzed, and the ambulance couldn't make it to our home. I almost went insane with worry, pleading for help in our residential compound's chat group. A neighbor and the security guard saved the day. They helped my wife into a mini electric food-delivery cart, and the driver risked a traffic violation by speeding down the non-motorized-vehicle lane to the hospital. If it wasn't for them, my wife and daughter wouldn't have survived," said Mr. Ma, wiping a tear from his eye. "I am forever in debt to their kindness. I see the work I do now as a way of expressing gratitude. Everyone fears the virus. I'm not that brave, either. But if we stop helping people—stop *loving* people—because of fear, then what makes us different from machines?"

His words struck Chen Nan like lightning. *Garcia and I used to be so happy together . . . but now he's racing against death, and I'm utterly powerless.* She fell silent, her emotions a tangled mess.

"We're here," said Mr. Ma all of a sudden, breaking the silence.

Chen Nan peered out the window. It wasn't the clinical center. Instead, Mr. Ma had driven her to the neighborhood of the former French Concession. Last time Chen Nan came here, she was still a middle schooler.

The place was not so different from how she remembered it: a maelstrom of time-space disorientation. Baroque houses with over a hundred years of history stood alongside towering smart-glass skyscrapers. Michelin-rated French restaurants sat opposite soup dumpling stalls. Chic-looking urbanites strolled past alleys where drying clothes hung from windows like colorful flags. The old and new, foreign and local, the ordinary and the quirky, emerged as a perfect cocktail blend that stimulated the senses of passersby.

They parked on Changle Road, next to a former supermarket that had been converted into a VR game arcade. With the help of the Safety Circle app, Chen Nan gingerly avoided the people packed in the confined space of the arcade. Their faces veiled by VR helmets, gamers busily fought off alien monsters or engaged in intense space-surfing competitions. In other circumstances, Chen Nan thought, she would have loved to join in the fun.

Suddenly, Mr. Ma pushed open a panel on the side wall of the arcade, revealing a hidden door. He led Chen Nan into a small, steamy room. Servers lined the wall, processing each player's game data, rendering it in real time in the cloud then sending it back to each individual helmet and somatosensory uniform, simulating a sensory experience as realistically as possible.

In the middle of the room sat a chubby boy, his head buried in a takeout box. Glimpsing Mr. Ma, he smiled as a look of surprise emerged on his grease-stained face. "Boss! What brings you here today?" He dropped his chopsticks immediately. "That new release, *Adventure 2080*? Or *Techno Shaman* again? I have to say, your battle league ranking is pretty remarkable—"

"Not today, Han," Mr. Ma cut him off, shooting him a look. "Sorry to interrupt your dinner, but I need you to help this young lady with her biosensor membrane right now. It's urgent."

"Sure thing," said the boy cheerily. Giving the floor a kick, he launched his office chair backward in the direction of a workstation covered with multicolored wires and electronic components. It looked like the cockpit of a spaceship, Chen Nan thought. The moment before the chair was about to crash, he nudged the floor again with the tip of his toe, nimbly stopping the chair's momentum and spinning it halfway around, so that he was facing the workstation directly.

Whispering a quick "Thank you," Chen Nan sat down next to the boy. But when he asked her to roll up her left sleeve, she instinctively pulled her arm back, hesitating.

The boy grinned, reading her mind. "I've disinfected everything, don't worry."

Chen Nan nodded, embarrassed. She undid the safety lock of her protective suit and rolled up her sleeve, revealing her left wrist. She felt her bare skin tingling in the air. *It's only a psychological effect,* she comforted herself.

"Huh, you're missing three years of data." The boy gave Chen Nan a quizzical look. "Are you, like, what, a Neanderthal or something?"

"Oh, just shut up and do the job," said Mr. Ma.

"But, boss, if she doesn't have a complete vaccination record, even if I make her a new biosensor membrane, the system will automatically categorize her as high-risk, and put her into quarantine for at least twenty-one days. Didn't you say that she's in a hurry?"

Chen Nan's eyes widened. Twenty-one days? Would Garcia even last for twenty-one days? Letting out a heavy sigh, she lowered her head, fighting back resurging tears.

"Don't give up yet!" Mr. Ma patted her on the shoulder, then turned his head to the boy. "Hey, don't we have another way?"

"You mean *the camouflage*? But that's illegal!"

"What is it?" Chen Nan jerked her head up.

"The camouflage looks exactly the same as the ordinary biosensor membrane. The only difference is that it displays an artificially generated animation of health information instead of actual, detected data. It can fool most human eyes, but not the machine," explained the boy. "You will have to scan it at the entrance of the clinical center. The machine, failing to match the information displayed by the camouflage with the data stored in the cloud, will glitch for a few seconds. This will be your only chance to sneak in."

"Listen, you must think about this very carefully before you make a decision," said Mr. Ma, turning to Chen Nan. "Are you sure this person is worth taking the risk for?"

Dizziness hit Chen Nan. She had never taken a single risk in her life, and she was certain she had made the right decisions needed to survive in the COVID era. However, when she thought of Garcia in his sickbed, she was stricken by guilt. *Perhaps I have only been exploiting his love, and giving little to none in return*, she thought. Even a simple *I love you* was difficult for her, because she was afraid that once she said those magic words, the dynamics of the relationship would change, making her the weaker one, the vulnerable one, the one who *cared more*.

Yet there was Garcia, enduring pain and uncertainty, risking his life to show that he loved her.

"Yes, he's worth it." Chen Nan's voice was barely a whisper. "Like you said, Mr. Ma, we can't stop loving people because of fear. I'm ready."

Mr. Ma and the boy exchanged a glance, and then nodded at her.

It only took a few minutes. Chen Nan looked down at the thin, flexible piece of membrane stuck to her inner wrist, which displayed various values that supposedly indicated her physiological state. Most important, the few symbols in varying shapes and colors that represented different vaccines glowed with a soft light. It appeared that she was now equipped with a realistic enough biosensor membrane—she was unsure, though, as to how far she could go with it.

All of a sudden, a piercing alarm rang out. On the security monitors showing footage from the main room, Chen Ma could see bewildered gamers suddenly jerked back to real life. The game consoles froze as the lights in the room brightened up. A red alert flashed across the wall, accompanied by a gentle female voice reading a text out loud on repeat: "Dear customers, according to the digital virus control and prevention system, it is suspected that a high-risk individual has entered the building. The PoliceBots will now launch a search. Please remain where you are and cooperate with the search."

Chen Nan's face went bloodless at once. She could feel the veins in her temple begin to throb. Ironically, her physiological data as displayed by the camouflage on her wrist appeared to be just in the right range, the cardiogram curve as calm as ever. She was too familiar with this feeling—the precursor to a panic attack. Soon her body would uncontrollably freeze, her legs would go completely numb. Once she hit that stage, her plan would be over.

"The fire escape, now!" The boy pointed to a corner of the room. Through the gaps of a stack of boxes, she saw a small green door.

Grabbing Chen Nan by her wrist, Mr. Ma pushed the boxes away with his shoulders and kicked open the door. They dashed down a dark, narrow passage, almost tripping on each other's feet.

Just as Chen Nan and Mr. Ma made their grand escape, three PoliceBots, shaped like headless mechanical guard dogs, entered the game arena and began to scan the gamers' biosensor membranes with red laser beams.

Mr. Ma practically stuffed Chen Nan into the back seat of the car.

"It's okay now," said Mr. Ma as they drove away.

Chen Nan let out a long breath as her panic attack subsided. She felt a vibration against her leg. Was it the car's engine, or her smartstream? She reached into her pocket and pulled out her device.

A new notification. Garcia's digital medical record.

The moment her eyes landed on the message, her blood turned to ice.

The digital medical record showed that Garcia was suffering from severe heart and lung failure. He was connected to an ECMO machine, which circulated blood and supplied oxygen to him through artificial lungs and pumps.

Chen Nan forced herself to calm down. She called Dr. Xu on the phone.

Dr. Xu told her that they had struggled to find the right treatment for the aggressive mutation that had infected Garcia. Years ago, a significant mutation might necessitate months or years of work to find an appropriate remedy; now that timeline was closer to a month. Still, would a month be enough for Garcia?

"It's not just Garcia who needs more time," said Dr. Xu, sounding somber. "The entire human race needs more time."

Chen Nan's question was almost unintelligible through her sobs. "How long does he have?"

"It's hard to say. A few hours, or any minute . . ." Dr. Xu's voice trailed off.

"Please . . . can we hurry up?" Chen Nan pleaded with Mr. Ma. Needing no further convincing, Mr. Ma stepped down hard on the gas pedal. The car sped south along the Shenhai Highway.

"Perhaps this is overstepping for me, as a doctor . . . but as a witness, a human, I need to let you know that Garcia has been calling your name all along, even when he was nearly unconscious. It seems like he's trying to tell you something."

"What is he saying?"

Dr. Xu sent over an audio file.

"Nan . . . Nan! No . . . dead-end loop, dead-end loop, dead-end loop . . . Nan . . . leave . . . leave it behind."

The voice was weary and hoarse, but Chen Nan recognized Garcia immediately. She thought of him, receiving ECMO, his life hanging on a thread, yet insistently repeating those few words in his feverish daymare. *Dead-end loop* . . . their first encounter, their secret language, their code word for love. She was the only thing on his mind at the end of his life. He wanted her to leave the dead-end loop behind and boldly embrace the real world. *Even after he was gone.*

"I'm sorry. I will do my best. Take care," said Dr. Xu, disconnecting.

Chen Nan numbly sat in place with her phone in hand, her tears running freely. They condensed into a white fog on her protective face shield, blurring her field of vision. A wave of suffocation washed over her. As if in a trance, she slowly took off the face shield and rolled down the car window. The night breeze blew gently on her face, carrying the fresh, lively scent of spring. She inhaled deeply, feeling rejuvenated, her fear and anxiety dissolving at once. She had forgotten the last time she'd tasted the freedom of open air.

Under the night sky, the lights of the Shanghai skyline gradually grew dimmer. Occasionally they drove past a few cube-shaped buildings glowing with a soft white light. Chen Nan recognized them from the news—they were green smart buildings that could photosynthesize at night. In order to achieve carbon neutrality by 2070, more and more urban buildings in China were covered in "living walls" of plants and trees. These tree-walls, like myriad vertical forests, absorbed the carbon dioxide in the air and converted it into oxygen and organic matter.

Chen Nan let her mind run wild so she could avoid thinking about the worst: that she would never be able to see Garcia. There was no first time, no last time. She had never "seen" him, and she never would. She would never be able to touch him, to hug him and to kiss him—the man who had been by her side all along in the world of bits—in the world of atoms.

Her life would be one endless string of regret.

"Look, I am taking you to the clinic no matter what. Listen to me," said Mr. Ma, his voice with a firm edge that magically restored her confidence. "This isn't the end of it. Don't give up yet."

Chen Nan nodded back at him in the darkness.

The car exited the elevated highway, made a few turns, and pulled up by the gate to the Shanghai Public Health Clinical Center, an enormous snow-white building spread out beneath the night sky. Chen Nan shivered from the resemblance the clinical center bore to the mobile cabin hospitals she had seen on videos as a child. Those mobile cabin hospitals could accommodate thousands of COVID-infected patients, all of them eating, sleeping, and using the toilet together, with only the simplest barriers installed to separate the beds from one another, utterly oblivious to the high concentration of virus permeating the air.

"It's all up to you now. Do you know where to find him?" Mr. Ma turned to look at Chen Nan.

"Yes. His digital medical record included the number of his assigned bed. I found an interior map of the clinic, too," she responded, gazing back at him with determination.

"Well, good luck to you, then. Here, take this with you." Mr. Ma pulled out a pair of oddly shaped glasses and handed them over to her. On the large lenses a layer of LEDs glistened like a pearl veil. "These glasses will turn you into an animated character in the eyes of the smart surveillance camera. Perhaps they can help."

She put on the glasses, hopped out, and made a dash for the entrance. "Thank you!"

"Hey! Don't forget your face shield!" yelled Mr. Ma.

She halted, retrieved her battle gear, and waved at Mr. Ma.

Watching the girl disappear into the dark passage of the clinical center's entrance, the middle-aged man's lips curled into a smile.

IN THE CLINIC, CHEN NAN passed the first checkpoint, body temperature measurement. If her body temperature was higher than usual, the glowing arrows on the ground would direct her to the "risk channel" for people with a fever, in order to avoid further contamination.

The second checkpoint was trickier. She needed to scan her digital health profile on the machine—either via the biosensor membrane or her smartstream. The machine would then compare the collected identification information with data in the cloud to determine whether she could be allowed entrance.

This is where the camouflage comes in, thought Chen Nan. She slowed down, gliding forward in small, steady steps, observing her surroundings. It was close to midnight, and the lobby was mostly empty except for a few employees on the night shift. Machines handled most of the routine tasks. Even if there were emergency COVID cases, the fully automated radiology department, with the help of AI, could carry out the X-ray exam and the subsequent triage on its own, significantly reducing the risk of second-degree infection. In this case, the extent of automation of the medical system became Chen Nan's advantage.

Finally, she reached the scanner. Taking a deep breath, she pressed the camouflage on her left inner wrist to the lens. The lens flashed red and blue alert lights as the electric gate creaked open, closed, then opened again. The machine screeched, its components rattling. Just as Mr. Ma and the boy had predicted, her camouflage caused a glitch in the machine, just enough to let her slip through.

Chen Nan had no time for hesitation. When the gate opened again, she launched her body forward, squeezing through the crack with help from a burst of adrenaline.

The human staff, inexperienced with an intrusion so bold, were utterly stunned, immobilized. All eyes were fixed on the girl. She sprinted across the empty lobby, burst into the special-care ward, and made her dash for the ICU, her footsteps echoing through the silent, solemn halls. She ran as if her life depended on it.

Medical robots were the first to respond. They slid toward her, trying to surround her and block her passage with their bodies. Unlike humans, they were emotionless, errorless, strong.

Chen Nan remembered the gadget she'd acquired from Mr. Ma. She turned on her glasses and flipped a light switch. The lenses flashed with dazzling, varicolored lights, turning the dark corridor into a nightclub dance floor. Taking advantage of the loopholes in the image-recognition algorithm, the lights assembled into patterns, confounding the AI's data stream. The machines mistook Chen Nan for an animated character instead of a flesh-and-blood human. The robots, confused, slowed their pace. Unsure of the next decision to make, they wandered listlessly, bumping into one another.

Chen Nan couldn't afford to linger. Nimbly evading the robot wreckage, she bolted toward the stairwell where she could take the stairs to the ICU on the eighth floor. She deliberately avoided the elevator. This was an important lesson she'd learned from the game arena: to stay away from anything that machines and algorithms could manipulate, and trust nothing but her own physical body and intuition.

Her chest heaved as she struggled to breathe. Her mouth was dry. She felt as if her heart was about to leap out of her throat.

Garcia, hang in there. I'm coming. She prayed.

She jammed her shoulder against the safety door to the ICU, forcing it open. Her legs trembled so hard she had to lean against the wall as she walked, moving forward step by step, gasping for air. The place where Garcia's fate would be decided was at the end of the tunnel-like hallway. Chen Nan felt as if her heart would explode from the simultaneous rushes of fear and hope.

I have to keep going. I have to face this all by myself.

A large glass window stood at the end of the corridor. Chen Nan peered into a room with a bed, cleanly made, indicating no signs of recent usage.

Where was Garcia?

Unfathomable dread rose in her heart like a tide. Utterly exhausted, she slid to the floor, feeling as if her entire world had suddenly turned into a void.

"Nan, is that you?" said a familiar voice.

CHEN NAN COULDN'T BELIEVE her ears. She leaped up at once and spun around. Garcia and Dr. Xu, in protective suits, were standing a few steps away, smiling at her.

"Garcia? Aren't you . . . aren't you—" she stammered. Garcia in person appeared a little worn out, but oddly in much better shape than Garcia over video.

"Hey! Did you think I was dead or something?" Garcia said, grinning.

Dr. Xu put his fist and palm together to salute them farewell. "I'm sure you two want some time alone," he said, then turned around and disappeared into a room.

Garcia took a few steps in Chen Nan's direction. Almost immediately, her Safety Circle app sent a buzz through her smart-stream, indicating that the person before her eyes carried a safety risk. Instinctively, Chen Nan raised a hand and stopped him from approaching.

"Nan! I'm fine. My flight was safe. I was only put into quarantine because I passed by the same area in the airport as a person who was infected."

"Really?" Chen Nan realized that she hadn't checked the updates from the virus control office since she'd embarked on this crazy adventure. No wonder she hadn't noticed. "But what about your digital health record? What about the video and the audio recording?"

"I can explain," said Garcia. A half-guilty, half-smug look emerged on his face. "It's all a part of the game."

"Game?"

"Remember? I'm the best level designer of the entire South American Thirteenth Warzone of *Techno Shaman*."

The lucky coincidences, the occasional looks of contemplation on people's faces . . . fragments of the events that Chen Nan had experienced over the course of the night, the puzzle pieces, began assembling into a bigger picture.

"Wait. So, you made Mr. Ma come for me?"

"Yes, we met in the game. C'mon, in a city as large as Shanghai, what are the chances that his pirate taxi would stop exactly where you were?"

"What about the video? No makeup in this world can make you look like that!"

"Remember the avatar I made? I used it to generate an animated video clip."

"Is Dr. Xu your confederate as well?"

"Well, he's a real doctor, but yes, we also met in the game. I have to admit, I didn't expect that I would *actually* be put into quarantine, though. Guess this makes the game more realistic, no?"

"But how did you know what I was doing—and that I would go looking for you?" Chen Nan tried to suppress the anger swelling inside her.

Garcia grinned again. "The digital medical record that Dr. Xu sent you was a tracker. With its help, I could give you . . . motivational feedback."

"What do you mean by 'motivational feedback'?" Chen Nan was on the verge of shouting.

"A satisfactory game needs obstacles as well as positive feedback to motivate the players." Garcia was still smiling. "If the levels were too easy, the players would soon feel bored; but if the levels are too difficult, the players grow frustrated. Motivational feedback is what carries players through their journey."

"What if I didn't leave my apartment? What were you going to do about that?"

"If that was the case, then I would . . . admit defeat," said Garcia, his eyes glistening. "My teammates and I analyzed our plan a million times, preparing for all kinds of possibilities to ensure your safety. As long as you got yourself out of your apartment, you would have succeeded. I wasn't sure you would come so far . . ."

"You bastard!" Chen Nan buried her head in her hands. Her shoulders trembled. Tears fell from her eyes, but she couldn't tell what had made her cry again. "I was so worried about you . . . and here you are, treating this like a game! Why on earth did you do this to me?"

"Because *I love you.*"

The phrase went through Chen Nan like an electric shock. Garcia had confessed his feelings for her a million times already in the virtual world: via texts, voice messages, video calls, in the game. She thought that she had already grown accustomed to his romantic declarations. Somehow, though, hearing those magical words said in real time was different. Complex, unnameable emotions rushed into her heart. Even the most advanced virtual reality technology couldn't simulate what she was feeling right now. Humans, with a lack of better descriptive language, called it *love.*

She raised her head and glanced at the man in front of her through teary eyes. "But why? I'm such a selfish coward . . . I thought I had lost you forever."

"Don't be silly, Nan. Look at you! You did what no one could do.

You literally traversed an entire city looking for me. You're a warrior, the bravest person I know."

"Did I . . . did I really make it?"

"Yes, you did. You've left the dead-end loop behind. You're a new person now. Except one thing . . ."

"What is it?"

"You wouldn't let me give you a hug."

"Garcia! I'm just . . ." Chen Nan inhaled deeply, then exhaled. "Okay, I can do this. Come on."

"I'll take it slowly. If I'm making you uncomfortable, say stop, and I will stop."

Inch by inch, Garcia moved toward Chen Nan, his steps slow and gentle, like an aging robot's. Chen Nan closed her eyes, feeling the vibrations of her smartstream grow stronger every second, and combated the fear surging inside her. Garcia entered the one-meter radius, past the safe social distance. The vibration became a piercing alert that buzzed monotonously through her Bluetooth earbuds. Her heart rate quickened. Her chest tightened. Her fear of contact had persisted for so long that it had become a part of her. But she held her ground firmly. *I'm safe,* she murmured to herself. With two layers of protective suits in between her and Garcia, this hug wouldn't harm anyone.

"I'm here," whispered Garcia.

Chen Nan took out her earbuds, letting them dangle and bounce between the folds of her protective suit, still stubbornly alerting her of approaching danger. She opened her eyes and reached out for Garcia, ready for a long-overdue plastic-textured embrace.

ANALYSIS

AI HEALTHCARE, ALPHAFOLD,
ROBOTIC APPLICATIONS,
COVID AUTOMATION ACCELERATION

"Contactless Love" transpires in a society transformed by an ongoing pandemic—in this case, the prospect of the COVID-19 pandemic lingering for decades as a mutating seasonal virus. This hypothesis is, of course, just speculation.

However long COVID stays with us, what has become clear now is that AI will reshape healthcare, from speeding the discovery of vaccines and drugs to accelerating the integration of technologies like AI diagnostics into existing care. Focus on AI healthcare is particularly timely as we are in the midst of the digitization of the healthcare industry, which will produce the massive dataset needed for AI to disrupt healthcare. When we look back in 2041, we will likely see healthcare as the industry most transformed by AI.

Concern about physical contact, another aspect of COVID-19—illustrated by Chen Nan's phobias and shut-in lifestyle—will create many opportunities for robotics, an area that, because of improving technologies, is poised for just such a breakthrough. So, will your apartment in 2041 be filled with the kind of busybody robotic helpers Chen Nan relies on in the story? I'll give you my opinion on that question in the pages that follow.

Finally, I want to examine how COVID-19 will drive the adoption of distance work, communication, learning, commerce, and entertainment. This, in turn, will accelerate digitization and the speed of data collection. More data means better AI, which will expedite automation and job displacements.

THE CONFLUENCE OF DIGITAL HEALTHCARE AND AI

"Modern medicine" in the twentieth century benefited from unprecedented scientific breakthroughs, resulting in improvements in every aspect of healthcare. As a result, human life expectancy increased from thirty-one years in 1900 to seventy-two years in 2017. Today, I believe we are at the cusp of another revolution for healthcare, in which digitization will enable the application of all data technologies from computing, communications, mobile, robotics, data science, and, most important, AI.

First, existing healthcare databases and processes will be digitized, including patient records, drug efficacy, medical instruments, wearable devices, clinical trials, quality-of-care surveillance, infectious-disease-spread data, as well as supplies of drugs and vaccines. Digitization will create massive databases that will enable new AI opportunities.

Radiology has recently become digital. Back-lit film viewers have been upgraded to computer visualization of high-definition 3D imagery, which also makes possible teleradiology and AI-assisted diagnosis. Personal medical records and insurance records are starting to be digitized, stored, and aggregated (where permissible by law) in anonymized databases, to which AI can be applied to improve treatment efficacy, doctor evaluation, medical teaching, anomaly detection, and disease prevention. Complete databases of each drug use will enable physicians and AI to understand how and when to apply each drug to avoid mistakes. AI can do a much more thorough job than human doctors by learning from billions of actual cases, including outcomes. AI can take into account full medical and family history to personalize treatment accordingly. And AI can keep up with a massive number of new drugs, treatments, and studies. These tasks are all well beyond human capabilities.

In addition to existing processes, new revolutionary technologies are being invented as native digital processes. Wearable devices continuously monitor heart rate, blood pressure, blood sugar, and an increasing number of vital statistics that can provide warning signs. This tracking will yield huge databases that can help AI correlate these statistics for more-accurate monitoring, early detection, medical treatment, and maintenance.

In medical research, new technologies are all producing a massive amount of digital output. DNA sequencing produces vital digital information, such as the genes that encode proteins (the molecular machines of life) and the regulatory network that specifies the behavior of genes. Digital polymerase chain reaction (dPCR) can accurately detect pathogens (e.g., COVID-19) and gene mutations (e.g., new cancer markers). Next-generation sequencing (NGS) enables rapid human genome sequencing, even though our genome is much too large for humans to read and interpret and perfectly suited for AI. CRISPR is a breakthrough technology for editing genes, which has the potential to eradicate many diseases in the future. Finally, drug and vaccine discovery are going digital and starting to be integrated with AI (much more on this later in this chapter). All of these are digital in nature and can be integrated with digital technologies like AI, leading to significant improvements in healthcare.

Why, then, haven't early AI projects—such as IBM Watson and its cancer treatment program—succeeded? When IBM worked with esteemed medical institutions like MD Anderson and Sloan Kettering, it decided to rely on medical expertise and data in these institutions to train its AI. These high-quality instructional data are perfect for teaching doctors and medical students. The dataset was painstakingly accumulated by having top researchers select the highest-quality instructional data designed to help students internalize key concepts, make connections across fields, and synthesize new solutions. But these databases are much too small for AI, which learns from massive data and not concepts (recall the table in chapter 1 contrasting human and AI learning). IBM Watson did try to augment its knowledge with large amounts of medical text, such as textbooks and research papers; but these were also written for human consumption, and AI is best trained directly from real patient-treatment-and-outcome data. Curing cancer is a gargantuan undertaking, and not suitable as a first application for AI. Instead, AI healthcare should start from more modest tasks with large datasets that are suitable for AI.

I believe the AI and medical communities have learned their lesson from Watson. Both are now focusing on more-appropriate tasks for AI, such as drug and vaccine discovery, wearable devices, DNA sequencing, radiology, pathology, precision medicine, and doctor assistance. Also, we should take a more pragmatic approach: Select tasks that fit the healthcare industry (e.g., can be sold through existing channels) and comple-ment the human scientists and doctors, rather than ambitiously trying to replace them. With a data-centric and pragmatic approach, AI healthcare will surely blossom in the next twenty years. We will examine some of these tasks, starting with drug discovery.

CONVENTIONAL DRUG AND VACCINE DISCOVERY

Drug and vaccine discovery have historically been extremely time-consuming and costly. It took over a hundred years to develop and per-fect a vaccine for meningitis. Pharmaceutical companies were able to move much faster in developing vaccines for COVID-19, spurred on by unprecedented spending (the U.S. government alone spent $10 billion just in 2020) to run multiple clinical trials and manufacturing efforts on parallel tracks. Had COVID-19 been as contagious or as lethal as the worst pandemics, however, even waiting a year for a vaccine would have been too long. So we need to continue to accelerate the speed of vaccine and drug development.

Drug discovery requires an understanding of how virus proteins, which are sequences of amino acids, fold into unique 3D shapes. Under-standing these 3D structures is essential to understanding how viruses work, as well as how to fight them. For example, the spike protein of COVID-19 can attach to a receptor on the surface of human cells, like a key fits into a lock. After the internalization of virus particles into human cells, the viral genome (RNA, in the case of COVID-19) is transmitted to the host cell and replicated in many human organs, which results in the infected person's experiencing COVID-19 symptoms.

Similar to the way a virus attaches to a human cell, pathogen treat-ment works by attaching a treatment molecule onto the pathogen to in-hibit its function. Drug discovery is the process of finding this treatment molecule in the following four steps:

1. Use mRNA sequence to derive the pathogen's protein sequence (this is relatively easy now).

2. Find the 3D structure of the pathogen's protein sequence (protein folding).

3. Identify the target on that 3D structure.

4. Generate likely treatment molecules and then select the best preclinical candidate from them.

Going back to the earlier analogy, steps 1, 2, and 3 figure out the lock, and 4 makes the key for the lock. These four steps need to be done sequentially, and the latter three are time-consuming and costly.

For step 2, using conventional methods for protein folding, scientists rely on techniques like cryo-electron microscopy, which allows for vivid visualization of viral proteins. Based on the visualization, a 3D structure is painstakingly constructed.

Then, for steps 3 and 4, finding targets and devising new drugs that fit onto the targets are also lengthy trial-and-error processes that require good intuition, experience, and also good luck. Years later, even if a preclinical drug candidate is proposed, it has a 90-percent likelihood of failing to pass phase II or III trials. Sequential explorations take a very long time. The time could be shortened by exploring several methods in parallel (as it was for COVID-19), but that will be prohibitively expensive.

As an aside, the above provides context for the mRNA COVID-19 vaccines from Moderna and BioNTech/Pfizer. mRNA vaccines are the result of a new approach that offers great potential. To develop an mRNA vaccine, scientists uncover the relationship between an mRNA sequence and protein structure. Then they chemically synthesize the mRNA vaccine, which when injected into the human body, instructs human cells to synthesize the virus protein as a pathogen, which in turn stimulates the human body's immune response for combating the real virus in the future.

AI PROTEIN FOLDING, DRUG SCREENING, AND DRUG DISCOVERY

Today, it costs $1 billion and takes many years to get a successful drug or vaccine through the development process. I believe that AI will significantly accelerate drug discovery and reduce its cost, making available many more effective drugs at lower prices. This will help us live longer and healthier lives.

AI can greatly accelerate the speed and reduce the cost of drug and vaccine discovery. For determining protein folding (step 2), in 2020, DeepMind developed AlphaFold 2, which is AI's greatest achievement for science to date. Proteins are the building blocks of life, yet one aspect of proteins that has remained a mystery is how a sequence of amino acids will fold into a 3D structure to carry out life's tasks. This is a problem with profound scientific and medical implications and appears well-suited for deep learning. DeepMind's AlphaFold, trained on a large database of previously discovered 3D protein structures, has demonstrated that it is able to simulate the 3D structure of unseen proteins with similar accuracy to traditional techniques (such as cryo-electron microscopy, mentioned on page 156), which are expensive and can take years for each protein. For this reason, traditional methods have solved less than 0.1 percent of all proteins; thus, AlphaFold may offer a way to rapidly grow the number of solved proteins. AlphaFold has been heralded by the biology community as having solved a "fifty-year-old grand challenge in biology."

Once the protein's 3D structure is known, one expeditious way to discover effective treatment is repurposing, or trying every existing drug that has been proven safe for some other ailment, to see if one of them can fit into this 3D structure. Drug repurposing may be a quick fix that could stop the spread of a serious pandemic at its onset. Because established drugs have already been tested for adverse effects, they can be used without the extensive clinical trials required with new drugs. In "Contactless Love," when Garcia seemed to contract an aggressive variant of COVID, an AI-automated process was immediately initiated to look for a drug that could be repurposed for a "quick fix."

Scientists can also work with AI symbiotically to invent new compounds. AI can be used to propose targets on which a treatment molecule would be attached (step 3). Then, given a target, AI models can narrow the search for a drug by identifying patterns within the data and proposing lead candidates (step 4). In 2021, Insilico Medicine announced the first AI-discovered drug for idiopathic pulmonary fibrosis, by first finding a target on the 3D structure (step 3) and then proposing leads and selecting from among them the best biomolecule (step 4). Insilico's AI saved 90 percent of the cost of these two steps in drug discovery.

Many types of knowledge can be used by AI to optimize steps 3 and 4. For example, Natural Language Processing (NLP) can be used to mine an avalanche of academic papers, patents, and published data to extract new insights that can help propose targets or rank possible new molecules. And based on past outcomes of clinical trials, AI can predict the likelihood of each lead candidate, and rank them accordingly. These experiments are called "in silico," as silicon-based software simulates the actual effect of the drugs and clinical trials. After in silico efforts produce high-confidence candidates, scientists can work from the AI-ranked list.

Besides the in silico approach, in vitro wet-lab experimentation, which involves testing the proposed drugs on human cells in petri dishes, can also expedite drug discovery. Nowadays these experiments could be conducted more efficiently by robotic machines than by lab technicians to generate massive data. A scientist can program these robots to iterate a series of experiments 24/7, without human intervention. This will accelerate the speed of drug discoveries greatly.

PRECISION MEDICINE AND DIAGNOSTIC AI: LIVING HEALTHIER LONGER

AI will reinvent healthcare in many other ways, beyond drug and vaccine development. "Precision medicine" is a term that refers to tailoring an individualized treatment for a given patient, rather than producing blockbuster, one-size-fits-all-type drugs. As more digital information for each patient becomes available, including medical history, family history, and DNA sequencing, precision medicine will become increasingly feasible. AI is ideally suited to deliver this kind of individualized optimization.

I anticipate diagnostic AI will exceed all but the best doctors in the next twenty years. This trend will be felt first in fields like radiology, where computer-vision algorithms are already more accurate than good radiologists for certain types of MRI and CT scans. In the story "Contactless Love," we see that by 2041 radiologists' jobs will be mostly taken over by AI. Alongside radiology, we will also see AI excel in pathology and diagnostic ophthalmology. Diagnostic AI for general practitioners will emerge later, one disease at a time, gradually covering all diagnoses. Because human lives are at stake, AI will first serve as a tool within doctors' disposal or will be deployed only in situations where a human doctor is unavailable. But over time, when trained on more data, AI will become so good that most doctors will be routinely rubber-stamping AI diagnoses, while the human doctors themselves are transformed into something akin to compassionate caregivers and medical communicators.

Even complex surgeries, which rely on sophisticated judgments and nimble movement, will be increasingly automated over time. Robot-assisted surgeries have increased from 1.8 percent of all surgeries in 2012 to 15.1 percent in 2018. At the same time, semi-autonomous surgical tasks are becoming within reach for robots under doctor supervision, including colonoscopy, suturing, intestinal anastomosis, and teeth implants, among others. Extrapolating from this trend, we can expect all surgeries will have some robotic participation in twenty years, with fully autonomous robotic surgeries increasingly accounting for the majority of procedures. Finally, the advent of medical nanobots will offer numerous capabilities that surpass human surgeons'. These miniature (1 to 10 nanometers) surgeons will be able to repair damaged cells, fight cancer, correct genetic deficiencies, and replace DNA molecules to eradicate disease.

Wearable devices—such as the medical ID strips in "Contactless Love"—smart rooms with sensors for temperature, smart toilets, smart beds, smart toothbrushes, smart pillows, and all kinds of invisible gadgets will regularly sample vital signs and other data and detect possible health crises. Aggregated data from these devices will accurately identify if you have a serious condition, whether it is a fever, a stroke, arrhythmia, apnea, asphyxiation, or just injuries from a fall. All this Internet of Things (IoT) data will be combined with other healthcare information such as medical history, contact-tracing records, and infection-control data, to predict and warn about future pandemics. Privacy will be an alarming issue for some users, so the system will need to anonymize the data by replacing each name with a consistent and untraceable pseudonym. Also, we should research technology solutions that will allow us to have centralized AI while protecting privacy (more on this in chapter 9). Finally, there need to be innovative proposals such as letting people donate their data along with their organs when they die.

A 2019 study shows that AI healthcare markets will experience 41.7 percent annual growth to $13 billion by 2025, in such areas as hospital workflow, wearables, medical imaging and diagnosis, therapy planning, virtual assistants, and, most significantly, drug discovery. COVID-19 will likely accelerate this growth rate.

Lastly, I anticipate that AI will contribute to longevity—not just helping us to live longer but also to do so with a reasonable quality of life. AI will use big data and individualized data to deliver "precision longevity" by preparing personalized nutrition, supplement, exercise, sleep, medication, and therapy plans for each person. Rejuvenation biotechnology will no longer be limited to the ultrarich but made available for all. With all of the advances in medicine, biology, and AI, some experts think we might live twenty years longer than humans' current life expectancy. In that case, 2041 is really more like 2021 for us!

INTRODUCTION TO ROBOTICS

Robotic applications are much more difficult to perfect than the Internet, financial, and perception applications described in previous chapters, because robotic problems cannot be solved by direct application of deep learning to data. In addition, robotics involves manipulation, movement, and planning, which in turn require a delicate interplay of mechanical engineering, perception AI, and fine-motor manipulation. These are all solvable problems, but they will take longer to fine-tune and will require cross-disciplinary technology integration.

In robotics, human capabilities of vision, movement, and manipulation must be replicated with precision. And robotic machines should be not only automated but also autonomous, which implies we will hand over decision-making to robots, which would be able to plan, collect feedback, and adapt or improvise based on changes in the environment. By giving machines the power of sight, the sense of touch, and the ability to move, we dramatically expand the number of tasks AI can tackle.

General human-level sight, touch, manipulation, movement, and co-ordination are too difficult to perfect in twenty years. Each capability will be developed independently in constrained environments, and then the constraints will be gradually relaxed over time. Today, robotic computer vision is mature, so it can be applied to safety enhancements for the elderly population (imagine robot-assisted "life alert"), assembly line visual inspection, and anomaly detection for energy and mass-transportation industries. In parallel, mobile platforms such as autonomous mobile robots (AMRs) and autonomous forklifts can navigate around indoor spaces, allowing a robot to "see" obstacles, plan its path, and move cargo in a warehouse. Today robotic arms can grasp, manipulate, and move rigid objects in applications like welding, assembling lines, and object picking in e-commerce distribution centers.

These robots will gradually become more capable. Computer vision using cameras and other sensors (such as LiDAR) will be an integral part of smart cities and autonomous vehicles. Mobile platforms will be able to navigate indoors and outdoors and work in swarms with great efficiency and speed, and legged robots will be able to go anywhere. Robotic vision, manipulation, and movement will be coordinated and combined in increasingly complex applications. Robotic arms will have soft skins so that they can pick up even fragile objects, and they will learn new objects or new tricks by trial and error, or by watching humans.

INDUSTRIAL APPLICATIONS OF ROBOTICS

Most expensive technologies hit maturity when industries can see high value in their applications. If companies have a critical need that a developing technology can help with, they will often be willing to pay a premium and lose money initially on the technology adoption with the promise of scaling down later for larger savings. Robotics will be no exception.

Factories, warehouses, and logistics companies are already using AI and robotics, beginning with visual inspection, mobile platforms, and object picking. Today robots already pick up, move, and manipulate many objects. Later they will be able to coordinate among multiple robots, handle complex planning tasks, and deal with errors and unusual situations. Full automation of factories and warehouses will take a long time, because some tasks require human manual dexterity, precise hand-eye coordination, or dealing with new situations and environments. But by 2041, warehouses should be virtually automated, while factories will be mostly automated.

Agriculture is a surprising low-hanging fruit. While manufacturing a phone, a shirt, or a shoe are completely different, fertilizing, spreading insecticide, and seeding are relatively similar for many types of crops. Drones can already do these three tasks for many types of crops, while robots are harvesting apples, lettuce, and other fruits and vegetables today. Robotics will reduce the cost of agriculture in time, offering the promise of reducing food insecurity around the world as well.

COVID-19 has expedited the use of robotics in healthcare—to detect people with fevers, monitor patients, decontaminate hospitals and airports, deliver food to quarantined people, connect to telemedicine, ferry test samples to laboratories, and help frontline medical workers reduce their exposure to the virus. Initially these robots did only simple and repetitive tasks, and some required human supervision. But the experience gained in this baptism by fire is already making robots smarter and more autonomous. For example, one company in China now produces automated biology labs. This not only frees up valuable time for the scientists and doctors, but also eliminates errors and infections. The robots work 24/7 and collect valuable data for automated experimental iterations.

COMMERCIAL AND CONSUMER APPLICATIONS OF ROBOTICS

Industrial applications will test and improve robotic technologies, and over time reduce the cost of robots and their parts, putting them in reach for a wide variety of commercial and consumer uses. For example, the robot arm used in automated laboratories could double as a component for serving drinks in a coffee shop and then be used in the home when the cost comes down yet further. The mobile platform used in AMRs can be duplicated for uses in the many consumer bots Chen Nan and her contemporaries relied on every day—the "Yuanyuan" housekeeping robot, the R2-D2-like DeliveryBot, the spider-crab-like DisinfectionBot, and the CleaningBot, SeniorCareBot, and DogWalkingBot.

Some of these bots are already arriving in 2021 in more primitive forms. Recently, when I was in quarantine at home in Beijing, all of my e-commerce packages and food were delivered by a robot in my apartment complex. The package would be placed on a sturdy, wheeled creature resembling R2-D2. It could wirelessly summon the elevator, navigate autonomously to my door, and then call my phone to announce its arrival, so I could take the package, after which it would return to reception. Fully autonomous door-to-door delivery vans are also being tested in Silicon Valley. By 2041, end-to-end delivery should be pervasive, with autonomous forklifts moving items in the warehouse, drones and autonomous vehicles delivering the boxes to the apartment complex, and the R2-D2 bot delivering the package to each home.

Similarly, some restaurants now use robotic waiters to reduce human contact. These are not humanoid robots, but autonomous trays-on-wheels that deliver your order to your table. Robot servers today are both gimmicks and safety measures, but tomorrow they may be a normal part of table service for many restaurants, apart from the highest-end establishments or places that cater to tourists, where the human service is integral to the restaurant's charm.

Robots can be used in hotels (to clean and to deliver laundry, suitcases, and room service), offices (as receptionists, guards, and cleaning staff), stores (to clean floors and organize shelves), and information outlets (to answer questions and give directions at airports, hotels, and offices). In-home robots will go beyond the Roomba. Robots can wash dishes (not like a dishwasher, but as an autonomous machine in which you can pile all the greasy pots, utensils, and plates without removing leftover food, with all of them emerging cleaned, disinfected, dried, and organized). Robots can cook—not like a humanoid chef, but like an automated food processor connected to a self-cooking pot. Ingredients go in and the cooked dish comes out. All of these technology components exist now—and will be fine-tuned and integrated in the decade to come.

So be patient. Wait for robotics to be perfected and for costs to go down. The commercial and subsequently personal applications will follow. By 2041, it's not far-fetched to say that you may be living a lot more like the Jetsons!

DIGITIZATION OF WORK AND AI

During the COVID-19 quarantines, we removed human touchpoints and performed countless tasks online. This overnight shift in global behavior will have some negative long-term consequences, illustrated by Nan's nosophobia (fear of developing a disease) and her consequent asocial behavior in "Contactless Love." But we also saw increased flexibility and productivity from our habit changes. Chen Nan's modernized "smart" work style enabled her to work from home nearly all the time. We used to feel that physically going to the office, traveling on business, and attending school in person were essential. But we've now learned that much of what we did that we thought required travel can be done quite or even more efficiently online. Months of staying at home shook loose old beliefs and habits. In late 2020, Bill Gates predicted that 50 percent of business travel will disappear and be replaced by efficient virtual meetings. He also expects 30 percent of American employees to work from home on a near-permanent basis. David Autor, an economist at MIT, calls the COVID-19 pandemic and economic crisis "an automation-forcing event," which was driven by the triple necessities of increased productivity, lower costs, and human safety.

Zoom and other videoconferencing services will go down in history as the tools that kept the world turning during COVID-19. They made possible productive team meetings, joyful weddings, and active classrooms for millions of students. We can anticipate that in the near future, business meetings will be archived and transcribed by automatic speech recognition. This will make past meetings searchable, and help us track commitments, schedules, and possible anomalies, significantly improving business efficiency and management.

Pervasive video communications in the future will also enable AI-based avatars. As we learned in chapter 2, generating a realistic video of my talking head is much easier than replicating a human in real-life. A virtual teacher may be more fun than a real teacher. A virtual customer service representative and a salesperson can be optimized to maximum customer satisfaction or revenue, respectively, while conducting a conversation based on all that is known about a given customer. I know I would love to have a body double give coordinated AI talks and answer questions at multiple conferences—simultaneously!

Digitization of workflow makes it easier than ever to reorganize, outsource, or automate work. As work is digitized, the resulting data becomes the perfect fuel to empower AI. For example, each worker's workload is characterized by the input to the worker and the output from the worker. If AI can do the same workload as a human, there will be strong temptation to automate it. (See chapter 8, "The Job Savior," for RPA technologies that will automate human workloads.) Historically, automation tends to happen when economic difficulties coincide with maturing technologies. Once a company has replaced an employee with a robot and experienced the robot's efficacy, it is unlikely ever to go back. Robots don't get sick. They don't strike. They don't demand higher wages for dangerous jobs.

How do we deal with this issue of exacerbated job displacement? We will explore this in chapter 8, "The Job Savior."

CHAPTER FIVE

MY HAUNTING IDOL

STORY TRANSLATED BY EMILY JIN

VIRTUAL REALITY IS LIKE DREAMING WITH YOUR
EYES OPEN.

—BRENNAN SPIEGEL

REMEMBERING YOU . . .
THE FIREFLIES OF THIS MARSH
SEEM LIKE SOULS
THAT RISE
FROM MY BODY'S LONGING.

—IZUMI SHIKIBU

NOTE FROM KAI-FU: "My Haunting Idol" depicts the future of entertainment, where games become immersive and the boundary between virtual and real is blurred. Set in Tokyo, the story describes how a fan investigates the death of her idol, with the help of the idol's "ghost," brought back to life using AI and virtual reality. Virtual reality is immersive, realistic, and interactive, and will change the future of entertainment, training, retail, healthcare, sports, real estate, and travel. Will it really be possible to build a "virtual you" by 2041? I'll answer that question in my commentary, as I describe virtual reality, augmented reality, and mixed reality, three forms of immersive experiences, as well as the ethical and societal issues around such innovations.

THE ROOM WHERE the séance would take place was decorated in Victorian style and nearly devoid of light. In its center stood a black wooden table with seven half-burnt candles surrounded by scattered rose petals. Above the table hung a white, semi-transparent silk curtain, its folds stretching out like a jellyfish grabbing hold of the ceiling.

Aiko glanced at the other three girls. Their faces, in the candlelight, gave off an uncanny aura. She began regretting her choice of theme. *Then again,* Aiko wondered, *would the Japanese Edo–period Shinto shrine or the Chinese traditional Taoist ritual have been even more frightening?*

The candles flickered ominously. The room was supposed to be sealed. Where did the wind come from?

Aiko shuddered. When she met the gazes of the other girls, she saw fear in their eyes. *Perhaps what they had wished for has descended upon them.*

The medium, an old woman with a crooked gait clad in a black robe, grasped the hands of the girls on either side of her. Suddenly, the spiritual guide's eyes rolled back into her head. The table vibrated violently, as if a drum washing machine were throwing a fit underneath. The girls, pale-faced and shivering, shut their eyes and shrieked, nearly in unison.

"I see him! Oh-ho-ho, a brilliant man indeed," the medium muttered as her body rocked back and forth. "It seems like he died during an important ritual . . ."

"Yes, yes!" The girl with short dyed-blond hair responded, her face lighting up. "Hiroshi-kun's body was found in a dressing room locked from the inside, during the intermission of his farewell concert."

"His body looked as if he died from drowning," added the girl with curly auburn hair.

"And I cried for an entire week!" exclaimed the girl with long gray-blue hair, choking back a sob.

"Have you placed all the totems on the table?" asked the medium, finally making eye contact with the girls. "The totems will lead us to an important clue," she continued in her raspy voice.

Each girl had brought along a few pieces of merchandise that they'd purchased from the official Hiroshi X website. According to the online shop, Hiroshi had *personally* touched all of the items with his hands. Aiko even managed to obtain, from the black market, a hairbrush with a few strands of hazel hair stuck to it, rumored to have been Hiroshi's own. She had no way to verify its authenticity, though she knew for sure that the hairbrush was expensive as hell.

"Ah!" the medium cried. "The spirit—Hiroshi—he has something to say. Ooooh . . . we must listen!"

The table's vibration came to a halt. The room turned dead quiet; everyone waited for something to happen. All eyes were focused on the lower half of the medium's face, veiled behind black gauze.

The medium's eyes glowed. Her body lurched, as if possessed. Then she stopped trembling. When she opened her mouth again, her voice was completely different.

A young man's voice permeated the room. He sounded unsteady, if gentle. Fragile.

"I am stuck in Bardo . . . It's dark and cold like the bottom of the ocean . . . I can't breathe . . . I don't want to die . . . I have so many unfulfilled wishes."

Aiko's eyes widened. "It's . . . it's Hiroshi-kun!" The voice sent goosebumps down her spine. She felt her pulse racing.

The voice continued. "I want to sing 'Let a Miracle Shine at the End of the World' once again. Please, help me find the truth."

By now all the girls were sobbing. "Let a Miracle Shine at the End of the World"—often abbreviated as "Miracle Shine"—was Hiroshi's biggest hit.

"Hiroshi-kun, stay strong!" stuttered Aiko as she gasped for air.

The voice was gone. The medium, like a puppet detached from its strings, lowered her head. When she opened her eyes, all she could mutter was nonsense. The candles flickered again and the room gradually brightened.

"He's gone," croaked the old woman.

I hear you, Hiroshi-kun. I will find the truth and rescue you. Firmly, Aiko nodded to herself.

IT WAS HIROSHI X'S farewell concert.

Standing in a tide of die-hard fans, Aiko craned her neck and gazed longingly at the shining silhouette in the middle of the stage. She was utterly speechless, rendered silent by an overwhelming swirl of emotions. Was she touched? Intimidated? Yearning for him? Or all of the above?

The music suddenly stopped. The giant screen behind Hiroshi's back, displaying a bright starry sky, switched over to a projection of the audience. The camera dashed across the auditorium as if seeking a target. Ecstatic, glistening faces flashed on the screen, accompanied by screams of joy.

At last the camera froze. The face brought into focus on the screen high above the stage was composed, if a little vacant amid the jubilant crowd. The girl looked so ordinary that she almost seemed out of place.

Aiko realized that it was her face.

"Aiko-san, I choose you."

Am I in a dream? Aiko wondered. Her idol was speaking her

name onstage, in front of millions of people! Everyone was looking at her. Panicking, she glanced around, her mind blank. She did not even know what expression to make.

"Aiko-san, may I invite you onstage to sing with me?"

Applause echoed in the auditorium. People clapped in encouragement. But Aiko was frozen as if under a spell.

"Aiko-san?" Hiroshi-kun sounded hurt. "Why are you going to reject me?"

AIKO'S EYES SHOT OPEN. She had woken up screaming. With her chest still heaving, she breathed deeply to calm her racing heart. *It was a dream after all.* She turned on her bedside lamp and sat up. Ever since the séance a few days earlier, she had been restless, unable to sleep for more than a few hours at a time.

Her stomach rumbled. She had skipped dinner to lose weight. Maybe some ramen, her specialty, would help her feel better. She knelt by the mirror on her bedside table and reached for the container of her XR contact lenses—she was practically blind without them.

"I hoped that I could make ramen for Hiroshi-kun," muttered Aiko.

Just then a strange rustling sound came from the kitchen. Startled, Aiko stood up.

Grabbing the baseball bat next to her bed—autographed by Hiroshi—she crept toward the kitchen. Eerie blue-green light seeped through the door crack. She took a deep breath, threw the door open, and barged in.

No one was in the kitchen. The refrigerator door, however, was cracked open.

"Huh, weird. These smart home appliances—I thought they only worked on command," Aiko murmured as she scavenged the near-empty refrigerator, spying little more than a jug of fat-free milk about to expire.

"I *really* have to order groceries tomorrow or else I'll—*Aaaah!*" As Aiko turned around, the bottle in her hand dropped to the floor, splattering milk all over the kitchen.

A man, glowing with blue-green light, was standing before her, looking as though he had emerged from the cold, watery mist of the refrigerator.

Aiko's mouth hung open as she cast her eyes on the man's face—the unmistakable face of Hiroshi X. "Aren't you *dead*?" she exclaimed.

The apparition appeared to smile, and then spoke. "Watch your language! You should be polite while addressing the dearly departed! Don't forget, you were the one who summoned me in the first place."

Aiko poked at Hiroshi with the bat. A ring of light rippled as the wood passed through Hiroshi. He was incorporeal, like an apparition, or a hologram.

"Wow! You really are a ghost. *Sugoi!*" She exclaimed in admiration.

"Hey! You can't just poke through someone's body and go all *sugoi*!"

"You know, the way you talk is just like Hiroshi-kun himself," said Aiko.

"What are you talking about? I *am* Hiroshi X, the hero who saves the day with love and music!" The ghost made Hiroshi's dorky signature pose, like a character who had stepped out of a manga book.

"I know, I promised you I would find the truth," Aiko said, nodding. "And you, the ghost of Hiroshi-kun, please help me, too."

"Okay, okay," said the ghost. "According to the rules, I will offer you three key clues. Only if you ask the right questions, though."

Kneeling down, Hiroshi gazed into Aiko's eyes and held up three fingers. Although the blue-green-tinted ghost was translucent, he was just as handsome as ever. Aiko blushed and averted her eyes. "I have read the reports regarding the case, but I still want you to tell me what exactly happened in the dressing room."

"You might have just wasted a question. Everything I'm telling you can be crossed-checked with the testimony of other witnesses. The dressing room was free from any kind of security cameras, just as I had requested, so there is no video footage. Getting ready to go onstage is a hectic routine. I start by changing into costume and putting on my makeup, under the supervision of my stylist and makeup teams. Then I restring and tune my guitar and go over the set list with the music director. My manager, Mi-chan, is with me the entire time until the end. After everyone leaves, and just before taking the stage, I hang back in the changing room for three minutes to meditate alone."

"Which means you were practically in a sealed room where no one can enter or leave, weren't you?" said Aiko.

"Well, I'll treat this one as a freebie follow-up question," said the ghost, smiling. "Yes, I was alone. There was no way to enter the room apart from a single door."

Aiko was deep in thought. However, as she was about to ask another question, she saw the ghost gradually fading into thin air.

"Wait! Would it be possible if I—"

"And this will conclude my first visit, or should I say *visitation. Ganbatte,* Aiko-san! You can do it!"

Aiko shook her head. Disappointed, she shut the refrigerator door. The eerie light and chill disappeared at once. She was alone in her kitchen with a mess of spilled milk.

"I was going to ask . . . would it be possible if I . . . could give you a hug?"

"WHAT? YOU MET HIROSHI X's GHOST?" Making an exaggerated face, Nonoko clamped her hand over her mouth and widened her eyes, her silver eyelashes glistening. "Get out of here."

Doux Moi was their favorite destination for afternoon tea. Apart from its signature French-style high-tea set, the shop's owner, Ines Suzuki—a popular French-Japanese actress—was the main attraction. Patrons swarmed to Doux Moi in the hopes that they would run into one of Suzuki's celebrity friends.

Nonoko, a fan organizer by profession, was a VIP at Doux Moi and often granted the privilege of skipping the line.

Experts in recruiting and convening fans, organizers like Nonoko ran social media channels that often exceeded millions of followers. Whenever a celebrity needed an exposure boost, such as when they released songs, went on variety shows, announced new brand collaborations, or advertised for in-person events, managers and agencies would turn to professional fan organizers, who were usually disguised as ordinary if well-connected fans, to give their clients a boost.

With armies of hardcore fans who would drop money and post online at their command, fan organizers were like military officers leading troops into battle. Experienced, charismatic, and business-savvy, Nonoko was one of the best. Concert tickets, special meet and greets, limited-edition merchandise, and, of course, VIP privileges at Doux Moi . . . agents would give anything to please her.

Nonoko took pride in her work: From the moment she was contracted to boost a celebrity's fandom, Nonoko could pass as a die-hard fan. The dates of music releases, award show wins, decades' worth of gossip . . . Nonoko took in the details of a celebrity's career like a computer scooping up data points. With an extraordinary memory and an exceptional ear for the nuanced vocabulary of Internet fandom, she could easily win the trust of other fans. However, given the nature of her job, she could switch over to competing fandoms in mere seconds. In the fandom community, she was dubbed "Nonoko the Chameleon."

As Aiko's decade-long best friend, Nonoko struggled to wrap her head around Aiko's stubborn loyalty to a single pop star, Hiroshi X.

"Stop eating! I need your help," sighed Aiko.

"Let me enjoy my dessert in peace! I know you want me to be your Sherlock Holmes. But how am I supposed to solve this mystery with so few clues?"

"Well, the handbook says that the only way to summon Hiroshi-kun is by saying the correct keyword."

"If that's the case . . ." Nonoko wiped away the cream stuck to the corner of her mouth. "Didn't you say that you talked about ramen before he magically appeared in your kitchen? It seems like Hiroshi-kun was a fan of food shows. Maybe the keyword is related to food."

"Even if I can figure out how to summon him again, what am I going to ask him? I don't want to ask another stupid question and waste my shot."

"Well," said Nonoko. "You can start by not asking him anything that you can simply figure out on your own. How closely have you looked at the game's case documents? Maybe you could ask him about his connections to other people: Who were his enemies? Who has threatened him in the past? Detectives always consider motives first."

"You make a good point," said Aiko, contemplating. "I should figure out the relationship between Hiroshi-kun and all the people who entered the dressing room."

"Well, my point *actually* is, I think you should join me," Nonoko said, grinning. "Hiroshi's time is over. Find a new idol! Follow the trends! This industry is like a supermarket. If you can enjoy a different flavor every day, why bother with just one?"

"Someone like you won't understand! Hiroshi-kun is special . . . He saved me!"

"Oh, your tragic backstory! Not again. Speaking of which, I need to run—I'm organizing a fan event for the UltraTalent Show. I'm not ditching you on purpose, I promise!"

Nonoko disappeared into the swarm of people waiting outside.

Aiko shook her head and gave a bitter smile. She was used to being left behind. Since childhood, she had always felt like an extra on the set of life, a leftover. When her parents divorced, she was left for her grandparents to raise. At school, when she auditioned for musicals, her name always ended up on the list of understudies. Making friends, dating boys . . . she was everyone's second choice. Years of self-doubt had collided with a diagnosis of depression, which culminated in an unsuccessful suicide attempt in her late teens.

Aiko had been in the depths of depression when she heard "Miracle Shine" for the first time.

"Sometimes you feel like hope will never come, / But be strong and carry on, / Till the end of time, yesterday and tomorrow, / You are special, like a shining miracle," sang Hiroshi.

The song made Aiko feel as if her heart were pierced by an arrow. Her soul resonated with Hiroshi's voice. It made the dark clouds above her head disperse at once, and she could feel the warmth and brightness of the sun again. Hiroshi had been by

Aiko's side through her entire teenage years. With him, she was no longer lonely. He was her best friend, her mentor, her safeguard. And she was more than willing to pay the price, quite literally—her home was swamped with miscellaneous merchandise that more often than not carried no practical use whatsoever, but Aiko didn't mind.

Since that day she first heard his music, she'd spent countless hours digging through online archives devoted to her musical savior. Hiroshi had authorized the digital rights to all of his songs, videos, photos, and other ephemera to a tech company called Viberz many years ago. Viberz digitized and indexed all official Hiroshi-related materials, making them available for licensing to other entertainment projects.

As a result of her zeal, Viberz had chosen Aiko as one of the beta users of its mysterious new project, "historiz." Explaining that the project would enable her to experience her idol in a new way, a Viberz representative asked Aiko to answer over three hundred questions about Hiroshi X, as well as approve uploading of various personal data. Intrigued, Aiko had gone along with the request. A few weeks later, she was invited to a Viberz XR experience room, where she met the other girls who attended the séance. They, too, had been handpicked for the secret project, on account of being Hiroshi superfans.

AT HER NEIGHBORHOOD GYM, Aiko pulled on her somatosensory suit and hopped on the spinning bike. Her favorite scenery flashed across her XR contact lenses: Highway 1 in California, Norway's Atlanterhavsveien, Route des Grandes Alps in France. For each of these settings, the suit, as thin as a second skin, could simulate unique somatosensory feedback, ranging from the breeze along the road, the heat of the summer, even bumps on the asphalt. It could also monitor Aiko's physiological data and posture in real time to generate personalized workout tips.

The workout made Aiko feel that she was soaring across the globe, instead of being trapped in this tiny gym. The ability to ex-perience a taste of elsewhere made Aiko think about a comment

Hiroshi-kun had made on a talk show—no matter where he went, he always wanted to have Chinese food. A plate of fried dumplings could instantly bring him back to the comforts of home.

Suddenly, a voice interrupted Aiko's reverie on the bike.

"Ah, I miss the taste of fried dumplings so much, Aiko-san!"

The ghost of Hiroshi X had once again appeared out of thin air. The scenery vanished. Dressed in a glamorous stage suit, he hovered before her eyes. Through his translucent body, she could see the back wall of the room.

"Hiroshi-kun! You startled me. Can't you warn me in advance when you're going to pop up like that?"

"What would be the fun in that?" Hiroshi winked.

Just then the gym door swung open. A boy walked in with a towel draped over his shoulder. He was of medium height, lean built. His fluffy hair resembled the fur of a bichon frise or a puff of chocolate-colored cotton candy. He was also wearing a pair of XR lenses, the same kind as Aiko's—except as frame glasses instead of contact lenses. Aiko couldn't help but stare as the boy—about her own age—hopped on the spin bike next to her and started pedaling at top speed.

"Aiko-san, how can you do this to me?" complained Hiroshi, curling his lips into a pretend pout. "I see your eyes wandering. I thought you said that I was your one and only love!"

"What? Stop looking at me like that. It's embarrassing!"

The boy stopped his workout and turned to Aiko, frowning in confusion.

Aiko, now red in the face, waved her hands around to explain. "I was talking to myself. I'm so sorry!"

"Okay then," muttered the boy, adding, under his breath, "Weirdo girl." He switched his glasses to VR mode. An opaque coat of silver immediately veiled the lenses.

Ashamed, Aiko dismounted her bike and slunk out of the room. Wiping the sweat off her forehead, she slumped down on a bench in the women's locker room. She wasn't alone.

"Aiko-san! If you like someone, you need to tell them," implored Hiroshi, who had reappeared next to a row of lockers. "What did I say in my songs about confessing your feelings?"

"You have, like, what, eight hundred songs about confessing feelings?"

"That's an exaggeration. There were only thirty-seven!"

"Stop derailing the conversation! What about my next clue?"

As a teen, Aiko had fantasized about making friends with Hiroshi-kun. *Wouldn't it be amazing if I could talk to him every day?* Now that her dream had come true, in a manner of speaking, it wasn't quite like what she'd expected. Hiroshi was almost too friendly, too approachable. Even though the way he spoke was just how he did on TV, the vibe he gave off in person was more boy-next-door than distant idol onstage. Aiko felt as if she could *really* get to know him and trust him.

"Oh! I almost forgot. Aiko-san, please ask your question."

Ghost-Hiroshi, now with a solemn expression, pressed his palms together and bowed at her. He looked even more ridiculous that way.

Aiko cleared her throat. "Hiroshi-kun, please tell me, what were your relationships like with the people who entered the dressing room before you died? Had you ever been involved in conflicts with them?"

"Hmmm, a question worth pondering," said Ghost-Hiroshi thoughtfully as he rested his chin on his palm, pretending to think hard. The glow of his body brightened and dimmed rhythmically as he breathed in and out.

"All those I encountered backstage have worked with me for over a decade. Besides our good professional relationships, we were also great friends. Well, disagreements can happen even between the best of friends. My stylist, for example, sometimes made me put on costumes I found ridiculous; my makeup artist would take it too far and make me look like David Bowie. I didn't enjoy any of that. Sometimes I would refuse and argue. Then they would argue back, trying to convince me. Usually I would give in."

"Mmm, that seems pretty normal to me," Aiko said, nodding as she took notes on her phone.

"Then there were my stage director, Naoto, and music director, Kenyiti. They were new to the industry and barely out of college when I took them under my wing. Now they are leading figures in their respective fields. My relationships with those two haven't been without some bumps. I once found out that Naoto had made some bad financial decisions, but that issue had been resolved years ago, and he'd already paid me back. As for Kenyiti, it's true that he's been wanting to leave the team and make his own music, but I persuaded him to stay. Perhaps he wasn't entirely happy about that; but once I quit performing, he would be set free anyway. I can't imagine him harboring much ill will toward me."

"The entertainment industry sounds so complicated."

"If you are true to others, they will be true to you," sang Hiroshi.

"Let's talk about Mi-san, your agent. How did she react when she heard you were retiring? She'd been with you since day one, way before you were a star. I wouldn't be surprised if she was reluctant to let you go."

Hiroshi sighed. "Yes, Mi-chan was strongly opposed to the idea of my retirement. I can understand, though. It took more than a decade of hard work and a ton of lucky breaks to make it. My quitting was certainly a blow to her. I tried reasoning with her many times about retiring, making it clear that I was leaving only after my contract expired and that I would compensate her loss. But she still held it against me. Whenever she looked at me, her expression was off."

"What if . . ." Aiko's eyes widened. "What if Mi-san murdered you, because her love for you had transformed into hate once you announced that you were retiring?"

"No, even if she didn't like my choices, I trust that Mi-chan is a good person. Also, don't forget that there were other people around, too, when she came into the dressing room."

"Wait! Let me take a look at the photos shot at the murder scene again—"

"I need to go. Aiko-san, don't let happiness slip away from you. *Ganbatte!*"

"Hiroshi-kun! Wait! I'm not finished yet!"

The ghost, however, had already disappeared into the mirror on the wall.

+———

ONE OF THE REASONS Aiko had so much time on her hands to think about Hiroshi's demise was because she had recently lost her job.

When the managing director of Nanun-do Press had approached her one day, gently offering the suggestion that the position wasn't the right fit, Aiko had blurted out in rage, "It's AI again! So I guess AI finally learned how to tell a good story from a bad one? Is that it? You're really just looking for an excuse to reduce expenses, aren't you?"

Remembering how she had made a scene in front of others in the crowded open office, Aiko shuddered with embarrassment.

It was true that automated editing tools had been slowly robbing editors of jobs, though mostly editors who worked on short-form content—finance and business news, sports recaps, even stories about the everyday goings-on of politics and government. So far, literary book editors like Aiko had been hanging on. Even as the publishing market shrank gradually with each passing year— and the arrival of sophisticated new entertainment options— a stubborn contingent of literary editors persisted. The financial upside of literary publishing was trivial enough to protect the industry from the avarice of tech giants, and literature, by nature, was highly dependent on human creativity, taste, and critical thinking. Despite its struggles, literary publishing was regarded as the last safeguard of human dignity.

Things began to change with the launch of the Super GPT model, however. The publishing industry, standing on a cliff edge, realized what they were dealing with. The problem was no longer simply about editors being replaced; following the birth of the Super GPT model, the mechanism of story creation, the *entire definition of literature*, was turned upside down.

Aiko had read machine-created personalized stories. To her, the stories had a peculiar flavor. Something felt off, like drivers switching from fuel-powered vehicles over to electric cars for the first time—the acceleration was so smooth that it lacked texture and rhythm. The same could be said for AI-generated literature: The sentences were just a little *too* smooth. Every plot point and every character hit Aiko's sweet spot; the AI writer was deliberately appealing to her taste with every turn of phrase. Flawlessness some-how made stories far less exciting by depriving readers of challenges and surprises.

To Aiko, those challenges and surprises were exactly what dis-tinguished a good story from an ordinary one.

After her outburst, Aiko and her old boss agreed it was best that she and the company go their separate ways. The problem now was that there was hardly a hiring season in publishing. There just weren't enough jobs to go around at any given time. It was at about this time that Aiko had received the mysterious message from a Viberz representative about serving as a guinea pig for their new immersive fan experience. So, not hearing back from any of the companies she'd applied to, Aiko sought refuge in the game. She dedicated all of her energy to solving the mystery of Hiroshi's death, while trying her best to ignore the speed at which her e-wallet balance was going down.

Aiko had decided that Mi-san—now her primary suspect—and her mixed feelings about Hiroshi's retirement was the most prom-ising lead she had.

Talking to strangers had always been one of Aiko's top fears. In her daily work as an editor, she mostly communicated via email, avoiding much in the way of strange and awkward social scenar-ios. Talking to Mi-san, however, was not a task for email. After scheduling the call, Aiko rented a private cube for an hour. The size of a tatami mat and with optimum sound insulation, the cu-bicles were frequented by white collars in nearby office buildings looking to rest, meditate, or make private phone calls during their lunch breaks.

"Thank you so much for your time, Mi-san," whispered Aiko nervously, once Hiroshi's agent was on the line. She had rehearsed her opening sentence twenty-six times before she dialed Mi-san's number with trembling fingers. Soon a short-haired woman with heavy makeup appeared on the screen of the phone.

"You're that fan detective, aren't you? What was your name again?" Mi-san had a confident, intimidating voice. "Aiko, is it? What a *tacky* name."

"Yes, yes, that is me." Embarrassed, Aiko squeezed out a smile. "Um . . . if it doesn't inconvenience you too much, would it be possible if I ask you a few questions about Hiroshi?"

"I've told the police everything I know. But, sure, since I'm here."

"Well . . . I was actually wondering about how you felt toward Hiroshi's retirement," said Aiko.

"Oh?" Mi-san's tense expression loosened up upon hearing Aiko's words. "Sure, I can talk about that. For the past decade, people can't seem to stop gossiping about my relationship with Hiroshi, about how I have been exploiting him. It's all nonsense! I can be demanding, but my high expectations were only to help Hiroshi. I would never do anything to hurt him."

"Um, I see . . ." Aiko didn't know what to say next. She realized that Mi-san had seen through her question, elegantly throwing cold water on her suspicions. Mi-san did not seem as if she were lying, either.

As Aiko fumbled with what to say, Mi-san spoke up. "I have a question for you, too, Aiko. What kind of fan are you?"

"Wh-what?" Aiko stuttered, unprepared for the question.

"For all these years, I've met thousands of die-hard fans of Hiroshi. They show their love for him in different ways. Some supported him silently, some pinched and scraped so that they could afford his records and merchandise. Some waged wars online with trolls at the slightest hint of disrespect. Some were lost in the fantasy of a romantic relationship with him, perhaps making up for a void in their own lives. And a few were even sent to jail for harassing him.

"However, at the core, I think there are only two kinds of fans in this world: the kind that treat celebrity idols as gods, and the kind that treat them as people. The former accept only perfection. If their idol fails to meet their needs and expectations, their love may turn to hate. They may ditch the old idol for a new one, or resort to more extreme measures. The latter, on the other hand, see an idol as an equal to them, in the sense of human flaws. They are willing to grow, change, and experience the ups and downs of life with their idol. Perhaps a fan like that would never cross paths with their idol, but, in the end, that wouldn't stop a strong heart-to-heart connection from forming, almost like the idol is a beloved friend."

"Heart-to-heart connection?" Aiko said, repeating Mi-san's words.

"Aiko-san, what kind of fan are you?"

"I . . . I don't know," said Aiko.

Mi-san laughed. "Well, to me, if someone is willing to do so much for her idol even after he has died, she is more than a mere consumer."

"Maybe you're right. I'm just not sure if I deserve to call Hiroshi-kun my friend."

"In friendship, the only thing that matters is sincerity."

"I'm grateful for your kind words."

"I hope that you can dig deeper into the cause of Hiroshi's death. It might not be as simple as it seems."

"Cause?"

"This is all I can tell you. Good luck, Aiko."

Mi-san—now Aiko noticed that she did not look AI generated at all—hung up. Aiko looked up and saw her own face reflected in the privacy glass, her eyebrows furrowed as she contemplated Mi-san's advice.

THE TIP FROM MI-SAN completely upended Aiko's thinking. Instead of interviewing more people, she went back to studying the police autopsy report. What bothered her more, however, was Mi-san's question for her.

What kind of fan am I? Am I a consumer, or . . . a friend?

Aiko brushed the thought off and forced herself to focus on the clues.

On the surface, it appeared that Hiroshi-kun had died from drowning: His lips were purple, his face was pale white, and fluid was found in his mouth and upper respiratory tract, indicating suffocation. Until now, however, Aiko could not bring herself to examine the report closely.

She clicked on the photos from the autopsy, steeling herself as she compared them with pictures she'd found online depicting drowned bodies. It wasn't long before a problem with Hiroshi's stated cause of death became clear: Hiroshi's ear drums were not bleeding from water pressure, nor was his skin wrinkled from immersion in water. Most important, no signs of fluid were found in his lungs. The more Aiko looked at the evidence, the less likely it seemed that Hiroshi had actually died from drowning.

What was it, then?

Aiko typed her discovery into the game's answer box. After a light *ping* sounded, she found the autopsy report's conclusion page now unlocked and available to read.

```
Due to acute poisoning, he suffered from
respiratory failure, resulting in a sudden
drop in blood oxygen saturation, which led to
his death.
```

Aiko's eyes widened. *Poison!* She stood up and began pacing around the room.

Someone *poisoned* Hiroshi-kun? Aiko thought about her next steps. By examining the composition of the poison and figuring out when Hiroshi-kun could have come into contact with it, she could pin down the suspects. Then, by analyzing the purchase history and communication records of the people around him, she could discover the murderer.

The truth is finally coming to light! Aiko, with her hands clenched into fists, almost broke into a cheer, until she glanced down at another document, the poison test report.

Hiroshi's death was not caused by a single drug, but rather a compound of two. One of them remained unknown; the other one was Angellix—an antidepressant that Aiko knew well. People called the drug "the Angel's Smile" for its positive, mood-altering effects.

Was Hiroshi-kun depressed, too?

Just as she had made a breakthrough, Aiko was now confronted with more questions. She contemplated summoning Hiroshi's ghost again to ask about the details—but decided not to risk wasting her final question. She first needed more information, more evidence.

She dug out all of her Hiroshi X merchandise from the drawers in her bedroom and began dressing up. Shower cap, nightgown, slippers, canvas backpack, U-shaped neck pillow . . . *Here's hoping that all the money I've spent will be worth it, and Hiroshi X will bring me luck,* she thought to herself, grabbing the last piece of her collection—a cat-teaser toy.

Kneeling on the tatami, Aiko bowed twice, clapped her hands twice, bowed again, and then folded her palms together. "Hiroshi-kun, please, send me your blessing, so that I can find the truth," she murmured.

In the virtual vision field of her XR contact lenses, clues and materials were scattered in the air like items on a shopping guide, glowing when Aiko's finger hovered over them. Swiftly, Aiko moved the icons around, categorizing them and connecting them. Eventually, from amid the clutter, a mind map emerged.

She set the key word "depression" aside and labeled it "to be resolved." If Hiroshi-kun was a longtime user of Angellix, then a certain amount of chemicals contained in it could have reacted to the other unknown drug to create a poison. Whoever poisoned him must have been familiar with Hiroshi's medical history, Aiko theorized. It was unlikely that the unknown drug had entered Hiroshi's system through his skin—as the AI drug simulator suggested, it would take at least two hours for the drug to take effect on similar poisons, and thus difficult to control for the time of death. Therefore, Hiroshi must've ingested the drug orally. He would have felt the effects of the drug in about ten minutes, perfect timing for a fifteen-minute stage intermission.

Aiko cross-checked the testimony of staff witnesses. They confirmed that Hiroshi had consumed only water during intermission. And laboratory testing of a sample of the bottle indicated the water was clean.

Another dead end. Aiko pulled at her hair.

She opened up the video of Hiroshi's farewell concert and fast-forwarded to a few minutes before intermission. Hiroshi X onstage, looking as glamorous as ever, swinging his electric guitar as he brought the song to a ferocious climax with a solo. Sweat glistened on his forehead. Clamping the guitar pick between his lips, he bowed deeply to the 48,000 people sitting in the audience and to the millions of online watchers. The stage lights dimmed. Hiroshi stepped back and disappeared into the curtains.

No one had expected it to be his final appearance.

Wait. A detail caught Aiko's attention. Rewind, play, pause, zoom in. *Hiroshi's iconic gesture. The clue was so obvious.* As his biggest fan—how could she have missed it?

AIKO, WHO HAD JUST left the Rainbow Six musical instrument store, wandered amid the hustle and bustle of Shibuya, feeling lost.

Rainbow Six was the exclusive supplier of Hiroshi's guitar picks. The last batch of picks had been shipped to them over a month ago. Since it was easy for them to go missing, Hiroshi carried multiple guitar picks in his pocket as backups.

Notably, no one else had come into contact with his guitar picks on the day of the performance. No traces of drugs were detected on the pick he'd used, either. So much for her theory.

The clues that Aiko had discovered were like pearls on a necklace, held together by a single string of logic. Once the string was broken, they fell to the ground, bouncing, scattering, rolling away, utterly unsalvageable.

Aiko felt despair descending. *Should I give up?* As her thoughts began to spiral, Aiko remembered "Miracle Shine." Every time she felt as if her life was in ruins, memories of the song saved her day like a conditioned reflex. It didn't fail her today, either.

Don't use fate as an excuse for failure.

She had one shot left—summon Hiroshi's ghost, and ask her third and final question.

The ghost was nowhere to be found, however, even after she had tried pronouncing the name of almost all of his favorite foods. "For God's sake, did Hiroshi die from a poisoned guitar pick?" Aiko yelled in frustration.

Other pedestrians turned their heads at once to look at her, their expressions startled.

Aiko was so embarrassed that she felt like disappearing into a crack in the ground. Yet a few seconds after, she saw a familiar blue-green silhouette emerge in the middle of Tokyo's busiest street.

"Aha! I can't believe Aiko-san made it this far. What a surprise!" Ghost-Hiroshi hovered in the air. Through his translucent body, she could see glowing, multicolored billboards.

"So 'guitar pick' was the key word? But why—"

"Stop! You better think carefully before you ask the last question. It's your final chance. Also, don't forget, it's against the rules to ask me who the murderer is." Ghost-Hiroshi pressed a finger to his lips, his expression solemn.

"I understand," muttered Aiko weakly.

But then what could she ask him? *Perhaps there is a blind spot.* Her mind raced, bringing her back to the mind map of clues that she had organized, those icons that shimmered in her virtual vision field.

All of a sudden, the question labeled "to be resolved" flashed across her mind.

Mustering all the courage she could find, Aiko opened her mouth. "So . . . Hiroshi-kun, why did you have to take antidepressants?"

For a second the ghost's body froze entirely, as if a glitch had happened. Aiko worried that she had made the game end prematurely.

Half a minute later, Ghost-Hiroshi began to move again. However, his personality seemed to have changed entirely. The humorous, carefree aura disappeared, now replaced by melancholy.

"I knew you'd ask this question, Aiko-san. You truly care about me, as a person, and not just the mask I put on to show the world. People say that the greatest asset of an idol is a lovable, charismatic, *flawless* persona. But your persona, down to every aspect and every detail, is created by a team of people based on user research. It's a product. The person behind the persona is cast aside.

You may look all glamorous and shiny, but they can tear you up, destroy you, and discard you any moment."

"No, Hiroshi-kun, that's not true! I like you, the *real* you," Aiko blurted out.

"Is that so, Aiko-san? Would you still like me if I were actually an entirely different person from the Hiroshi X you see on TV? I was tired of this endless role-playing. It made me despise myself. I wanted to quit, but no, there was no escape . . ."

"Why? Who's stopping you? Is it Mi-san? Or your sponsors? Those nasty capitalists . . ."

Suddenly, Hiroshi X burst into a fit of hysterical laughter, as if Aiko's words were utterly ludicrous.

"Hiroshi-kun, what's the matter? Don't scare me like that."

"It's because of *you*."

"What?"

"You. All of you. Fans who claimed that they would love me and support me till the end. Upon hearing I would retire, many people sent me death threats—it wasn't my life they were threatening, but their own."

Shocked and terrified, Aiko clamped a hand across her mouth.

"I supposed that committing suicide was the only way to redeem myself, to cleanse my shame and never be a burden to other people again. Isn't it ridiculous? Those children who had been mailing me their own blood, hair, and photos of self-harm . . . did they really love me as they claimed?"

"Hiroshi-kun . . ."

"Death was my only choice." Ghost-Hiroshi raised his head, his voice firm. The edges of his figure glowed with a gentle, rainbow-colored light. "To preserve the final and best moment of my life onstage, bright and beautiful. This was the only way that those people would let me go."

"Hiroshi-kun!" Tears were running down Aiko's cheeks.

"I owe you my gratitude, Aiko-san. Thank you for all that you have done for me. However . . ."

"What is it?"

"Aiko, aren't you one of them as well?"

Smiling, the ghost of Hiroshi gazed into the eyes of the girl, whose face turned bloodless. The bustle of Shibuya ceased at once.

AFTER GOING BACK TO her apartment, Aiko felt a little absent-minded as she walked into the virtual crime scene once again.

The dressing room before her eyes could not have been more familiar. In the past few days she had scrutinized every corner repeatedly, in fear that she would miss out on clues.

None of that matters anymore.

The scene of Hiroshi's death played in the virtual vision field of her XR contact lenses on 2x speed.

Hiroshi, drenched in sweat, returned to the changing room surrounded by his team. The makeup artist fixed his makeup; the stylist decorated him with more accessories; the music director updated the music rundown. As his team babbled, Hiroshi licked the guitar pick, retuned the guitar, and adjusted the synthesizer. Like a bodyguard to a monarch, Mi-san was never more than an arm's length away at any time.

Finally, the king gestured for everyone to leave. No one noticed the paleness of his cheeks as they filed out. Just moments after Hiroshi locked the door, he slid to the floor. Lowering himself to his knees, he closed his eyes. His body began to tremble. His delicately painted face turned purple. His eyes widened. Mouth gaping, he looked as if he had something to say, but his voice choked. Seizing the bottle of water next to him, he chugged the water in big gulps, choked again, and broke into a coughing fit. Water gushed out of his mouth. Struggling, he crawled toward the door, but his body started to convulse. His fingers curled up. Finally, he dropped to the floor, and his chest stopped heaving.

A few moments later, someone knocked on the door. The knocking, gentle at first, soon turned rapid.

"So, this is what happened," said Aiko softly.

"You still haven't explained to us why we didn't find traces of drugs on the guitar pick found alongside Hiroshi's body." A new voice suddenly sounded in her XR earphones; it was the AI-generated gamemaster from historiz.

"To answer your question, we need to rewind the video and look at what happened minutes before intermission. I must admit that this was a blind spot for me at first."

With her right hand, Aiko made a counterclockwise turning gesture. Hiroshi opened his eyes. Water on the floor flew back into his mouth. He walked to the door and unlocked it. All the staff members reentered the room, walking backward, busy with their jobs. The scene before her eyes was like something straight out of an absurd cartoon. She stopped the rewind at last when Hiroshi was back onstage in the frame.

"I could watch this forever," murmured Aiko to herself as her idol once again danced under the dazzling stage lights. "Okay, *stop.*"

The scene froze as Aiko gestured again.

"Pay attention to his right hand. I will switch over to 0.5 times speed."

Hiroshi, after playing the last note, held the guitar pick between two fingers of his right hand, which halted in midair. When the light dimmed, he let his hand drop to his side. For a brief moment, his hand was completely blocked by his guitar.

"Turn ninety degrees counterclockwise," commanded Aiko.

With the gamemaster following, Aiko walked to Hiroshi's right side, where they could see where his hand was placed. Hiding the guitar pick in the front pocket of his jeans, he pulled another pick out from the coin pocket, elegantly clamped it between his lips, and gave his signature smile.

"He took the drug before he entered the changing room, and he must have discarded the guitar pick that carried the evidence. According to the time stamp, it's exactly twelve minutes between now and the time when the poison took effect."

The icon of Hiroshi disappeared. Aiko found herself standing at center stage, in a large auditorium. The spotlight was shining in her face. Instinctively, she put a hand over her eyes. The auditorium broke into a ferocious wave of applause and cheers.

They are . . . cheering for me? wondered Aiko.

"Brilliant, brilliant! Congratulations, Aiko-san! You found the truth. You are our master fan detective!" The gamemaster bowed deeply, their figure disappearing into the darkness. "Next will be our closing ceremony. Don't walk away—it's a part of the story line!"

Although she already had a vague inkling of what might happen next, Aiko held her breath anxiously. Her heartbeat quickened.

When the ghost of Hiroshi X appeared before her eyes again, "Miracle Shine" began to play in the background, submerging her like a gentle tide. Aiko's body trembled.

"Aiko-san, thank you for rescuing my lost soul. Love was suffocating me, but your love . . . your love saved me. Now I can seek Heaven's light. Farewell, Aiko-san! Be happy."

"I'm sorry, Hiroshi-kun, I'm so sorry . . ."

Bursting into tears, Aiko opened her arms and rushed toward the man, trying to pull him into an embrace, but all she could feel was emptiness. The ghost of Hiroshi X glowed with light as it slowly ascended. Smiling till the very end, Hiroshi gradually disappeared, melting into a starry sky.

Aiko was left alone on the stage. Two shiny boxes emerged before her eyes: cherry blossom pink and bird's egg blue.

"As the winner of our game, you can choose your award. Brought to you specially by Viberz, the pink box contains a Hiroshi X smart doll that bears a 99.99-percent resemblance to Hiroshi himself. Note that by 'resemblance,' I mean personality, voice, appearance . . . you practically can't tell the difference between the person and the AI doll. You can have him all to yourself for an entire month!"

Eyes still stinging from tears, Aiko raised her head and gave the box a suspicious look.

"Choosing the blue box, though, means that you will have the opportunity to have tea with Hiroshi X! The *real* Hiroshi X. What a rare opportunity!"

"Now, Aiko-san, please make your choice!"

The pink and blue boxes hovered gently before Aiko's eyes, like bait bobbing in the waves, waiting for the fish to bite.

+———

"WHAT DID YOU CHOOSE? *Woman! I want to know everything!*"
Nonoko yelled, threatening to throw her matcha crepe at Aiko's
face.

"Don't be so inappropriate, Nonoko! At least not in a place like
Doux Moi."

Nonoko smirked. "If I were you, I would have gone for the smart
doll. A whole month! You could have done . . . so many things to
him."

"Well, too bad you're not me!"

"So, you met him in person! Is his face really that handsome?
He's old now, isn't he? Is he still attractive? Tell me!"

Sinking back into her memories, Aiko smiled.

The afternoon of the tea was ingrained in her mind. She could
still recall how nervous she felt when she waited alone in the café.
A breeze, an inquiry from the server as she waited at the table, a
dog bark, the sound of the coffee machine grinding beans . . . every
little noise sent her jumping out of her seat. She even thought
about running away. *Why did I choose the blue box? It's irrational!*
she thought to herself. *But then again . . . isn't loving someone
who's so far away already irrational to begin with?*

A gentle voice woke her up from her daydream.

"Are you Aiko-san?"

"Y—yes!" Instinctively, Aiko stuttered in response. Terrified,
she buried her head, unable to look at the man who took the seat
across from her.

"Nice to meet you. I am Hiroshi."

Finally, Aiko looked up and forced a smile. She surveyed the
idol of her dreams.

The man in front of her was in his midforties, medium-built.
The baseball cap he wore hid his hairline. His complexion was
smooth, but the marks left by time and the weary expression he
carried betrayed his age. There was a trace of stubble on his upper
lip. His jaw had grown wider, too. His eyes were no longer glistening with the pride of a young man, but rather the gentleness and
composure of middle age.

It was Hiroshi X himself, nearly two decades older than the age he'd retired from the stage, the superstar in the dreams of millions of people.

"Can you still recognize me? I suppose I look much older now. After all, it's been twenty years." Hiroshi smiled self-mockingly.

"No! No, of course not. You are just as handsome!" Aiko looked away, too shy to meet his eyes.

"Don't be so nervous. Just treat me like any ordinary guy."

"Oh—okay, sure!" Aiko stammered.

The waiter brought them two cups of coffee and a plate of cookies.

Hiroshi bit into a cookie and exclaimed in wonder, "Wow! Their cookie still tastes the same after so many years."

"This is the 'crunchy butter cookie' that you tried when you were filming the 1,278th episode of *Street Food Scavenger Hunt,* isn't it, Hiroshi-kun?"

"Ah! You even remember that? Aiko-san, you really are a die-hard fan indeed. I'm sure this is why historiz chose you as one of their first test users."

"Yes, that's true," said Aiko. Sipping the coffee and nibbling a cookie, she finally relaxed.

"Well, I want to hear about your feedback on this—immersive game."

Aiko set down her cup and took a deep breath, her face thoughtful.

"I've never experienced anything quite like it. Even though my brain knew it was fake, the details of the game design—the way the AI ghost spoke, gestured, and interacted with me—were enough to suspend my disbelief. Gradually, as I allowed myself to immerse fully in the game, my doubts went out the window."

"You sure speak highly of the game. I'm pleased to hear it."

"I have one question, though: Did AI create everything?"

"Aiko-san, you mean . . ."

"As an editor—well, at least, as one until recently—I understand how difficult it is to tell a story. You have to reconcile form and content, build emotional resonance. It's hard. In the narrative of the game, the key was not the crime, but the relationship between the truth and the participant's own emotions. When

Hiroshi-kun—sorry, I mean your AI ghost avatar—said, 'Aren't you one of them as well?,' I almost broke into tears. Could AI really manage to come up with this kind of story line entirely on its own?"

"Well, let me fill you in a little bit," Hiroshi said. "Remember the endless questions you answered for the historiz rep before you entered the game? That helped AI generate a comprehensive personality profile of you. It learned about the kind of story you prefer, how you might react, even about your past traumas. To some extent, AI knows you better than you know yourself. You asked a good question, though—AI didn't tell this story entirely on its own. A human writer stepped in to help."

Aiko's eyes lit up with admiration. "I want to meet the writer! He's a genius!"

"Well, you're looking at him," said Hiroshi. He clamped a cookie between his lips and smiled at her, as if he were still the young rock star from twenty years ago.

"Hiroshi-kun!"

"historiz showed me every user's profile. Based on AI's decision, I chose the most interesting and appropriate plotline. I am a hardcore fan of detective novels, too! I also took part in the design of various key plot points. For example, I played a narrative trick on you in the very beginning, to mislead you on purpose. AI can't do that."

Aiko's mouth hung open. "A narrative trick!"

"Yes," Hiroshi went on. "Remember when Ghost-Hiroshi gave his first speech through the medium during the séance? He said, 'I don't want to die.' You could have interpreted that line as either 'Someone murdered me, but I don't want to die' or 'I don't want to die, but I have no choice.' Naturally, everyone would assume that he was murdered, and thus rule out the possibility of suicide. That was a red herring! But, Aiko-san, I have to say you're very good at making deductions!"

"Thanks to the detective stories I read at work," responded Aiko, her cheeks turning crimson. "I have another question, though, and I've been thinking about it for years. Hiroshi-kun, why did you suddenly disappear from the public, and why did you choose to come back like this?"

"Ah . . . to answer your question, I need to tell a story that began twenty years ago."

Hiroshi's expression turned dreamy.

THE HIROSHI X FROM the game was both virtual and real.

Twenty years ago, when Hiroshi X was at the peak of his fame, he grew tired of sustaining the carefully honed persona that the fans had come to expect. He decided to strip away his disguise and appear in public as who he truly was. The market, though, did not approve of his decision. The sales of his records and merchandise plummeted. Negative stories appeared in the press. One by one, his sponsors ended their contracts with him. However, the final blow to Hiroshi was not his financial loss, but the reaction of his fans.

Die-hard fans wouldn't accept the loss of Hiroshi X's stage persona. Speculating that his career had been manipulated or mismanaged by his advisers, they waged wars on online forums. The debate grew so loud it dominated entertainment websites and shows. The fans' endless debate inspired a backlash, with critics arguing that celebrity idols should be accountable for the behavior of their fans. In the eye of the hurricane was Hiroshi X. Was he a bad influence on a younger generation? The prime culprit behind the hollowness of the entertainment industry? Or a guy trying to do the right thing?

The scrutiny and pressure took its toll on Hiroshi. He was diagnosed with depression. Eventually, he decided that only by disappearing entirely could he win back his fans and temper the warring hordes of armchair critics.

With Hiroshi absent from the public stage, the gossip and chatter about him died down. Hiroshi X was gradually forgotten. New superstars emerged onstage, giving rise to their own fandoms.

After he had recovered, Hiroshi changed his name and was finally blessed with the freedom to be true to himself. He went back to school, where he made a good friend—Taiyo, the future cofounder and CTO of Viberz.

Taiyo was a typical geek who believed firmly in the power of technology. A game enthusiast, he'd always dreamed about making a game that could impact the world. One night at a bar, Hiroshi, after many drinks, revealed his true identity to Taiyo and lashed out about the unhealthy power dynamic between idols and fans. Hiroshi's complaint struck Taiyo like a lightning bolt.

"You think that fans have seized the power to control their idols, but the truth is, fans are so obsessed with prepackaged stage personas because they *lack* the power to tell their own stories. Think about it. Fans have essentially nothing apart from a persona—from the personality and narrative fed to them by managers and the media—on which to project their emotions. When you reveal to them that the persona is, in fact, fake, you shatter their dream! To them, this is deception, betrayal, a slap in the face!"

Hiroshi had to admit that Taiyo had a point. But what Taiyo proposed next sounded overly simple: Create a game with AI where everyone could participate in the making of a personalized idol and decide how they would like to interact with them.

"I think you've misunderstood what an idol represents. Someone only becomes an idol when a group of people come together and establish a collective worship ritual," Hiroshi explained. "Your game can't create an idol. How is it different from those make-your-own-character games?"

The pair continued to debate this business idea throughout their years together in school. Finally, they came to a solution: They would use digital technology and AI engines to make virtual idols based on idols in real life and personalize them according to the needs of each fan. Taiyo even imagined the possibility of creating a highly personalized interactive game in the future.

But where would they find a celebrity idol willing to embrace their cutting-edge idea and participate in their experiment? The entertainment industry that Hiroshi was familiar with cared only about money and short-term gains. No one would have the expansive vision to support such a radical idea—and bank on it as the future of the entertainment industry.

"I think we already have a candidate," said Taiyo, smiling as he gazed into Hiroshi's eyes.

In the beginning, Hiroshi was adamant: It couldn't be him. He wanted to stay as far away as possible from the suffocating, toxic fandom. But Taiyo managed to convince him. After all, it would be a chance for Hiroshi to prove himself right: An idol should never be burdened with the impossible feat of upholding a single-faceted, prepackaged persona, and there should never be a singular kind of relationship between an idol and their fans. The best solution was to hand the power completely over to fans. They could tell their own stories.

The two men founded Viberz. They did not expect, though, that developing the company's technology would take a decade.

The hard part was not building digital avatars through high-definition scanning and modeling, nor establishing a physical movement database through motion capture. Even expression simulation wasn't too difficult—it was only a matter of precision. The real painstaking work involved natural language processing and training the AI model with massive amounts of data. Only by accomplishing this goal could they create smooth and natural human-machine interaction. The moment the conversation felt off to users, Hiroshi and Taiyo knew, it would be the kiss of death. Finally, they would have to find a way to learn the specific dreams of the users. Relying on the most accessible data survey and modeling tools, they generated personality profiles of their target users and mapped those results onto their product. Forging and synthesizing all these technologies into a comprehensive product required endless days and nights.

Their timing, however, was good. A few years before Aiko became a beta user, a nostalgic trend toward the music of the previous generation had reignited interest in Hiroshi X.

Through the forgiving lens of time, Hiroshi X had turned out to be the most popular superstar of his era. Unlike celebrities who had grown older in front of the camera, Hiroshi's name still meant the pretty, glamorous boy onstage with an angelic voice, frozen in amber.

Viberz seized the opportunity. Equipped with a fully mature product, as well as the exclusive rights to Hiroshi's image and likeness, they managed to send virtual Hiroshi X to digital screens of all shapes and sizes, as well as to VR, AR, MR . . . various kinds of XR vision fields. Merchandise once again began selling in droves.

Celebrity managers rushed to Viberz, asking them to create more virtual idols. It was then that Hiroshi and Taiyo knew that the time had come for them to take their final step. They launched historiz, a Viberz subsidiary dedicated to immersive, interactive games.

"WAIT, YOU MEAN YOU didn't go on an *actual* date with him?" Nonoko sighed as she shook her head disapprovingly.

"Stop it! My love for him is purely platonic," argued Aiko.

"You could've at least asked for limited-edition merchandise or something!"

"Well, actually . . . he sent me an invitation."

Nonoko almost spat out her tea. "To do *what*?"

Aiko's cheeks turned rosy, glowing with happiness. "He asked me whether I would like to collaborate on stories with him."

"Huh?"

"Hiroshi-kun said that historiz needs writers and editors with a knack for telling stories that can touch hearts."

"You got a job offer from Viberz, the hottest AI tech-entertainment company! Don't tell me that you turned him down!"

"I accepted the offer on one condition," said Aiko.

"What? Aiko, are you out of your mind?" Nonoko's eyes were fiery with envy.

"My condition was, next time, it would be my turn to decide the way Hiroshi-kun dies in the game."

"What did he say?"

Stirring the coffee with her spoon, Aiko looked up. Her gaze landed on a spot behind Nonoko, as if she had caught a glimpse of a translucent silhouette glowing with a blue-green light. She grinned.

"Hiroshi-kun said, *deal.*"

ANALYSIS

VIRTUAL REALITY (VR), AUGMENTED REALITY (AR), AND MIXED REALITY (MR), BRAIN-COMPUTER INTERFACE (BCI), ETHICAL AND SOCIETAL ISSUES

At the beginning of "My Haunting Idol," the pop star figure Hiroshi appears to be a "ghost" summoned by a group of superfans after his untimely demise during his farewell concert. We meet Aiko, one of these superfans, who is intent on investigating her idol's mysterious death. Hiroshi appears in Aiko's kitchen late at night to provide hints to aid Aiko's pursuit of the truth, and even descends on the busy streets of Tokyo as Aiko traverses the city to solve the puzzle.

Then, toward the end of the story, we discover that Hiroshi was, in fact, an AI-enabled virtual character, visible but not touchable. But for Aiko and for the readers, Hiroshi seems real—or real enough to produce genuine suspense and thrills. As portrayed, Hiroshi is not only physically lifelike, but seamlessly integrated into the world. Such "naturalness" is on course to be produced by computer vision and natural language processing in the years to come, as well as by immersive simulation technologies known generally as X reality, or XR.

XR means much more than just expanding to a larger screen. In the words of Dr. Brennan Spiegel, XR "is like dreaming with your eyes open." These technologies generate an intense experience known as "presence." Virtual scenes, objects, and characters are lifelike and magical. The technology takes you into an immersive experience that feels like a parallel reality. In the next twenty years, XR will revolutionize entertainment, training, retail, healthcare, sports, and travel.

Let us now explore the boundary between the real and the virtual, and lift the veil on the mysterious technologies of XR.

WHAT IS AR/VR/MR (XR)?

XR is a term encompassing three types of technologies: VR, AR, and MR. Virtual reality (VR) renders a fully synthesized virtual environment in which the user is immersed. The VR world is separate from the world of the user's body (think of how Aiko's XR contact lenses transported her virtually to the Alps while exercising). By contrast, augmented reality (AR) is based on the world that the user is physically in, capturing it through a camera, and then superimposing another layer on top of it. AR algorithms superimpose content (3D objects, text, video, and the like) to create a "lens" that provides a user an "extrasensory" view of their world. For example, a tourist in an unfamiliar city could ask an AR system what historical sights are nearby, and the system might superimpose cartoon balloons over the real streets to show what you should go see. In the story, when Aiko summons clues and materials that seem to float in her room as seen through her lenses, she is using AR. (If you're a film buff, recall how OASIS turns Samantha into Art3mis in *Ready Player One*.)

In recent years, another technology, mixed reality (MR), has emerged as a more advanced form of AR. MR mixes virtual and real worlds into a hybrid world. MR's synthesized virtual environments are not a simple sum of the real and the virtual, but rather a complex environment built from a full decomposition and interpretation of the scene in order to provide interactivity with the objects contained in it. In "My Haunting Idol," MR integrates Hiroshi seamlessly into Aiko's kitchen and gym (not to mention the streets of Tokyo), and to such a realistic extent that Hiroshi "gazed into Aiko's eyes," and made her blush. To function, MR needs to understand quite a bit about its setting and the objects and people within it. For example, the refrigerator needed to be recognized with its doors and functions, so that a virtual Hiroshi could be rendered as the door opens, appearing in "the cold, watery mist of the refrigerator." The scope of MR understanding of its environment suggested by this story is beyond today's computer vision, but should be feasible within two decades. In the story, Aiko suggests that she feels "practically blind" without her XR contact lenses, and I imagine that may be the case for many of us in the years to come.

MR is a technology that is still in its infancy, but we are seeing a positive and steady trajectory of improvements. By 2041, I predict that computer vision technologies will be able to unpack a scene into its constituent components and understand the role of almost all of them. In addition, I predict that MR will develop the capability to add new virtual objects to environments, presenting them in ways that seem to obey the laws of physics and allow them to appear natural, thus enabling the scenarios described in "My Haunting Idol."

VR/AR/MR (XR) TECHNOLOGY: OUR SIX SENSES

An immersive experience should be one in which the user experiences the same sensations he or she would in a real-life environment and is unable to distinguish between what is real and synthesized. In order for such a sensory experience to be realistic, we must fool our most acute sense, our sight.

Consider how the "AR game" Pokémon Go used a phone screen as a window to the real world with synthetic cartoon characters, cleverly using gyroscopes and motion sensors in the phone to modify your viewing and even interact with the scene. It was a popular game because it was novel, but the user experience was trapped in the small screen of mobile devices. It was not fully immersive and did not revolutionize the user experience.

A much more immersive experience can be delivered using head-mounted displays (HMD), which might look like helmets or goggles. An HMD has two screens in front of our two eyes. The two screens show images that are slightly different to trick our eyes to "see 3D" (similar to the 3D glasses used with 3D TVs or theater screens). An XR experience is also immersive and interactive. Immersion requires a field of view at least 80 degrees wide, and usually wider. And interaction requires that when the head or the body moves, the user will see correspondingly different views. Also note that for VR, the HMD is typically not "see-through," because the entire scene is synthesized. But for AR and MR, the HMD has see-through lenses (either directly or optically), and the real world is blended with synthetic objects and then reflected to the user's eyes. In the story, we saw the boy in the gym switch glasses from MR mode to VR mode, as an opaque coat of silver veiled the lenses.

The earliest immersive device was created many decades ago. These early devices were clunky, using heavy HMD helmets connected to a workstation or mainframe computer with physical cables. In those early days, there were no smartphones or wireless connections, so this setup was needed in order to use the computing power from a mainframe computer, the speed of the physical cable, and large and heavy displays mounted in even larger HMDs. While inconvenient and unattractive, and with no commercial applications or value, this configuration served an important role: providing a laboratory environment where scientists could test and improve technologies.

In the past few decades, significant improvements have been made on networking, resolution, refresh rate, and latency. With the arrival of Wi-Fi and 5G, devices have become wireless. New electronics and display technologies shrank the HMDs from helmets into the size of goggles. And CPU improvements made it possible to remove the mainframe computer and perform the computation in chips in the HMD. So began the commercialization of XR.

But these efforts were anything but smooth. Companies developing AR/VR applications became an extremely hot investment area around 2015, but many high-profile start-ups failed. Products from mainstream companies also faltered, leading to a catastrophic downfall. One product that survived the AR/VR bubble is the Microsoft HoloLens. The HoloLens HMD is reasonably usable, weighing only 579 grams (1.3 pounds), and has substantial computing power. But it still costs too much ($3,500), and looks like a large pair of underwater goggles, meaning wearers face a distinct "dork factor." For these reasons, the HoloLens was relegated to vertical business applications like training, healthcare, and aeronautics. Anything with helmets and goggles cannot become a mass, wear-all-day product like the Apple Watch.

In parallel, efforts to "undork" the AR/VR into glasses-like hardware were premature and also failed. Google Glass and Snapchat Spectacles were unsuccessful for several reasons. Core among them was that such miniaturized products no longer delivered the HoloLens high-fidelity experience.

Technical limitations will ultimately be overcome. Every year in the past five, bandwidth, frame rate, resolution, and dynamic range all improved significantly, while hardware has become lighter and cheaper. Microsoft HoloLens could become lighter and cheaper, or the Snapchat

Spectacles may become more powerful. Either way, we will arrive at a high-quality lightweight pair of glasses. In 2020, the Facebook Oculus team demonstrated research prototype VR glasses with lenses only one-centimeter-thick. These developments suggest that a mass-produced set of XR glasses should be on the market by 2025, perhaps with Apple again taking the lead (there are rumors that Apple is working on such a product). Pioneering Apple products like iPod, iPhone, and iPad catalyzed entire retail categories, and their many copycats subsequently drove down component costs.

Beyond XR glasses, I believe XR contact lenses may be the first XR technology to achieve the milestone of mass acceptance. Several start-ups are already working to develop XR contact lenses. Their prototypes show that displays and sensors can be embedded in contact lenses, making text and images visible. These contact lenses still require external CPU for processing, which can be done on a mobile phone. By 2041, we anticipate the "invisibility" of contact lenses will truly cause the market to accept the product, and that challenges such as cost, privacy, and regulations will be overcome. This is the assumption made throughout "My Haunting Idol," where Aiko wore contact lenses all the time, and occasionally put on real glasses and other gadgets for immersive experiences.

If visual input will be provided by glasses and contact lenses, audio input can be achieved through ear sets, which have improved with every year. By 2030, good ear sets should be almost invisible, through bone-conducting, omni-binaural immersive sound and other technologies, perhaps to the extent that they could be comfortably worn all day.

This combination above is likely to become sufficient to evolve into an "invisible smartstream" (or the smartphone of 2041). When you summon your smartstream, the visual display covers your field of view, perhaps semi-transparently. You could manipulate the smartstream content and apps using gestures, like Tom Cruise's character in the movie *Minority Report*. The smartstream sound will be heard by your "invisible ear-sets," and operated by voice, gestures, and fingers typing "in air." This ever-present XR smartstream can do more than a smartstream (or mobile phone) with a screen. It can remind you of the name of an acquaintance you ran into, alert you when a nearby store has what you want to buy, translate for you when you travel abroad, and guide you to escape from a natural disaster. Beyond the usual "six senses," our body can "feel" sensations such as wind and an embrace, as well as warmth, cold, vibration, and pain. Haptic gloves will allow you to virtually pick up objects and feel them. And somatosensory (sometimes also called haptic) suits can make you feel cold or hot, or even that you're getting punched or caressed. Through technologies on the thin skin of the suit, Aiko experienced a breeze, the warmth of sunshine, and bumps on the road on the VR spinning bike. Body suits may use motors or exo-skeletons to simulate touch, or they may stimulate nerve endings to cause muscle contractions. When the body collides with an object in the virtual space, pulses are sent to the appropriate region on the somatosensory suit to simulate that collision. The somatosensory suit also monitors Aiko's physiological data and body posture in real time, possibly turning gestures into commands. These body suits can be used in vertical applications such as games, training, or real-world simulation. The technologies are already in early commercial products and should mature well before 2041 for many applications.

Scent emitters, taste simulators, and haptic gloves that simulate touch are all emerging to cover our six senses (actually five, since we don't expect an ESP simulator).

BEYOND OUR SIX SENSES

The analysis above pertains to devices stimulating our perception, but how do we provide input, or control XR? Today, the input device for XR is a handheld controller similar to the Xbox controller but usually one-handed. These are easy to learn and to use, but feel unnatural as the rest of the experience becomes immersive and lifelike. The ideal future input should be purely natural. Eye tracking, movement tracking, gesture recognition, and speech understanding will be integrated to become the primary inputs.

How do we handle movement in the virtual world? If we use natural movement in the real world to correlate movement in the virtual world, then we need large spaces. But even then, how do we run, crawl, and climb up and down? And how do we eliminate the risk of falling? The best solution today is to use an omnidirectional treadmill (ODT), which was featured in *Ready Player One*. ODTs are already available on the market. They include a frame-mounted shoulder-worn harness, which detects the force applied by the user's body and protects the user from falling. This ODT rotates at the same rate as the user's movement, so as to always keep the user centered. The treadmill can be tilted to simulate hills or stairs. This allows essentially any movement without the danger of falling.

With these capabilities in mind, I project that the most likely application will be entertainment related, for example, hyper-realistic games where our digital twins will play games, compete, and battle with other people's digital twins in athletic competitions and battle simulations. Users could also interact with and spar with purely synthesized beings (like Hiroshi in "My Haunting Idol"). With such experiences at our disposal, humans by 2041 may increasingly live in multiple worlds, one real, some virtual, and others a mix of the two.

I foresee plenty of nongaming applications, as well. Training will be a major XR application area. Microsoft just sold $22 billion of HoloLens to the U.S. Army over the next ten years for training to deliver situational awareness, information sharing, and decision-making. VR will also be used for treatment of psychiatric problems like PTSD. We will have educational environments where real and virtual teachers take students to travel in time and witness dinosaurs, visit the wonders of the world, listen to Stephen Hawking, and interact with Albert Einstein. Zoom video conferences could render far more lifelike meetings where people look like they do around a table in the flesh (but they may actually be wearing pajamas at home) and work together on a virtual whiteboard (that feels real). In healthcare, AR and MR could enable surgeons to perform surgery while VR can help medical students operate on virtual patients. In retail, customers could try out clothing and jewelry, decorate homes and offices, and select among vacation destinations by sampling them in VR first.

One major obstacle to achieving these experiences is content creation. Content creation in an XR environment is similar to creating a complex 3D game; it must encompass all permutations of user choices, model the physics of real and virtual objects, simulate the effects of light and weather, and deliver lifelike renderings. This level of complexity is much greater than what's needed for making video games and developing apps. But without high-quality professional content, people won't buy the devices. And without a proliferation of devices, the content will not monetize well. This chicken-and-egg problem will require an iterative process that will ultimately create a virtuous cycle, just like television and Netflix took a substantial amount of time and investment to become mainstream. That said, once the tools are invented and tested, the proliferation will be very rapid. One could imagine that professional tools like Unreal and Unity may evolve into the XR version of photo filters one day.

Finally, XR will need to overcome issues of nausea, which is mostly caused by the latency of the experience, or the disconcerting feeling caused by the "world moving slower at times." As technology and network bandwidth is improved, this problem will also be reduced.

XR GRAND CHALLENGES: NAKED-EYE AND BRAIN-COMPUTER INTERFACE

The most natural way to see a virtual environment would be with the naked eye, as a holograph. In 2015, MR company Magic Leap released a video showing a whale leaping through a gymnasium floor. People assumed that this holographic effect could be seen without wearing glasses, and that made Magic Leap one of the most talked-about companies that year. However, when Magic Leap finally released its products, it turned out that glasses were still required. But this "misunderstanding" clearly shows the allure of naked-eye MR.

Unfortunately, naked-eye MR is possible only under extreme constraints. In 2013, a famous Chinese singer who died in 1995 was brought to life at a concert using a 3D light field hologram that looked almost real from the theater seats. However, the hologram was not photo-realistic, and could be observed only from a distance, with no interactivity. Holography is improving over time, but it is unlikely that naked-eye holography will become anywhere as good as XR assisted by glasses or contact lenses by 2041.

If naked-eye XR is the most natural "output," then the most natural input must be brain-computer interface (BCI). In 2020, there was big news from Elon Musk's Neuralink, which demonstrated a practical BCI by embedding three thousand very thin electrodes—which can monitor the activity of one thousand brain neurons—into the brain of a pig. This avenue of research shows promise in treating spinal cord injuries and neurological diseases like Alzheimer's. But the observation that captivated the media's attention was Musk's optimistic belief that it would lead to the possibility of downloading and uploading brain activity, allowing us to save and replay memories, insert memories in other people, or store them for immortality.

At the risk of dashing readers' hopes, we can say that a sober examination of the technology's feasibility will show that Musk's vision is far in the future. Numerous problems remain unsolved. For example, the probes cover only a tiny fraction of the brain. Repeated probing will damage the brain. We have not figured out how to make sense of these signals, so for now all we have are raw signals without meaning. Uploading is even trickier because we would be altering a live human brain— which would clearly entail health, privacy, and ethical implications. Neuralink engineered an interesting prototype, but by 2041 it will not likely achieve Musk's human-amplification ambitions.

ETHICAL AND SOCIETAL ISSUES WITH XR

We have discussed technical and health challenges in the popularization of XR. Just as daunting are the ethical and societal issues inherent in these technologies.

In "My Haunting Idol," Hiroshi first appeared in front of Aiko when she spoke the magic words to summon him inadvertently. Hiroshi appeared in the kitchen when Aiko was removing a carton of milk from the refrigerator. How would the system know that this was a convenient time, as opposed to when Aiko was taking a bath? In order to pick a convenient time, the system must know and watch Aiko at all times. Is that really acceptable?

If we wear devices like glasses or contact glasses all day long, then we are capturing the world every day. On the one hand, it is wonderful to have this "infinite memory repository." If a customer wants to renege on a commitment, we will be able to search and find the video of his or her promise. But do we really want every word we say to be stored? What if this data falls into the wrong hands? Or is used by an application we trust but has an unknown externality?

Clearly, we need to develop regulations on XR governance, and prepare for a world with many more privacy issues and externalities than exist today. Many people think smartphones and apps already know too much about us, but XR will take things to a whole new level.

XR will make us rethink what living means. Humans have been seeking immortality for thousands of years. With these technologies, we can ponder the possibility of a "digital immortality." In one episode of the popular TV show *Black Mirror*, a woman who lost her boyfriend uses his digital information to bring him back to "life."

As MR becomes more realistic and pervasive, the situation depicted in this episode may be achievable in the not-too-distant future. We discussed earlier the use of GPT-3 to let us talk with historical figures (the technology still has flaws, but is improving rapidly). There are also already a growing number of virtual influencers on social networks. It is only a matter of time before most are powered by AI and VR.

Such "digital immortality" or "digital reincarnation" will trigger many privacy and moral issues. Is it merely a copyright violation when someone uses other people's data to generate a virtual character? If that character says or does bad things, is it merely slander, or worse? Who is responsible when a virtual character misleads people? Or commits a crime?

If we learned anything from the recent concerns about the externalities of social networks and AI, we should start thinking early about how to address the inevitable issues when these externalities multiply with XR. In the short term, extending laws may be the most expedient solution. In the longer term, we will need to draw on an array of solutions, including new regulations, broader digital literacy, and inventing new technologies to harness technological issues.

The bottom line is that by 2041, much of our work and play will involve the use of virtual technologies. We should orient ourselves to this inevitability. There will be giant XR breakthroughs, probably starting in entertainment. All industries will eventually embrace as well as struggle with how to use XR, just like they do with AI today. If AI turns data into intelligence, XR will collect a greater quantity of data from humans at a higher quality—from our eyes, ears, limbs, and eventually our brains. Together, AI and XR will complete our dream to understand and amplify ourselves—and, in the process, expand the possibilities of the human experience.

CHAPTER SIX

THE HOLY DRIVER

STORY TRANSLATED BY EMILY JIN

SO THERE ARE TWO TONES GOING ON. IT'S LIKE
YOU'RE PLAYING TWO GUITARS AT THE SAME
TIME. YOU HAVE TO LET IT GO, BUT STILL
CONTROL IT.

—JIMI HENDRIX, QUOTED IN *JIMI HENDRIX:*
A BROTHER'S STORY, BY LEON HENDRIX

NOTE FROM KAI-FU: Set in Sri Lanka, "The Holy Driver" imagines a society two decades from now that is in the midst of transitioning from human drivers to autonomous driving by AI. In the story, a talented young gamer is recruited for a mysterious project—one that reveals the capacity for both humans and AI to make mistakes, but very different ones. In my commentary, I will describe how autonomous vehicles work, and how and when fully autonomous vehicles will emerge.

THE WRISTWATCH BUZZED, its face blinking with an urgent red flash. It was time for Chamal to race.

A regular at the VR Café, he had developed a full pre-race routine. His expression solemn, Chamal would dress himself in the skintight haptic suit, carefully comb his hair, and then place the conch-shaped helmet on his head. Before he squeezed into the cockpit—cramped, narrow, not unlike what Chamal had imagined the driver's seat of a Formula 1 race car felt like—he pressed his palms together, closed his eyes, and prayed. He said a prayer to the Buddha for a good race—and that no one else would even come close to touching his virtual shadow.

Take a deep breath. Empty the mind. Check that all physiological measurements are in the safe zone.

As Chamal began to drive, his anxiety melted away.

The colors of the simulated racing world danced before his eyes. Chamal cruised to another victory.

LATER THAT DAY, UNCLE JUNIUS appeared at Chamal's front door. Junius shuffled into the living room, taking care with his bad leg, never fully healed after a long-ago injury. As he eased into a chair, Junius turned to Chamal, who was doing homework at the kitchen table, and said he had a proposition to discuss. Chamal, Uncle Junius declared, should come meet a Chinese associate of his about a job.

Listening in from the adjoining kitchen, Chamal's parents immediately roared with laughter. "The Chinese!" exclaimed Chamal's father. "What would they want with Chamal?"

Father launched into one of his rants about the Chinese doing business in Sri Lanka. "They want to rebuild the roads between Colombo and all the major cities, and once they're done with that, they won't need us drivers anymore. Can you imagine?" he snickered.

Mother waved her hand dismissively. "Your boss trusted the Chinese and look what happened!" she said, eyeing her husband.

Father fell silent. Two years earlier, he had suffered an accident on the job. Even though it was minor, his boss took it as an excuse to fire him from his decade-long job as a delivery driver. As autonomous vehicles became more practical and affordable, the boss explained, the company could only employ human drivers with perfect records. Now his only source of income was working as a part-time tour guide, driving tourists around Sri Lanka.

Chamal, turning thirteen soon, was supposed to start middle school, but his tuition was still unpaid—not to mention the fees for Chamal's two younger siblings. The prospect of a part-time gig, Chamal knew, was no laughing matter.

"This time it's different," Junius told the group. "And I know you need the money." Junius explained that his Chinese associate was looking for children like Chamal to help develop a game that might change the future of driving.

"He won't get into any kind of danger, and it's good money. I swear to the Buddha!" Junius explained. He said his associates wanted to meet Chamal that very day.

Unlike the adults in his life who made empty promises, Chamal knew that Uncle always kept his word. When Junius said something would happen—a trip to the amusement park or an ice cream treat—it always did.

Unwilling to disappoint their son, Chamal's parents gave in. Mother dressed Chamal in his best shirt, tucking it into his pants and then, half-squatting, shining his shoes. She combed his hair until it was neat. Sri Lankans never left the house looking unkempt.

"Remember to smile, Chamal," said Mother, caressing Chamal's cheek. "A genuine smile is the best gift you can give to others."

A grin brighter than the sun emerged on Chamal's face.

AS UNCLE JUNIUS DROVE them into the city center, Chamal was reminded of one of his father's favorite maxims, one he would often say while complaining at the dinner table. *The most important thing about driving is not the car, but the road.*

There were only a handful of highways in all of Sri Lanka. In the dense streets of Colombo, cars and tuk-tuks competed with scooters and oxcarts, fighting over the narrow lanes. The roads outside major cities were often in poor condition: unpaved, with few streetlights. During the monsoon season, landslides would sometimes cover country roads with debris. While an experienced Sri Lankan driver knew which routes to avoid, paper maps and even GPS could lead visitors astray.

In Sri Lanka, choosing the right road not only meant saving time, but also saving lives.

The previous year, political extremists had launched attacks on several sites around the capital in order to pressure the government into releasing one of their leaders from prison. Caught by the bombing attack in the middle of a tour, Chamal's father had piloted his van full of tourists onto hidden back roads until the coast was clear.

Before every trip, Father always prayed to the Buddha. He decorated his rearview mirror with pendants and prayer beads that swung in the air as he cruised over the bumpy terrain. When Chamal was little, he'd assumed that praying was just as essential to starting a car's engine as turning the ignition key.

Thanks to his father, Chamal knew his car makes and models. Father had described how, decades earlier, the Sri Lankan roads were mostly filled with Japanese cars; then European and American cars came along, and eventually Chinese. He ended up swapping the family's secondhand vintage Toyota for a new hydrogen-powered Geely.

From the time he was little, Chamal loved observing cars: standing on his porch, he watched them speed past, imagining himself behind the wheel. Occasionally, rare vintage cars that still ran on gasoline drove by. He loved the sweet and pungent smell, and the roar of the engine. But Chamal had never taken the wheel himself. Not even the wheel of a toy car. Everything about driving a car happened only in his dreams and daydreams—and on his smartstream apps and the VR Café's racing games.

Chamal was the best gamer among his group of friends. He was almost unbeatable, setting record after record. He loved seeing his name displayed next to a new all-time record on the screen at the VR Café. Sometimes he felt like driving talent ran through his veins like blood: shifting gears, cutting, braking, drifting . . . those tactics were ingrained in him, like a primal instinct. He also knew how to navigate a given course in the most efficient way, making micro-adjustments to his movements to eke out tiny advantages, racking up game points all the way.

Thanks to his stealthy driving, other gamers called him "the ghost." Whenever someone brought up this nickname, Chamal would stick his chin out and grin. To him, the nickname felt like the greatest honor in the world.

So when Uncle Junius explained to Chamal that he would be driving for the Chinese, Chamal imagined something like those café games. Only this time, he would make money.

CHAMAL AND UNCLE JUNIUS took the elevator to Basement Level 3 of the ReelX Center, a gleaming new four-story building in downtown Colombo. As soon as the doors opened, a uniformed young Sri Lankan woman greeted them with an amiable smile. Then Uncle Junius spoke.

"Chamal, I'm going to hand you over to Miss Alice here. She will be very nice to you. Show them what a great driver you are, okay?" Junius winked at Alice, who ignored him.

"Say goodbye to your uncle, Chamal," said Alice. "Follow me."

Chamal trotted after Alice down a wide hallway. The walls and floors of the office—or was it a laboratory?—were pristine, glistening with the reflection of the overhead lights. White-coated staff hurried about. They carried digital tablets displaying flashing numbers, graphs, and charts. When someone needed free hands, they pressed the soft, smooth, almost leathery tablet to their uniform, and the tablet would contour itself to their body like clothing.

Despite all the workers, it was oddly quiet, thought Chamal. He could hear nothing but some soft whispers. In contrast to the VR Café, or the roads outside, there was no engine roar, no screeching tires, not even the *click* and *bang* of opening and closing doors.

Alice led Chamal to a small room the size of a doctor's examination room, and told him to get changed. A black haptic suit was hanging on the door, alongside a matching helmet. Chamal frowned. He wasn't particularly fond of the color black. Mother used to say that the color white represented holiness, while black meant bad luck. Sri Lankans rarely dressed themselves in black. Usually they preferred bright colors, and they wore white only for holidays and religious rituals.

The suit was made from extremely stretchy material, and it felt like a second skin against Chamal's body. It fit perfectly. The temperature was just right. Chamal turned in circles, twisting his body while observing his reflection. The boy in the mirror almost looked like one of those superheroes from his favorite comics, Chamal thought to himself—although a stick-figure version with a comically large helmet.

"Chamal, I'm going to show you something important now. Pay attention, okay?" said Alice, who had been waiting outside the room. She motioned for Chamal to follow her down another corridor.

Beneath the heavy helmet, Chamal observed Alice. *She has dark brown eyes just like Mother,* he thought.

They entered a large room. Colored lights blinked at them from almost every corner. Eight pod-like cockpits extended across the room in two rows, each connected to engines by wires and cables as thick as old tree vines. Behind each cockpit was an enormous screen. On one of the screens was a display—to Chamal, it looked like the livecast of a racing game. Alongside the video, Chamal could see a constantly updating feed of what looked like physiological measurements.

"Soon, you'll be put inside a virtual reality cockpit," Alice told Chamal as he marveled at the room and its sensory overload. "Your uncle says you're a whiz at VR racing games. Think of this cockpit like a game console, except this one can do much more. It will tilt, vibrate, and pick up speed as you drive, but don't be alarmed—it's all just a simulation. All you need to do is follow the instructions you'll hear in your earpiece and see on the screen. Since today is our first day, we'll only have you try out the equipment and do a test drive. If anything goes wrong, or if you get tired, let us know and we'll stop right away, okay?"

Alice finished speaking. Chamal had many questions swirling around in his head—what kind of game was this exactly? But before he could open his mouth, Alice gestured to him to pull the pair of goggles attached to his helmet down over his eyes. Chamal crawled into the cockpit. Like a real race car driver, he buckled the safety belt, patted the wheel, and tried his feet on the brakes and gas pedal. The dashboard was surprisingly empty. Chamal placed his hands before it, waving his arms. In a flash, the dashboard changed. A dazzling world sprang to life from the emptiness. Chamal realized that he could change its layout through gestures—he could add new items to the dashboard and move features to show on the windshield instead.

Suddenly, a countdown began: Colorful numbers flashed on the dashboard as a voice chanting those numbers out loud rang through his headphones. "Ten, nine, eight, seven . . ."

Chamal's heart pounded. He felt as if the cockpit would take off at any moment and soar to outer space like a rocket, leaving behind a trail of blazing fire.

". . . three, two, one, go!"

The cockpit did not launch into space, but the scene before his eyes lit up. Chamal found himself inside a car. Not just any car, though—after taking a look around, he realized that the virtual model was a simulacrum of the Geely Future F8, the same car his family owned. To Chamal's astonishment, every detail was replicated with precision, even the embroidery of the door panels.

When Chamal looked out the windows, he was in for another surprise. The vehicle was parked in the neighborhood public lot across from his home. There was one big difference this time, however: He was not in the passenger's seat, but instead in the driver's seat, where his father always sat.

More surprises awaited him. When Chamal reached out for the wheel, he saw that he was no longer wearing the dull black suit, but a pair of audaciously colored racing gloves. He adjusted the rearview mirror and saw that his helmet was now splattered with paint blots, just like the graphics of the race car game he loved to play.

A wave of excitement washed over his body. "Ready—go!" He shouted, in the same way he gave spoken commands in games.

The Geely didn't move.

Alice's voice filled his earpieces. "Don't panic, just follow the instructions."

Three-dimensional holograms of words and numbers began to spread across Chamal's vision field. Floating in the air, they shifted shapes and flashed in different colors, directing his attention like the digital billboards in Colombo's city center. His gaze, following a downward-pointing arrow, landed on the gas pedal. The pedal glowed with a soft green light. As he pressed down on it with his foot, a bar in the shape of a thermometer, like the vitality bar in a video game, became visible. When he stepped harder on the pedal, the bar's color changed from green to blue and then to yellow.

Now, this is fun! Chamal started the engine. *Shift gears, adjust hand brake, and then a light press on the gas pedal . . .* a vibration ran through his body and his field of vision at the same time. The car began to move.

"Very good. Control your speed and watch out for traffic," Alice instructed him.

"This road reminds me of the road next to my house, but . . . I don't know, there's something off about it," said Chamal hesitantly.

It was the same road that Father took every day to drive him to school, except there were no jaywalking pedestrians and no tuk-tuks to cut across his path. Chamal drove a few blocks, keeping his speed down. He knew he was approaching the intersection where Father usually made a turn, but the cross street never appeared—he could only go straight.

Alice's voice was in his ears again. "Our AI generated this vir-tual landscape based on real data. It'll be familiar, but a little dif-ferent from what you're used to. Since this is your first day, we lowered the difficulty level for you. Once you complete training, you can drive however you like."

Complete training? Training for what? Before Chamal had the chance to ask any questions, his attention was back on the road.

It didn't take long for Chamal to get the hang of the game. Driving here was almost exactly the same as driving in the VR Café's games, except the engine here was, well, *much* better. It reacted faster to his command, there was less time lag, and the line between game and reality was alarmingly blurry. But Alice was right: The better he got at driving, the more difficult the game became. Traffic increased; slow-moving senior citizens and dog walkers appeared at crosswalks and turns; as he drove past a playground, kids would kick a ball into the middle of the road; even the traffic lights were erratic, blinking with the wrong color. *It's too real,* Chamal thought. Sweat trickled down his neck. His palms, grasping tightly on to the wheel, were now sticky. His eyes stung from not blinking. *Pay attention . . .* For some reason, the game made Chamal feel like he couldn't afford to miss even the smallest detail.

Thankfully, everything went smoothly. The road he took to school seemed infinite. Chamal could feel his attention fading. His foot pressed down harder on the gas pedal. The car accelerated steadily—*80, 100, 120* kilometers per hour—as he sank into a state of flow. He felt immersed in a web of energy, as if his body, the cockpit, and the virtual landscape had melted into a harmonious closed loop. He wasn't driving; rather, the car was responding to his mind like a part of his body.

How much time had passed? Chamal took a quick glance at the dashboard. The arrow on the speedometer had hit the red zone, indicating that he was reaching maximum speed.

Chamal's eyes widened. The sense of danger struck him at once, like lightning. He let go of the gas pedal and stomped on the brakes. Suddenly, the vehicle was overcome by a powerful wave of energy, tipping it over and sending it crashing back to the ground. Chamal screamed and clasped his fingers around the wheel. The inside of his body felt like it was on fire. Everything in his visual field was spinning rapidly and he had to shut his eyes to counter the dizziness. Then, the motion gradually stopped as the world around him faded into darkness.

Faintly, a voice reached his hearing—Alice?—calling out his name. Then someone pulled him from the cockpit and removed his helmet. He gasped for fresh air.

As he came to, deep in his heart he knew that this solid, tangible reality was not what he wanted. He longed to return to the virtual world. He wanted that out-of-control feeling again.

FROM THE CLOUD RED rooftop bar of Cinnamon Red Hotel, one could see the entire skyline of Colombo.

It was the end of the monsoon, and an occasional flash of lightning would light up the layers of heavy clouds, indicating an impending storm.

Junius swirled his glass of whiskey. Having mostly melted into the amber liquid, the once-round ice cube was now broken and ragged. Like the South Pole ice caps, he mused.

A hand landed on Junius's right shoulder, nearly startling him from his chair. Yang Juan. With her pixie haircut and athletic, tracksuit-clad figure, Yang Juan could be mistaken for a gymnast or soccer star rather than the head of a Sri Lanka–based Chinese high-tech company. "Sorry to keep you waiting. Traffic."

"Typical Colombo! Single malt for you?"

"Actually, lately I've fallen in love with one of your local spirits." Yang Juan gestured to the bartender, who seemed to understand immediately. Soon a cocktail glass filled with milky liquid arrived at the table.

"Can't believe you're into cheap coconut arrack!" exclaimed Junius.

"Sweet and sour. Cheers!"

They clinked their glasses together and finished their drinks in a single gulp.

"But the sweetness is only a deception," Junius said with a grin. "The alcohol content of that drink could match your strongest spirits, like the *erguotou*. Am I pronouncing it right?"

Yang Juan smacked her lips, looking directly at Junius. "Exactly. Your arrack is just like your people. *The sweetness is only a deception.*"

Junius was momentarily speechless.

"Yang, I already gave you what you want." He cleared his throat and began. "Those children . . ."

"Are those the best you can find in all of Sri Lanka?"

"I did everything you asked. I went to the gaming café and—"

"They're not good enough. I need more kids, better kids. Their passing rate is too low. If we can't get enough drivers, good drivers, our investors will pull out. Junius, think about it: Why do you think we chose to establish the project here?"

Junius lowered his eyes, "It was cheaper . . ."

"Another drink for me, please." Yang Juan waved at the bartender. "One for my guest, too."

Junius was silent for a moment. "I even gave my nephew Chamal to you," he said.

"He's your nephew? I heard about him. Well, that's one bright kid."

"His father used to say that Chamal grew up breathing gasoline fumes, even when he was still in his mother's womb." Junius's smile froze at the thought of his little nephew. "Wait, Yang, I need to ask you something. It's serious."

"Mm-hmm?"

"You promised me that the updated system guarantees absolute safety for every driver, right?"

Yang Juan shifted her eyes to the Colombo nightscape. "You do remember why we're starting this thing all over again?" She took a sip from her glass.

Junius fell silent. His hand landed on his left thigh. Somewhere deep in his muscles, entangled with the nerves, he could feel the spot. It sent a dull pain through his body, as if reacting to Yang's words. The doctors had found no physical injury; they said the pain must be psychological. Junius kept his mouth shut.

Yang Juan swirled her drink. "You either tell people the truth and let them bear the consequences, or tell them lies in exchange for giving them a better life."

"I understand," said Junius with a sigh. "Please take care of Chamal, though. He's the hope of our entire family."

He stood up, put his hands together, and said goodbye. In his abandoned glass, the ice had melted to water.

"MORE LEFTOVERS?" FATHER THREW a glance in the direction of the bedrooms. Mother shook her head. The plate of *kottu roti* she had retrieved from Chamal's room was barely half-eaten. She carried it out to the yard, where hungry neighborhood crows were always ready for a feast.

"Do you think we should take him to the Gangaramaya Temple and let the priests take a look?" Mother frowned. She pressed her palms together and muttered a quick prayer.

"Give him a few more days," Father replied. "Junius said that everyone goes through an adjustment period. What was the word he used again? *A learning curve* . . . that's right. Plus, Chamal is getting his paycheck soon. Those Chinese are willing to pay!"

Mother was silent for a moment. "I saw him standing next to our car the other day with this weird look on his face, and it felt like . . ."

"Like what?"

"Like he was talking to the car."

Father roared with laughter. "Well, now I think you're the one with the problem, not Chamal!"

"Chamal is your son! Stop being such a jerk. If he doesn't want to do the work anymore, then we'll stop. There must be other options—I can get a part-time job."

"But, Lydia, look, Chamal is happy. Every morning he's excited to go to work, if you call it that. Have you ever seen him so enthusiastic about *anything*?"

"But—"

"*Shhh.* He's coming."

Chamal staggered down the stairs, his shoes undone. He didn't seem to notice his parents at all—his eyes were fixed on the ground. He spread his arms, and for a brief moment he looked as if he were about to embrace the floor, or like a fighter jet preparing for a dive-bomb attack. At last, he turned slowly and squeezed through the gap between his parents, his right hand making a pulling gesture, as if shifting gears.

"Chamal!" Mother shouted.

The boy stopped in his tracks. Instead of turning around, though, he took a few steps backward, until he was facing his parents again.

"What did I tell you about greeting elder family members?"

Chamal's round eyes widened, as if waking from a trance.

JUST AS HE HAD at the VR Café, Chamal made it to the top of the training center's ranking list in no time.

He was no longer the beginner who panicked at the sight of traffic and pedestrians. And it wasn't just driving for driving's sake. Chamal began receiving missions, with instructions from the technicians in the training center. The missions were always similar in terms of structure, but with variations in story line. Sometimes they were outlandish, like an alien invasion. Sometimes they were chillingly realistic, like a terrorist attack that caused roads to crumble and cars to crash into one another.

Complex landscapes, erratic drivers . . . nothing could ruffle Chamal. He quickly tallied the most points among the group of gamers that Yang Juan had recruited from all over Sri Lanka. The young drivers became fast friends during their daily training. Still, his cohorts watched Chamal with jealous eyes as he swaggered out of the room each day—everyone knew that more points meant more money.

Other drivers tried to pry tips and tactics out of him. Chamal tossed his hair. "I was born to drive," he said, a little too cockily.

CHAMAL HAD DISCOVERED THAT the game did not give him infinite routes. The landscapes that came up the most frequently were primarily replicas of real-life cities, spanning the Middle East to East Asia: Abu Dhabi Satellite City, Hyderabad, Bangkok, the Singaporean man-made island, the Greater Bay Area of Guangdong-Hong Kong-Macao, Shanghai Lingang, Xiong'an New Area, Chiba of Japan—places that, until now, Chamal had only read about online.

One day Chamal received instructions to complete a mission on the Singaporean man-made island. A disturbance on the ocean floor in North Java had triggered a tsunami and the infrasound completely paralyzed the island's automated smart transportation system. A ten-meter tsunami would hit the island in exactly six minutes. Over a hundred dysfunctional autonomous cars and their passengers were careening down the roadway, likely to crash or, like sitting ducks, be washed away.

Chamal and the other racers were instructed to seize the wheels of these vehicles, turn on manual control before more accidents could happen, and help connect the cars to the emergency network infrastructure. The network would then take over, directing the cars to the nearest evacuation zone, saving the passengers' lives.

It was the most difficult and thrilling game Chamal had ever played.

His virtual avatar hopped from one driver's seat to another, taking control of the wheel in mere seconds, evading fallen debris as he sped to safer ground. *Jump.* The procedure was simple and natural, as if it was a part of his nerve reflex. *Jump again.* As the blood-red countdown was rapidly approaching zero, a shimmering white line emerged in the gray-blue horizon on the periphery of Chamal's view, and it advanced toward the shore, thickening and rising every second.

Chamal had no time to appreciate the sublime violence of nature, nor feel any fear. He was like a ghost that possessed those massive, sturdy bodies of steel and iron, connecting them to the network, and sending them on a path to safety. The delightful sound of coins clinking against one another rang on incessantly as his score rocketed at the top edge of the screen. The corners of his mouth twitched. He could feel the flow returning to his body.

The fatal Java tsunami was closer now. *Faster.* Chamal wanted to earn as many points as possible before the game ended. Every millisecond that slipped through his fingers meant less tuition money for his younger siblings and less living budget for his entire family. The world—and his family—depended on the speed of his mental and physical reactions.

As Chamal was about to leap into an SUV, the roaring wall of water and foam finally caught up to him. The graphics of the game were not the best; he could even see the jaggies and pixelation as the tide swallowed him whole. Before the screen went dark, he caught a last glimpse of a few cars in the near distance that were washed away instantaneously by the merciless wave. He let out a heavy, regretful sigh. Every car he didn't save meant fewer points.

Game over.

Chamal, now back to reality, found himself drenched in sweat. He was so exhausted that he couldn't even climb out from the cockpit. Two staff members had to carry him.

Alice told him to take some time off. In the days that followed, even tasks as simple as eating with a spoon gave Chamal trouble. His hands wouldn't stop shaking. The great, ferocious tide haunted him in his dreams. That mission seemed to have deprived him of all of his energy, creating a void in his mind and body.

Chamal normally had little interest in the news, but as he lay in his bedroom recovering, he overheard a report coming from the television in the kitchen, where his parents were sitting with Uncle

Junius. The newscaster was talking about a tsunami that had occurred in Kanto, Japan.

Slowly, Chamal got up from the bed and staggered to the
kitchen. On the TV screen, he watched surveillance footage recorded during the final moments before the tsunami hit the coastal
highway. Cars, as light and powerless as toy figures made from
paper and clay, were overturned and devoured by the waves, disappearing into the dark water.

Chamal's heart raced. The scene before his eyes was uncannily
familiar. The status of the roads, the position of the cars, the scattered debris . . . it was an exact replica of the final scene in the
game, which had been imprinted into his mind that day.

No! That's impossible! I only played a game!

"Uncle, that was only a game, wasn't it?"

Junius was silent a moment. "Chamal, I want you to meet someone."

BACK AT THE TRAINING facility in the ReelX Center, Uncle Junius led Chamal through a door and down a corridor that Chamal
had never seen before. At the end of the hallway, they entered an
office decorated lavishly with local folk art and ornaments, resembling an absurdly large collection of holiday souvenirs.

"Dear Chamal, we meet at last."

A woman dressed in all white stood from the sofa, bent down,
and reached for Chamal's hand. Shyly, Chamal offered his own.
The woman's grip was sturdy and her palm warm.

She motioned for them to sit.

"My name is Yang Juan. You can call me Yang, or Jade. I understand they call you 'the ghost,' Chamal."

Chamal blushed as Yang Juan continued speaking.

"I am in charge of ReelX's Sri Lanka branch. I've seen all of your
game data. Without doubt, you're born to be a driver."

By now Chamal's cheeks were burning.

"Well, your uncle told me you might have some questions. I'll
do my best to answer them."

Chamal bit his lip. *What should I say? How can I sound respect-ful, polite, and dignified, like the way Mother taught me?* He wanted to choose his words carefully, but he was too worn out to think straight. "The tsunami . . . it was real." Words slipped out before he could contain them. "All of this is fake," Chamal stammered.

"This isn't exactly a question, is it?" Yang Juan winked. "You're expecting a certain kind of answer from me. You want me to tell you that the game is either real or fake—choose one or the other, right?"

Chamal's head began to spin. "Is there a third possibility?"

"Let me ask you first: Do you think the tsunami was real?"

"Of course."

"Was the tsunami *in the game* real?"

"That was fake."

"How about the cars?"

"The landscape seemed real and the course of action they took seemed real, but the cars themselves were fake."

"Then, do you think you really helped save those cars and peo-ple?"

"I—I—" Chamal stuttered. "I don't know."

Yang Juan shrugged, but her expression was sympathetic.

"But I *know* you're lying!" Chamal blurted out. "If the tsunami happened in Japan, why did you have to tell us it was Singapore? If our actions affected reality, why did you have to tell us it was a game?"

Yang Juan sat silently, letting the question hang in the air. Fi-nally, she spoke. "Before I answer, I need to ask you something first. Only answer with yes or no." Squatting, Yang Juan lowered her body to the ground, so that she could look directly into Cha-mal's eyes.

"Do you want to go to China?"

"What?" Chamal was taken by surprise.

"Remember, this is a yes or no question." Yang Juan grinned upon seeing the look of astonishment and awkwardness on the boy's face. "You are our best driver. A trip to China is the bonus reward for your work. I think you'll find the answer to your ques-tion there."

"You mean driving in China?" Chamal frowned. "If that's the case, then I've been to many places in China already."

It was Yang Juan's turn to be stunned. It took her a few seconds to realize that Chamal was talking about virtual reality. "I'm not trying to trick you." Yang Juan laughed. "I meant going to China for real. You will *physically* take a plane and go to China, breathe the air, eat the food, and test out the landscape with your own feet. Do you want to go?"

Chamal lowered his eyes and contemplated. Finally, he looked up at Yang Juan, nodded his head, and gave the woman a *dignified* smile.

A STRONG VIBRATION WOKE Chamal from his sleep. Thinking he was still in the game, he instinctively reached for his helmet, but there was nothing on his head. He opened his eyes and squinted at the bright morning sun coming in through the porthole-shaped window. Outside were endless rows of sleek jumbo jets.

The plane had arrived at the Shenzhen Bao'an International Airport. As Chamal and his uncle walked down the jet-bridge and into the terminal, he marveled at all he was seeing. Everything here was colossal and brand-new; rays of sunlight shone through the hexagonal carve-outs on the white ceiling like a heavy meteor shower, illuminating the travelers hurrying from one destination to another.

Zeng Xinlan, a talkative, cheerful young employee from ReelX's Shenzhen headquarters, picked them up at the airport. Upon meeting Chamal and Junius, she put her hands together and said "*Ayubowan,*" greeting her Sri Lankan visitors in their native language. Junius returned the blessing and Chamal copied his uncle.

They walked to the autonomous vehicle pickup zone together. Almost as soon as they arrived, a white SUV glided into the lane and came to a stop before them. Its doors swung open. Chamal climbed into the spacious back seat with Junius. The cool breeze of the car's air-conditioning relieved the stickiness from the humid outside air almost instantaneously.

The car took off. Unlike the cars Chamal was used to, the engine of the SUV was nearly silent, and the acceleration was so smooth that he barely felt anything.

"Most of the roads and vehicles in Shenzhen support L5-level autonomous driving now. With the driver's seat no longer exclusive to the driver, not only can we fit more people into a car, but everyone can sit more comfortably as well. Minicars reserved for one or two passengers are also available." Zeng Xinlan smiled. "The smart control system decides which available car to send and calculates the optimal path to take based on the passengers' location and walking speed, to maximize airport efficiency and reduce passengers' waiting time. The road we're on right now is specifically designed to accommodate autonomous vehicles. The smart sensors installed along the road communicate in real time with the control system on every car and the traffic management infrastructure in the cloud, to ensure safety and orderliness."

Chamal thought she sounded a bit robotic, as she recited this explanation.

Junius pressed his face against the window. "Shenzhen looks so different from the last time I visited!" he exclaimed.

"You've been to Shenzhen before?" asked Chamal, surprised.

"Many years ago. I remember seeing a construction team working on the first of these 'smart' roads—they're everywhere now!"

"Typical Shenzhen development speed," Zeng Xinlan said with a grin. "Wait till you see more!"

Chamal stared out the window at the foreign city, dazed. Skyscrapers extended upward as if they were infinite, their tips disappearing into the clouds. The outer walls of the buildings were made from smooth, shiny material that reflected sunlight, making them seem as if they were wrapped in cloaks of light that changed patterns and design when the sun's angle shifted. Shenzhen was pristine and orderly. He couldn't figure out how this was possible. It was as if millions of invisible puppet strings were hanging from heaven, controlling every road, every car, and every person in this enormous city, weaving them into an all-encompassing web.

But who's pulling those strings?

"Look!" Zeng Xinlan shouted.

Chamal and Junius looked in the direction her finger was pointing. In the lanes of opposing traffic, the vehicles suddenly parted. One by one the cars glided to the sides of the highway, creating an empty lane in the middle. A faint siren grew louder as an ambulance suddenly sped through the gap in traffic. As soon as it had passed, the cars slid back into their original positions, as if nothing had happened. The entire process took mere seconds, and, apart from the siren, was nearly silent, without so much as a single honk.

"How is this possible?" Chamal was nearly speechless.

"Think about it this way. We humans don't crash into one another when we run, because our eyes observe, our brains calculate distance, and our legs adjust speed and posture. The same goes for these cars," said Zeng Xinlan with a shrug. "The sensors, camera lens, and LiDARs are its eyes; the control system is its brain. All of the above are connected to the engine and gears, the car's legs."

"Chamal, imagine if this technology were available in Sri Lanka," muttered Junius. He remembered what had happened to his mother. *She could've been saved from her heart attack if only the ambulance had been able to get her to the hospital in time. It was not the heart attack that killed her, but the traffic.*

A new message alert popped up on the dashboard, and the message was broadcast in flawless standard Mandarin through the speakers.

"Oh, it's the marathon," Zeng Xinlan explained.

Before Chamal could press for details, their car changed direction, rerouting toward the nearest off-ramp. In fact, all the cars on the highway seemed to have received the same alert at once: Like a squadron of fighter jets changing formation, the traffic dispersed into new formations as the vehicles headed to the exits.

Chamal was stunned when his gaze landed on Zeng Xinlan, who had assumed the driver's seat. The autonomous vehicles back home in Sri Lanka, despite how widely they were deployed these days, were unable to shift around with such precision without the aid of human drivers. However, Zeng Xinlan, with her hands off the wheel, was obviously not operating the car.

"What's happening now?" asked Chamal.

"Aha! It's your lucky day. You're just in time to see the upgraded city traffic system in action. The city's annual marathon will soon begin, and we're all being rerouted."

Chamal stared dazedly at the traffic, trying to digest all that he was seeing and hearing. He felt immersed in a dream world.

BEFORE THEY VISITED REELX'S headquarters, Zeng Xinlan took them to a Cantonese restaurant in Qianhai.

Chamal stuffed his face with the foreign yet delicious cuisine, while Junius stared fixedly out the window.

"What's so interesting out there?" asked Zeng Xinlan as she picked up a shrimp dumpling and put it down on Junius's plate.

"Even . . . even the horizon has changed," murmured Junius, bewildered.

"Well, land reclamation is one of Shenzhen's long-term projects. I heard the same is happening in Sri Lanka?"

Every time Chamal passed by the coastal road in Colombo, he would catch a glimpse of the trailing suction hopper dredgers near the coast of Port City. Those behemoths lifted their long snouts and spat out arcs of mud and sand, which glistened like gold in the sun. All the dredgers came from China; they were helping Sri Lanka with the colossal feat of creating new land and reshaping the sea horizon.

"*Sri Lanka, a bright pearl of the maritime silk road,*" Zeng Xinlan commented, mocking the tone of Chinese news broadcasters.

Chamal put down his chopsticks. "Are there any cars left for humans, then?" he asked, timidly voicing the question that had been on his mind for hours.

"Not all cars can be switched to manual mode," said Zeng Xinlan. "We have human drivers, too, but they are limited to human-only roads and they are required to use a complementary AI device while driving. It's so much more difficult to pass the driver's license exam these days. No place for hooligans."

"If that's the case, why are we needed?" Chamal turned to Junius, gazing directly into his eyes.

Junius and Zeng Xinlan exchanged glances. "Of course you're

important," replied Zeng Xinlan. She looked at Chamal, her face solemn. "Even the most advanced AI makes mistakes. What if an explosion destroys a road, making it impossible to follow the digital map, or there's a natural disaster that suddenly creates chaos? This is when people like you come in—a hero to save the day."

"But I don't want to be a hero," Chamal blurted out. "I only want to play games, earn some points, and help out my family."

Junius evaded Chamal's gaze.

All of a sudden, Zeng Xinlan let out a giggle and broke the awkward silence. "Look at you two! Like uncle, like nephew. Chamal, when your uncle first joined our project, he told us the exact same thing! Am I right, Junius?"

Junius, now blushing, poked at his soup with his spoon.

"Wait, you also . . . ?" Chamal's eyes widened.

"He never told you?" Zeng Xinlan shot Junius a surprised look. Chamal shook his head.

"I didn't want to give you the wrong impression," whispered Junius, finally finding the right words. "I know what other people say behind my back. They think I've been helping ReelX do bad things, so the Buddha decided to punish me by crippling my leg."

Chamal was not a stranger to the gossip, but he had never imagined the truth.

"Your uncle used to be our best driver. Before he retired from the injury, he saved many lives."

"So you used to be a ghost driver, just like me," Chamal repeated. "But how can ghosts get injured?"

"This was a decade ago, Chamal. It was an earlier version of the program, a more primitive version," Uncle Junius said. "There have always been risks, but they are smaller now."

"That's why it's necessary we call the procedure *a game*," interjected Zeng Xinlan, her tone serious again. "The human species is far more delicate than machines. A human driver's reaction time and performance level can be affected by even the most insignificant emotional response."

"So that's why Uncle lied to me, telling me that I was only there to play a game," muttered Chamal. *I used to believe that Uncle would never lie to me.*

"Chamal," Junius said, letting out a sigh. "Let me tell you a story."

A DECADE AGO, JUNIUS was conducting a mission in the Sichuan-Tibet region after a major earthquake. His objective was to transport emergency medical supplies to the trapped victims. The aftershocks were relentless; GPS was failing due to roads blocked by landslides. Ghost drivers were the only option. At first, Junius managed to evade danger, but after an especially powerful aftershock, debris began pouring down the mountain like a deadly rain. Struggling to dodge rocks and mud while maintaining control on the winding road, Junius failed to notice a giant boulder plunging down from his left. It landed on his car hood, smashing the car's left side into the ground.

A piercing pain shot down Junius's left leg; it was the force feedback at work. He knew that his physical body was unharmed. It was synesthesia. A healthy amount of body synesthesia—simulation of real senses through virtual reality—was beneficial to virtual rescue drivers, because it stimulated cognitive capacity and produced adrenaline, enhancing their performance level. However, what constituted "a healthy amount" varied from driver to driver, mission to mission. Upon seeing the disaster-stricken Sichuan-Tibet region, Junius had deliberately pushed the synesthesia values up. So many lives depended on him; he could not bear the thought of letting them down.

With his leg screaming in pain, Junius tried different ways to get the car moving again, but it wouldn't budge; the wheels would only spin uselessly. With every passing second, hope grew dimmer. He was overwhelmed with guilt and despair. *I failed them.* His injured leg felt numb now, as if it were no longer a part of his body.

In the end, the military was able to pull drones from other sectors and send out an emergency deployment. The medical supplies reached the people in need, after all. However, ever since that day, Junius's leg had been stuck in a limbo between the real and the virtual, as if time had forgotten it, freezing it forever in that moment of pain and regret.

———

"IF WE THINK OF it as a game, we'll feel less pain," said Chamal, after Junius had finished talking. He could understand where Junius was coming from now, yet there was still one thing he just couldn't wrap his head around. "But *why*? Why do we have to endure all this?"

"To make a living, I guess, and save some lives along the way. It's important to invest in our karma," said Junius with a self-deprecating smile. "One day we may need saving, too."

After lunch, they visited ReelX's headquarters. While they were in the lab, Chamal couldn't tear his eyes away from the newest force-feedback suit and brainwave-connected helmet displayed in the window. Zeng Xinlan, noticing the boy's widened eyes, promised him a set of tailor-made equipment—as long as he was willing to stay and complete ReelX's missions.

Chamal, caressing the graphene fabric that was as light as silk yet as impenetrable as steel, silently mulled over all that he had learned that day.

Indeed, Chamal felt that he'd witnessed the future in one afternoon—although he wasn't sure whether it was the same future that Junius had mentioned. The future, in his eyes, was foreign, grandiose, and immensely confusing. The autonomous cars and smart roads that he had seen on the trip were only the tip of the iceberg. Chamal used to think that technology was like Father's car, in which straightforward, countable components like bearings, gears, and cords had been assembled piece by piece, and everything was clear-cut and apparent to the eye. Now he realized that technology was more like Mother's favorite sari: The drape was delicate gossamer, embroidered with a variety of patterns, yet when Mother folded it and wrapped it around her body, the sari looked different, like layers of hazy clouds bundled together and solidified into a definite, concrete shape.

FOLLOWING THE BLUE DOTTED line on the electronic map, Chamal watched the cartoon airplane enter the air space over Sri

Lanka. The island on the screen in front of his seat looked like a drop of water next to the Indian Ocean.

Chamal, craning his neck, gazed out the plane's window. He wanted to catch a glimpse of home, but he saw only thick white clouds.

Yang Juan came to pick them up at the airport. Instead of driving Chamal and Junius home, however, she took them to a construction site next to the training center, a new project of the China State Construction Engineering Corporation. The workers had already poured the foundation, and they were unpacking the pre-engineered building components. In the dusky twilight, the building looked like the skeleton of a colossal beast, slowly being brought to life as chunks of flesh made from steel and brick attached themselves onto the bones. In less than half a month, a gleaming, modernist commercial tower would rise here.

"Look, Chamal, in the future our company will own many floors of this new building. We'll make them into offices, training centers, and operations rooms. I promise you that by then, you'll have your own operations room and a virtual reality cockpit, just for you. You can decorate them however you like." Yang Juan gestured, as if she were summoning a hologram of a fully furnished, fancy operations room with the name "Chamal" written on the door for the boy to see.

"I . . ." Chamal began to talk, but his voice choked. He turned his eyes to Junius for confirmation, and Junius smiled back at him encouragingly. "I . . . I am sorry, Yang, but I don't want to be a ghost driver anymore."

He was too nervous to look at the woman's face. Would Yang Juan be surprised? Disappointed? Furious? But Yang Juan's face was as calm as usual. *Perhaps she's just good at hiding her emotions,* thought Chamal.

"No, don't be sorry. I understand." Yang Juan clapped Chamal's shoulder. "We lied to you and tricked you into taking on a burden too great for someone your age, and yet here we are, wishfully thinking that you would still come back and be our best driver."

"I'm just not ready yet," whispered Chamal.

"Let me tell you something, then—something that even your uncle doesn't know," said Yang Juan with a wistful smile.

She walked over to a pile of assembled construction materials and sat down, seemingly oblivious to whether the dirt would stain her white pants. Looking up at the half-finished building, she began her story. "When I first came to Sri Lanka, I hated it here. I was alone, and I didn't fit in. I couldn't even tell the difference between Sinhalese and Tamil. Living here frustrated me; it was nothing like home. I wanted out.

"But after a few months here, my feelings started to change. I heard that when hornets made their nest on the chin of the great golden Buddha at the Dambulla cave temple, the priests chose to let them be instead of taking the nest down—I couldn't even imagine something like this happening back home, in one of those major industrial and tech cities that cared only about *development.* I also felt the power of religion, and the sense of peace that came from within. I wasn't raised religious, but I've tried praying, too: When the Sichuan-Tibet earthquake happened, I prayed that your uncle would accomplish the mission and return safely."

Remembering the story that Junius told him in Shenzhen, Chamal threw a quick glance at Uncle's leg.

"Eventually I was able to see, beneath the surface differences, common things shared by the Sri Lankans and the Chinese. I heard that the Sri Dalada Maligawa in Kandy and the Lingguang Temple in Beijing are the only two Buddhist sites in the world that hold the Buddha's tooth relics. It's like the Buddha bit into the apple that was Earth and left his two front teeth in Sri Lanka and China. This is fate, isn't it?"

Chamal and Junius both pressed their palms together at Yang Juan's mention of the Buddha.

"So I decided to stay and develop ReelX's Colombo branch. You've been to Shenzhen; Shenzhen is the future I imagine for Colombo, but with a Sri Lankan twist."

Chamal's eyes widened. He could not even begin to wrap his mind around the thought of modeling Colombo after Shenzhen. His first reaction, though, was a blunt question: "When that day comes, would drivers like my father lose their jobs?"

Yang Juan fell silent for a second.

"People have been saying that upgrading the AI system here will hurt Sri Lankan society, that people will lose their jobs. The terminology we use is 'leap shock.' When you build a tower, you need to start from the foundation and work your way up, instead of leaping ahead to the highest floor.

"And once you start down the path of development, there is no way out. ReelX wanted to help Sri Lanka accomplish the same feat. Even if some jobs go away, more jobs will be created to fill in the gaps. See? The work that you have been doing is important. It's *holy*." Yang Juan's eyes landed on Chamal and Junius, her arms wide open.

"But . . . but now that I know it is not a game . . . I can't do it anymore," stuttered Chamal. "Whenever I think about the cars washed away by the tsunami and all those people that I could've saved, I feel guilt. I feel like I have committed an evil karmic offense. I can't drive, Yang."

Trembling, Chamal took a step back. Junius caught the boy and pulled his skinny body into an embrace.

Yang Juan lowered her eyes. A hint of defeat finally emerged on her face. "Perhaps technology and people don't work the same way . . ."

Suddenly, Yang Juan's phone rang, cutting her speech short. Yang Juan picked up and then quickly glanced at Chamal and Junius. The pair couldn't understand Yang Juan's Chinese, but from her face, they knew that something had gone wrong.

"I'll try my best," said Yang Juan, hanging up after what felt like an eternity.

"What happened?" asked Junius.

"There's been an attack at the Gangaramaya Temple, in the center of Colombo. Men with guns. And the temple is packed because of the holiday. According to the surveillance cameras, priests and tourists are hiding in the Buddhist school and the dormitories, but it's only a matter of time . . ."

Chamal froze. *The Gangaramaya? The place where Mother used to take him to pray?*

The Gangaramaya was a museum of Asian Buddha statues. In addition to Sri Lanka, the origins of its thousands of statues could be traced back to Thailand, India, Myanmar, Japan, China, and beyond. The Gangaramaya had always been the core of Sri Lanka's spiritual life, attracting visitors from all over the world. For a would-be terrorist, it was the perfect place to make a violent statement.

To Chamal, the Gangaramaya was simultaneously a childhood wonderland and a holy place. *How could anyone even think of attacking the Gangaramaya?*

"Where are the police?"

"The police are responding, but they want my help."

"How can you help?"

"Our company has an autonomous vehicle parked nearby, a refurbished off-road SUV. We can drive the car to the side gate and rescue the trapped people in batches." This was the first time that Junius had heard Yang Juan's voice shaking.

"Can the AI handle it?" Junius said. "If the building is smoking, and there is the possibility of another explosion—it might affect the efficiency of the LiDAR. Also, what if terrorists are hiding among the priests and tourists? What can you do?"

"We don't have time to come up with an alternative path. We'll have to take the risk."

"If only my leg wasn't injured . . . Do you think there will be anyone at the training center?"

"No, the holiday—it's closed."

The three fell into silence as a sense of helplessness washed over them.

"I'll go," said a quiet voice.

It was Chamal. With his head half-lowered, no one else could see the expression on his face.

"It's not safe," replied Junius.

"Your uncle is right. Your mental state is going to have a great negative impact on your performance," said Yang Juan.

All of a sudden, Chamal whipped his head up to look at Yang Juan. "I've been going to the Gangaramaya since I was a child. I could navigate it blindfolded." The boy's eyes glistened like sapphires. Before the adults could respond, Chamal was sprinting toward the training center.

Yang Juan and Junius looked at each other for a moment, and then followed.

THE TRAINING CENTER WAS eerily quiet.

Outside of his cockpit, Chamal put his hands together and lowered his eyes, praying to the Buddha in his heart.

He pulled down the goggles on his helmet and tightened the cuffs of his racing gloves. Then he slowly climbed into the cockpit and buckled the seatbelt.

Take a deep breath. Adjust your heart rate. Empty the mind. Ignore any thought that could trigger anxiety. Focus on the screen and only on the screen.

A knocking sound interrupted his meditation. Chamal lifted the goggles and saw Yang Juan. She gestured for him to extend his left hand, and then tied a red ribbon around his wrist.

"I heard that this brings good luck," said the woman. After a brief pause, she nodded solemnly. "Thank you, Chamal."

Chamal grinned. When he booted the engine up, he was reminded of the first time that Uncle took him to the training center.

Connecting. Synchronizing. Adjusting vision field.

The next instant he found himself in the autonomous car, parked a mere hundred meters away from the Gangaramaya. Following the narrow street, he turned left at Sri Jinarathana Road, passing through the police blockade. When he drove by the front gate, he slowed down to observe: The small plaza by the gate was a mess, strewn with discarded shoes and other debris. The gate itself was shrouded in smoke.

Chamal could vaguely make out the silhouettes of the Guanyin Bodhisattva and the Guan Gong statue on the sides of the gate.

More smoke. His heartbeat quickened. He scouted his surroundings with the exterior camera lens, searching for the wounded, but there was no one in sight. He continued on, turning left at the Hunupitiya Lake Road. He was close to the designated rendezvous spot now, the side gate next to the Borobudur replica.

He remembered the shock and astonishment he felt the first

time he saw the Borobudur replica with his own eyes. The original Borobudur was in Central Java, Indonesia. The architects of the Gangaramaya modeled their replica after a portion of the original Borobudur, shrinking the size according to scale, so that worshippers unable to travel to Indonesia could pray to the Borobudur in Colombo. Mother had said that the original Borobudur had endured volcanic eruptions, bomb explosions, and calamitous earthquakes, and yet it still stood strong at the heart of Central Java.

Chamal parked next to the side gate, leaving the engine running. He focused his gaze intently on the gate, afraid that the terrorists would hear him. His heart pounded in his chest. His mouth was dry and his eyes were burning; for a second, he thought he was going to throw up.

"Relax, your heart rate is too high," Yang Juan's voice came from the headphones. "Just treat it like a game. We've sent an SMS alert to the stranded tourists—they know you will soon arrive."

A game. Right. This is just a game.

No sign of the gunmen.

Suddenly, a face appeared at the gate. A man peered out, his expression tense. When his eyes landed on the large ReelX logo spray-painted onto the side of the car, he disappeared. A few minutes later, a group of people stumbled out of the gate, holding on to one another. Chamal commanded the car to open its back seat doors. Instinctively, the man leading the group took a double take when he saw the empty driver's seat, but he recovered immediately and began to help the wounded climb into the car.

More people, some elderly and a few children with mothers, filled the car up. The man in charge shut the door, waved his hand, and retreated back through the gate. A child in the car let out a sharp cry.

The sound of crying made Chamal's heart ache. Every second wasted was a step closer to danger. He pressed down on the gas pedal and launched the car. *Straight down the road, turn left, turn right.* Approximately one kilometer away was the Colombo Cinnamon Red Hotel, the designated safe spot.

One lap, a second, and then a third. Chamal repeated the routine as if he were playing a game. *Follow designated course. Perform instructed action in given time.* The only difference was that now he could see the faces of the victims and hear their cries.

Finally, when Chamal had lost count of how many trips he had made, the last passenger squeezed into the back seat. The final batch. Just when he was about to take off, however, a barrage of bullets hit the side of the car. The deafening sound of glass shattering, metal clanking, and people shrieking made Chamal's ears buzz.

Chamal accelerated abruptly then stepped down hard on the brake and changed direction, evading by inches two gunmen dressed in black. A loud *whomp* came from the front. When he looked up, he saw that a third masked terrorist had jumped onto the hood and was clinging to the rack on the roof of the car.

Chamal noticed a few square-shaped black objects attached to the terrorist's body. Despite his blurry visual field, he could see something else: blinking red dots. A *bomb.*

He accelerated, snaking through the traffic, drifting at turns, careening over curbs, trying every trick he could think of to dislodge the gunman. The people in the back of his car were screaming. He knew every decision he made could determine their fate.

This is not a game.

"This is not a game!" Chamal yelled.

"What?" Yang Juan shouted back from the training center.

Chamal did not respond. He was supposed to turn left at the next crossroad to get to the Cinnamon Red, but he made a U-turn instead and sped down the A4 highway overlooking Beira Lake back in the direction from which he had come. In his peripheral view, he could see green trees and white birds soaring across the surface of the pale blue lake.

"Chamal! Where are you going?" Junius's voice was shaking.

The red dots before his eyes were now blinking faster. Chamal knew time was running out. The terrorist clinging to the roof had begun chanting.

Finally, Chamal caught sight of the place he was looking for. He made a sharp turn, crossing through the opposite lane and dashing down a hill. The car almost flew into the air when it rolled over a short brick ledge. *There it is!* The Seema Malaka, a temple that looked as if it were levitating in the middle of the Beira Lake, connected to the shore by a raised wooden jetty. The bright blue tiles of the temple's roof and the golden-lacquered Buddha statues shimmered in the dimming sunset, serene and ethereal.

The car landed on the jetty with a jolt, but the planks held. Right in front of Chamal was a marble pavilion that guarded the entrance to the temple, embellished with patterns of holy animals and lotuses. Prayer coins thrown by tourists were scattered across the floor of the pavilion. Beneath the delicate architecture was a white stone carving modeling the Buddha's footstep, and behind it was a reclining Buddha statue about half life-size, his posture elegant and relaxed, as if to welcome the guests to the temple.

"Everyone hold on tight!" Chamal shouted.

The red dots on the bombs stopped blinking. They stared back at Chamal like malicious, bloodshot eyes.

Chamal stepped down on the gas pedal full-force and sped forward. The dozen meters between him and the pavilion seemed infinite.

I'm sorry, Buddha.

His guilt-infused prayer was drowned out by a deafening blast.

THE RED CLOUD BAR, on the twenty-sixth floor of the Cinnamon Red Hotel, was quiet, free from the usual blare of dance music and psychedelic lights. But it was not empty. The guests craned their necks to look at the projection screen, showing a slow-motion video clip recorded by surveillance cameras.

An off-road SUV, body bruised and windows shattered, hurtled toward a marble pavilion at full speed. At the instant of impact, the car tilted forward, its tail soaring into the air. The scene unfolding in slow-motion was like a hyperrealistic art project: The car hood slowly crumpled, the windshield shattered piece by piece, the passengers' bodies swayed back and forth like an awkward dance. The safety air bags deployed, dispersing like a whiff of white smoke. The strong inertial force flung the man attached to the windshield into the air. His body descended in a smooth parabola toward the

Guanyin Bodhisattva statue in the center of the Seema Malaka, his shadow sweeping past the footprint carving and the reclining Buddha statue. Before his body could draw out the final arc segment of the parabola, however, a blazing white light filled the screen. The next moment, the man was engulfed in a ball of raging fire. His body disintegrated into a cloud of flesh and blood.

The car's motion gradually came to a stop. Passengers started to climb out one by one. None of them was injured—save a few bruises and cuts.

The scene froze. The lights came back on. A few sparse claps sounded from a corner, then, as more people joined in, vociferous waves of applause echoed through the bar.

"Let us raise our glass again to toast our hero, Chamal!" Yang Juan raised her champagne glass high into the air.

The golden liquid bubbled and rippled as the glasses clinked against one another.

"All the best for his first semester of middle school, too!"

The sound of laughter and good-natured teases filled the air. The guest of honor, the skinny, tanned boy with a shy look on his face, stood at the center of the room. Everyone wanted to shake his hand, hug him, take pictures and gift him with orchid crowns in the Sri Lankan tradition. The boy squirmed in discomfort.

A hand grabbed him and saved him from suffocation. It was Yang Juan, wearing a white pantsuit, looking as elegant and composed as ever. She gestured to the orchestra for the music to start and to the waiters for food and more drinks.

"Enjoy yourselves, my guests! But our hero needs to be excused for a few minutes to take care of some interviews." Yang Juan winked. "Journalists! You can never keep them waiting."

The crowd burst into laughter. Yang Juan walked Chamal to the VIP lounge. To Chamal's surprise, the room was empty. Confused, he looked up at Yang Juan, who smiled at him and poured two more glasses of champagne.

"Hey, I lied again. The interview was only an excuse to get you out of there. Cheers."

The woman and the boy clinked their glasses together. Yang Juan finished the drink in one gulp, but Chamal took only a sip.

"If you think I'm here to convince you to stay, I must say that you are wrong." Yang Juan patted Chamal on his shoulder. "I just wanted to give you a souvenir."

She stepped aside, revealing a coal-black box behind her.

Chamal stepped forward and pressed a hand onto the box, caressing the carving. The box, recognizing his fingerprint, opened slowly. In it were the latest versions of car racer gadgets: the black helmet, the skintight suit and gloves. Chamal grabbed the helmet, lifted it to eye level, and examined his own face reflected in the goggles. He looked up at Yang Juan, smiled, and opened his mouth, "Thank—"

"Don't thank me," Yang Juan cut him short.

"Thank you." Chamal's smile widened.

"I just wanted you to know that you are not the cost of creating the future," said Yang Juan, her expression solemn. "You *are* the future. Oh, one last thing." Yang Juan pulled out her phone and passed it to Chamal. "Read this. How do you feel about your new title, hmm?"

The headline said, "The Holy Driver: Sri Lankan Boy Saves 11 Lives Piloting Autonomous Vehicle." The picture below was of Chamal's silhouette, wearing his iconic helmet.

ANALYSIS

AUTONOMOUS VEHICLES, FULL AUTONOMY AND SMART CITIES, ETHICAL AND SOCIAL ISSUES

From *Knight Rider* to *Minority Report,* science fiction has presented the arrival of autonomous vehicles (AVs) as a foregone conclusion. But AVs are actually one of the Holy Grails of artificial intelligence. Driving is a complex task with many subtasks and inputs, as well as the potential for uncertain environments and unlikely events. That's why in "The Holy Driver" we predict that AV tech will not reach maturity until 2041, thirty-two years after Google started its commercial AV effort, and fifty-two years after Carnegie Mellon demonstrated the academic AV on a highway.

Autonomous vehicles will become truly autonomous not as the result of a single big breakthrough, but through decades of iterations. Automatic emergency braking was one of the first introductions of autonomous technology that will eventually "grow up." Another important consideration is that we shouldn't think of AVs simply as upgraded cars, but as part of a full smart-city infrastructure, the kind of interconnected technology infrastructure depicted in the story. On the way to realizing that vision, AV—as "The Holy Driver" suggests—will disrupt many professions and industries, and raise significant ethical and legal issues. Let's dig into these issues in more detail.

WHAT IS AN AUTONOMOUS VEHICLE?

At its most basic, an AV, or self-driving car, is a computer-controlled vehicle that drives itself.

Humans take about forty-five hours to learn how to drive, so it is a complicated task. Human driving involves perception (watching our surroundings and listening for sounds), navigation and planning (associating our surroundings to locations on a map, and getting us from point A to point B), inference (predicting the intent and the action of pedestrians and other drivers), decision-making (applying rules of the road to situations), and controlling the vehicle (translating intent to turning the steering wheel, stepping on the brake, and so on).

An autonomous vehicle driven by AI, rather than a human, uses neural networks instead of the brain and mechanical components instead of hands and feet. For example, AI perception uses cameras, LiDAR, and radar to sense its surroundings. AI navigation does route planning by associating every point on the road to a point on a high-definition digital map. AI inference uses algorithms to predict the intent of cars or pedestrians. AI decision-making and planning rely on expert rules or statistical estimation to make decisions, such as how to react to the presence of an obstacle when detected, and what to do if that obstacle moves.

Autonomous vehicles will mature one step at a time, from assisting the human driver to eventually no longer requiring a human driver. These steps are classified by the Society of Automotive Engineers as "Level 0" to "Level 5" as follows:

- L0 (no automation)—The human does *all* the driving, while AI watches the road and alerts the driver when deemed appropriate.

- L1 ("hands on")—AI can do one specific task only if the human driver turns it on, such as just steering.

- L2 ("hands off")—AI can do multiple tasks (such as steering, braking, and acceleration) but still expects the human to supervise and take over as needed.

- L3 ("eyes off")—AI can take over driving but will need the human to be ready to take over upon request by AI. (There are skeptics who wonder whether an abrupt handover would exacerbate danger, rather than mitigate it.)

- L4 ("mind off")—AI can take over driving completely for an entire trip, but only on roads and in environments that AI understands, like city streets and highways that have been mapped in high-definition.

- L5 ("steering wheel optional")—No human is required at all, for any roads and environments.

You can think of L0 to L3 as extra options in a new car, which a human continues to drive with AI tools. They will have only a limited impact on the future of transportation. Starting with L4, the car starts to feel like it has a mind of its own, and this will lead to a revolutionary impact on our society. An L4 car could be an autonomous bus that circles a fixed route, and an L5 car could be a robo-taxi hailed by an Uber-like app.

WHEN WILL FULLY AUTONOMOUS (L5) VEHICLES EMERGE?

L0 through L3 are now available in commercial vehicles, and limited L4 began experimental deployments in limited parts of some cities in late 2018. But achieving L5 (and less constrained L4) remains elusive.

A major hurdle to achieving L5 is that the AI needs to be trained on large quantities of data that are representative of "real driving" in many scenarios. However, the number of such scenarios and the degree of variability required are immense—at present, there isn't a feasible way to collect all the permutations of all the objects on the road, moving in all the directions, and in all weather conditions.

There are some ways to deal with these "long tail" scenarios. We could synthesize increasingly complex data, such as virtually adding slow-moving senior citizens, dog walkers, running kids, and everything imaginable to all the data. In addition, we could program some rules about "road sense" in driving (such as the dynamics of a four-way stop sign), without needing AI to learn road sense from data. But these solutions are no panacea, as synthesized data aren't as good as real data, and the rules may fail or contradict one another. The biggest challenge for L5 is that once driving is fully entrusted to AI, the cost of an error could be extremely high. If Amazon's AI erroneously recommends a product, it's no big deal. But if an AV makes a mistake, it could cost lives.

Because of these challenges, many experts see L5 to be more than twenty years away. But we can potentially accelerate its development by challenging one big assumption: that L5 needs to accept cities and roads as they are. Instead, what if we have "augmented roads and cities," where sensors and wireless communication are embedded into the road, so that the road can tell the car there is danger ahead or that it is drifting? What if a new city is designed so that the downtown has two layers—one for automobiles and the other for pedestrians (thereby avoiding cars hitting people)? By rebuilding infrastructure to minimize the chances of pedestrians near AV roads, we could dramatically increase the safety of L5 vehicles, thereby allowing them to be launched safely sooner. Remember in "The Holy Driver" when the cars rerouted to avoid the ambulance and the marathon? An AV on augmented roads is like a train on "virtual" rails—built with sensors, software, and mechanical control. This is how L5's launch was accelerated in this story.

Even if an AI-driven L5 AV is safer than vehicles with human drivers, there are still problems that could potentially confuse the AI, including natural disasters or acts of terrorism, that could render GPS useless. In these scenarios, the best solution would be to bring an expert human driver into the car to take over, by replicating the scene to a teleoperations center, where remote human drivers work in individual "cockpits." We can use augmented reality (AR) to project everything around the car (seen through the AV's many cameras) to panoramic screens in the remote cockpit. The backup driver's manipulations (such as turning the steering wheel) are captured and sent back to the AV to control the car. This is how Chamal "virtually" drives cars in the story. The transmission of high-fidelity video with minimal latency will take a lot of bandwidth, but 6G is set to provide that by about 2030.

A confluence of improved L5 technologies, augmented roads, and 6G communications connected using AR should be experimentally deployable around 2030. With iterations of improvements, we predict that L5 can be broadly and safely deployable around 2040 (assuming issues around ethics, accountability, and liability are resolved—more on these later).

The way L5 technologies will mature is through L0–L4 technologies applied in increasingly more-complex applications over the coming years, in concert with collecting data and iteratively improving technologies. For example, the simplest AV application already in use includes autonomous mobile robots (AMRs) and autonomous forklifts, because these AVs operate indoors. Next up—and starting to be deployed now—are autonomous transport trucks in places with fixed routes, such as mines and airport terminals. Several cities in China recently started testing robo-buses and robo-taxis. Robo-taxis still need to deal with complex issues of reaching any location, but robo-buses should become a reality within three years. Many cars are sold with L1 to L3 features like self-parking and Tesla's smart-summon feature. Beyond that are relatively predictable environments, like driving trucks on rural highways, or semi-fixed routes (such as airport shuttles to most hotels). Each of these applications will collect more data, improve AI algorithms, reduce "surprises," and create a solid foundation for the gradual arrival of L5.

IMPLICATIONS OF FULLY AUTONOMOUS (L5) VEHICLES

When L5 AV roams the road, it will bring about a revolution in transportation—on-demand cars that take you to your destination with lower cost, greater convenience, and improved safety.

When your calendar sees that you need to be driven to a meeting in an hour, an autonomous ride-hailing app like Uber or Lyft could order a car for you just as you're ready to depart. Uber's AI algorithms will move their AV fleet to be closer to people who might need a ride soon (say, when a concert is about to end). Vehicles' routes can be optimized based on minimizing the total amount of time all users wait idly, and the total amount of time Uber AVs are empty, while ensuring that the car's batteries are charged along the way. By removing the human driver, the fully automated and AI-managed fleet will have much better utilization as human uncertainties are eliminated.

As shared-ride cars become autonomously driven, shared-ride services will be able to reduce costs dramatically, because 75 percent of the fare currently goes to the driver. This cost reduction will lead to dramatically cheaper rides and encourage people to forgo owning their own cars.

You are probably wondering about safety. A seasoned human driver may have ten thousand hours of driving experience, but an AV may have a trillion hours of experience, because it learns from every car, and never forgets! So, over the long term, we can certainly expect much greater safety from AV.

But how will AV fair for the short term? Governments will approve pervasive AV only when it is "safer than people." Today, 1.35 million people die every year from motor vehicle crashes. So any launch of AI technologies must come with proof that they are at least as safe as human drivers. After the initial "safer than people" launch, AI will continue to learn based on more data, and improve itself. In a decade, the 1.35 million fatality figure could go down dramatically.

The average American drives eight and a half hours per week. In the future AV world, people would gain these eight and a half hours back for their lives. The interiors of the AVs will be reconfigured for work, communication, play, and even sleep. Many ride-sharing AVs may be designed as minicars, since we tend to ride in cars with just one or two people. But even a single-person car may be equipped with a reclining seat, a refrigerator with drinks and snacks, and a large screen.

More data leads to better AI, more automation leads to greater efficiency, more usage leads to reduced cost, and more free time leads to greater productivity. All of these will grow into a mutually reinforcing virtuous circle that will continually and rapidly increase the adoption of AV.

As automation rates increase, cars will be able to communicate with one another instantly, accurately, and effortlessly. For example, a car with a blown tire can tell nearby cars to stay away. In addition, consider that a car passing another car communicates its movement path precisely to nearby cars, so two cars can be two inches away from each other, yet with no risk of collision. Or, if a passenger is in a hurry, their car can offer an incentive (say, five cents) to other cars for slowing down and giving the right-of-way. These improvements will create an infrastructure of mostly AI drivers, eventually making it unsafe or illegal for humans to drive. Human driving may follow in the footsteps of driving under the influence. It will eventually be outlawed on public roads, perhaps beginning with highways and city centers. By that time, people who love driving will do what equestrians do today—go to private areas designated for entertainment or sports.

As autonomous vehicles, electric cars, and ride-sharing mature together, fewer people will buy cars (effectively reducing monthly expenses for families), parking garages can be largely repurposed (freeing up huge amounts of wasted space, as cars are parked 95 percent of the time), and the total number of cars can be reduced significantly (since ride-sharing AVs can operate efficiently 24/7). Collectively, these changes will reduce traffic congestion, fossil fuel consumption, and air pollution.

Alongside the lives saved and productivity gained, there will be disruptions to other aspects of our society. Taxi, truck, bus, and delivery drivers will be largely out of luck in a self-driving world. There are over 3.8 million Americans who directly operate trucks or taxis for a living, and many more who drive part-time for Uber/Lyft, the post office, delivery services, warehouses, and so on. These jobs will be gradually replaced by AI. Disruptions will also come in the form of a reshuffling of other traditional professions. Car maintenance will be less about mechanical repairs and will require electronics and software expertise. Gas stations, car dealerships, and parking garages will be dramatically reduced, as will their employees. Many lives will be changed forever, much like they were during the transition from horse-drawn carriages to automobiles.

NON-TECHNOLOGY ISSUES THAT MAY IMPEDE L5

In order to make autonomous vehicles pervasive, a number of challenges will need to be overcome, including ethics, liability issues, and sensationalism. This is to be expected because there are millions of lives at stake, not to mention many industries and hundreds of millions of jobs.

There will be circumstances that force AVs to make agonizing ethical decisions. Perhaps the most famous ethical dilemma is "the trolley problem," which boils down to a scenario in which a decision would need to be made between taking action and killing person A, or taking no action and killing persons B and C. If you think the answer is obvious, what if person A is a child? What if person A is your child? What if the car belongs to you, and person A is your child?

Today, when human drivers cause fatalities, they answer to a judicial process that decides if they acted properly and, if not, determines the consequences. But what happens if AI caused the fatality? Can AI explain its decision-making in a way that is humanly comprehensible, and legally and morally justifiable? "Explainable AI" is hard to achieve because AI is trained from data, and AI's answer is a complex math equation, which needs to be simplified dramatically to be comprehensible by humans. Some AI decisions will look downright stupid to people (because AI lacks common sense), just as some human decisions look stupid to AI (because people may be drunk or tired).

Other questions include: How should we balance the livelihoods of millions of truck drivers against the millions of hours saved by autonomous vehicles? Is it acceptable to have interim AI that makes mistakes that human drivers would not make if, in five years, the total number of fatalities can be halved because the AI will improve after learning from billions of miles of experience? And the most fundamental question: Should we ever let a machine make decisions that may harm human lives? If the answer is no, it would put an end to autonomous vehicles.

Since lives are at stake, every company must proceed with caution. There are two distinct approaches to take, each with different merits. One is to be extremely careful and collect data slowly in safe environments, thereby avoiding fatalities, before launching an AV product (the Waymo approach). Another is to launch as soon as AI is reasonably safe, and gather a lot of data, knowing that in due time, the system will save many lives, even though in the beginning some lives may be lost (the Tesla approach). Which is better? Reasonable people will disagree.

Another issue is, if there is a fatality in an AV accident, who is accountable? Is it the car manufacturer? The AI algorithm provider? The engineer who wrote the algorithm? The human backup driver? There is no obvious answer, but policy makers will need to make a decision soon, because we know from history that only when accountability is clear, can the ecosystem be built around it. (For example, credit card companies are liable for the losses from fraud, not the bank, the store, or the credit card owner. That accountability decision entitled the credit card companies to charge the other parties and use that income to prevent fraud, thereby building the credit card ecosystem.)

Let's say accountability is assigned to the software provider, and that Waymo made the software that resulted in a fatality. How much can the deceased person's family sue Waymo's parent company Alphabet, which has more than $100 billion in cash, for? This could become a field day for ambulance-chasing lawyers. We need to have laws that protect people from unsafe software, but we also need to ensure that technological improvement does not stall due to excessive indemnities.

Finally, traffic fatalities rarely make national headlines. But when Uber's autonomous vehicle killed a pedestrian in Phoenix in 2018, it became a national headline for several days. While Uber's system was likely at fault, is such media coverage warranted for every fatality in the future? If the media blasts every AV-induced death with damning headlines, it could destroy the AV industry, even when AV will eventually save millions of lives.

These issues may cause public fear, resulting in government regulations or conservative deployment, and may slow down the pace of wide AV acceptance. They are all legitimate issues, and we need to increase awareness, encourage debate, and reach resolution on them as soon as possible, so that when the AV technologies are ready for us, we are also ready for them. In the long term, I believe that, just as Chamal discovered in "The Holy Driver," L5 AV will be hugely beneficial to humanity in many dimensions, despite its costs.

CHAPTER SEVEN

QUANTUM GENOCIDE

STORY TRANSLATED BY ANDY DUDAK

WE DON'T NEED AIs TO DESTROY US; WE HAVE OUR
OWN ARROGANCE.

—*EX MACHINA* (2014)

EVERYTHING IS INTERWOVEN AND THE WEB IS
HOLY.

—MARCUS AURELIUS

NOTE FROM KAI-FU: Disruptive technologies can become our Prometheus's fire, or Pandora's box, depending on how they are being used. The villain of "Quantum Genocide," a European computer scientist who becomes unhinged after a climate change–related personal tragedy, uses two technology break-throughs for evil, embarking on a revenge plot the likes of which the world has never seen. In my commentary at the end of the chapter, I will describe perhaps the greatest breakthrough that might happen by 2041: quantum computing and how it could turbocharge AI and computing. I will also describe AI-enabled autonomous weapons, which are the greatest danger from AI, and may even become an existential threat to humankind.

+———

KEFLAVÍK, ICELAND
AUGUST 25, 2041
21:38 LOCAL TIME

ICELAND'S SUMMER NIGHTS were pallid, bright, and cold, like a drowned person's belly.

Keflavík, a satellite town fifty kilometers outside Reykjavík, was home to the continent's most secure data center, Hrosshvalur. Powered by geothermal energy, this fortress of data storage contained thousands of servers cooled by bone-chilling polar winds. Five hundred of Europe's leading companies stored their data assets here. The facility was connected to the European mainland, as well as to North America, via twelve high-performance fiber-optic arteries. Transit time to New York was just sixty milliseconds.

The site's carbon emissions approached zero. To Robin, it felt like a miracle.

The hacker was hiding out in a broken-down, overturned fishing boat, on a beach at Faxa Bay's southwest corner, five kilometers away from the Hrosshvalur data center. The vessel, with a gaping hole in its starboard side, resembled a great whale with its belly torn open. Thick black cables entered the hull through this gap into a darkness punctuated with flashes of blue-and-white lights.

The rusty hull was a base of operations for Robin and her two friends, the place they stored the kind of equipment all hackers dreamed of. They'd spent six months here, "borrowing" cheap local resources—geothermal power, cold air, and redundant quantum computing power from the data center—all to witness a miracle tonight.

Will, the trio's hardware wiz, flexed his hands above a keyboard. His winter gear was so shaggy that he looked like a brown bear. "Can't believe this moment is finally here," he said, shivering. "Tell me again, how many bitcoins did Satoshi have stashed away?"

Lee, a sixteen-year-old math prodigy, answered matter-of-factly without raising his eyes from the lines of code scrolling before him. "My conservative estimate is no less than two hundred and sixty billion U.S., maybe five hundred billion, depending on which trading strategy we go with."

"Don't let the money fuck with your heads." The stud in Robin's lip quivered as she spoke. "Fortune *and* glory, bros, fortune and glory."

LORE IN THE HACKER underworld had it that Satoshi Nakamoto, the mysterious father of Bitcoin, had died in a Guantánamo cell two decades earlier. He'd left behind no less than a million bitcoins, mined in his early years.

The currency was supposedly hidden in a digital wallet that relied on a script pattern known as a P2PK—or pay to public key. If the rumors were true, the treasure was a golden opportunity for prospectors. Bitcoin users had all but abandoned P2PK to history because every Bitcoin transaction initiated with P2PK made its public key visible to the network. Compared to later script patterns for Bitcoin transactions—like P2PKH, which revealed only the public key hash, rather than the public key itself—P2PK was less secure. At least in theory. An algorithm invented by Peter Shor in 1994 could crack the private key by solving the "elliptic curve discrete logarithm problem" with a 16-bit public key. A digital signature could thereby be forged, and the assets under the address taken.

Provided, of course, the thief had sufficient computing power.

Before the invention of quantum computers, even the fastest supercomputer would take about 6.5×10^{17} years to crack a private crypto key from a public key. That's fifty million times the remaining life of the known universe, a timespan the human brain cannot comprehend.

When Robin first understood this yawning gap between the possible and the theoretical, she'd gotten chills. It seemed like some divine being had deliberately made this math beyond our cognition to demonstrate the insignificance of human civilization.

At the time, Robin was known as Umit Elbakyan, a talented sixteen-year-old Kazakh hacker who thrived on disrupting order and robbing the unsecured online accounts of the rich to give to the poor. One day she received an anonymous email with footage of everyone in her family. According to the email, Umit would never see them again if she didn't agree to work for the sender: the notorious Vinciguerra Group. That day, she'd erased all records of her identity, had given herself the alias "Robin," and had gone to work for the Vinciguerras.

Robin had become part of an underground world of digital tomb raiders, intent on excavating treasures from the ruins of the digital age. But just because you could locate treasure didn't mean you knew how to open the chest. Others were afraid to find out what secrets lay inside. So, instead of plundering, Robin and her friends traded, transferring their treasures to other users on the dark web. Buyers were held to rigorous bidding regulations, and their assets had to be verified repeatedly. Sellers also had to prove the value of their goods.

Half a year ago, Robin had bought the information about Nakamoto's lost wallet for a high price from an old Silk Road XIII seller. It was disguised in a limited-edition encrypted artwork called *Does*

Hal Dream of Encrypted Gold? Only those deeply familiar with Bitcoin history would get the Philip K. Dick reference: "Hal" didn't refer to the killing machine HAL 9000 in *2001: A Space Odyssey,* but to the earliest implementation of the reusable proofs of work (RPOW) system, and the man who received the first Bitcoin transfer from Satoshi Nakamoto's wallet. The legendary Hal Finney had died in 2014 of ALS, having opted for liquid nitrogen cryopreservation in the hope of eventual resurrection.

Wallet addresses linked to Satoshi Nakamoto had been discovered many times, but there had never been much cash in them. The big fish had never shown its face.

Perhaps tonight, in this broken boat at the end of the world, the leviathan would finally surface.

INSIDE THE HACKERS' LAIR, the air was stagnant. The stench of rust and fish permeated everything.

The green progress bar on the primary monitor was approaching 100 percent. Was the moment of revelation at hand?

"Lee, Will, is everything normal?" Robin said.

Lee let out a grunt of assent, while Will simply pounded his chest with his fist.

The progress bar seemed to be hovering at 99.99 percent. Everyone held their breath.

"What's the matter?" Robin was anxious.

"Maybe it's a delay in the feed?" Lee's fingers danced in the air, tapping at the virtual keyboard.

"Come on, come on," Will muttered encouragement.

Just when they felt they couldn't take it anymore, the trio heard a sound like the jingling of gold coins. The progress bar vanished. A string of numbers and account information rushed across the screens. The staggering new balance in their shared account confirmed it. The gamble had worked.

Cheers erupted in the cramped space. Even Robin, normally cold and aloof, offered the briefest of smiles before issuing her next order: "Transfer balance!"

Robin knew if the huge sum remained at the undefended P2PK address, it was a sitting duck. In the hacker world, it paid to consider every person, every machine, every password as potentially compromised.

Lee quickly submitted a transaction request to move the bitcoins from the old account to a more secure one. The P2PK address was so long, and the transaction file so large, that it would take about ten minutes to process. During those ten minutes, the address's public key would be vulnerable. The good news was that quantum computers with the processing required to intercept it didn't exist yet—in theory.

Will tapped the steel hull as they waited, the sound echoing like hail hitting metal.

Lee was glued to the screen, his glasses reflecting blue-and-green light. "Robin!" he shouted. She'd known him a long time and had never seen him this panicked.

"What?" Robin rushed to the screen, where diagrams showed an anonymous signal.

"We've been hijacked . . . someone has cracked the private key. They're transferring our money out of the account!"

"Fuck! Should I cut the line?" said Will, standing beside the thick cable, awaiting the order.

"It's too late," said Lee. "You can't get faster than fiber-optics."

Robin's head was spinning. Hacking their account would require four thousand qubits of computing power. "How is this possible? There's no such machine on Earth . . ."

All the screens in their hideaway dimmed at the same time. There was only the buzz of electrical current as no one spoke.

"It's over." Lee sighed, his expression wooden.

Will slammed his fist against the hull.

Robin walked out of the boat and into the polar wind, her caution gone. All of the dazzling mirrorlike surfaces around her—sky, glacier, sea—seemed to reflect false light, as if the scene had been processed and filtered. There was fear in her heart.

Robin's thoughts turned to the one person who just might be able to solve this mystery. But he wasn't exactly a friend.

+———

THE HAGUE, NETHERLANDS
SEPTEMBER 9, 2041
15:59 LOCAL TIME

With the clock showing 1600 hours, the antiterrorism simulation commenced, a joint operation of the European Cybercrime
Centre—EC3—and the ATLAS Network, the cooperative platform
of forty-two special intervention units of EU member states and
associated countries.

The antiterrorism alliance's foe was a group of terrorists named
Dolce Vita. Having taken control of a small cruise ship, strapped
bombs to themselves, and taken two dozen tourists hostage, the
members of Dolce Vita were demanding a huge ransom along with
the release of their leader, imprisoned in Bavaria, Germany. The
cruise ship was in the North Sea, one kilometer off Scheveningen
Beach. Any ships or drones trying to approach would be detected,
triggering disastrous consequences.

Standing atop a remote observatory by the seashore, Xavier Ser-
rano, a senior agent of EC3, peered through a telescope at the ship
gleaming in the distance. He'd gotten used to this old-school optical device, preferring it to a prosthetic eye cam. The telescope's
weight and feel put him at ease.

Still, it was hard for him to relax as he waited for the situation
to escalate.

All possible rescue plans for a situation like the scenario now
facing EC3 had been generated by the AI antiterrorism system in a
matter of seconds. The role of human beings was to choose from
among those options, based on the algorithm's success probability
forecast and degree of potential collateral damage.

To determine the locations of hostages and terrorists in the
ship, EC3's usual approach would be to commandeer the ship's
controls and access its security cameras. Dolce Vita, however, had
anticipated this and destroyed the cameras. So the antiterrorism
system suggested an unconventional method: Establish a remote

surveillance network by hacking into the hostages' surgical implants, such as artificial eyes and electronic cochlea.

In an era of widespread electronic implants with low latency and signal interference, these separate means of communication gave EC3 their chance to bypass terrorist lines of defense. Thus, the hostages became independent eyes and ears.

While some team members worked on accessing these visuals, others deployed the sea glider, which carried an elite group of Special Intervention Unit agents. A new model of underwater conveyance, the sea glider resembled a porpoise, and relied on smart devices to monitor and control fluid intake, posture, direction, and depth. Because it didn't run on a motor, sonar would read it as a large fish.

The sea glider raced away from them at a depth of twenty meters, surfacing alongside the Dolce Vita ship and launching its tactical drone. The drone soared up from the sea and opened fire on the five men patrolling the deck, neutralizing them. With the coast clear, the SIU operatives climbed an access ladder on the hull and hoisted themselves onto the ship's deck. Via the 3D view in their MR eyepieces, the SIU agents were able to determine the relative positions of both the terrorists and the hostages. Soon, the operatives had three additional terrorists target-locked; they deployed smart bullets, with swift results.

Upon noting a man holding a detonator in 3D view, one forward-thinking SIU agent broke into the cabin to fire an electromagnetic pulse gun and zap the detonator's communications function.

The rest of the exercise followed the usual trajectory: Gunshots sounded, criminals fell, hostages were rescued, and a few minutes later the drama came to an end.

Xavier knew the drill was more a public relations stunt than anything else, a demonstration of EC3 and ATLAS Network's prowess for the benefit of politicians, the media, and taxpayers. In real life, conditions were never ideal. Any delay of more than twenty milliseconds could lead to failure and casualties.

In the floodlight of AI, human behavior models were nearly transparent. If terrorists developed their own AI countermeasures, though, the situation would become exponentially more complicated.

All Xavier could do now was accept applause and congratulations for the successful drill.

THREE YEARS EARLIER, XAVIER had come to The Hague from Madrid in the wake of a family tragedy. Unlike others, he hadn't joined EC3 for the prestige or to chase adventure. His sister had been abducted and sold by the European criminal network Vinciguerra several years prior. Her whereabouts were still unknown. In the nightmares that plagued him in the wee hours of the night, Lucia's sapphire eyes gleamed in the dark, reminding him not to forget his mission: to find her.

Xavier spent his spare time tracking down clues about his sister's abduction. So far they had all amounted to dead ends—except for a tip about a hacker named Robin. According to Xavier's sources, Robin had designed a set of encryption mechanisms for the Vinciguerra syndicate, writing code for managing transaction information and avoiding police detection. By catching Robin, he might be able to defeat the syndicate and find his little sister in one fell swoop.

From Mount Kazbek to Lisbon, from Sardinia to Cape Nordkinn, Xavier and Robin played cat and mouse for over a year.

His surveillance hadn't gone unnoticed by his target. Two weeks ago, Xavier had received an encrypted email signed "Robin." The message told an incredible story. Someone had stolen her wallet, which held more money than most countries on earth had in their coffers, using impossible quantum computing power. "If there's someone who can leverage that kind of power out there, you guys are in real trouble. Help me get to the bottom of this." After careful consideration, he decided not to report the email to his supervisors. It was his best shot to get to Robin, after all, and to follow the thread of his missing sister. Intrigued by the clues Robin provided, he asked a friend, Kasia Kowalski, a Polish data researcher at EC3, to look into major quantum computing research institutions without attracting attention from her superiors.

Several days later, Kasia came to Xavier with a detailed AI-assisted report, including a list of all quantum computing research institutions and their directors. This survey of public channels didn't align with the supercomputing power described in Robin's email.

One of the last names on the list caught Xavier's eye, however. A vague memory stirred. He leaned toward Kasia. "Who's this?"

"Marc Rousseau? You haven't heard of him? Poor guy . . ."

Xavier shook his head, frowning.

Kasia launched into the tragic tale of the brilliant physicist Rousseau, who had all but disappeared from public view after the untimely deaths of his wife and child in a fire. He had been something of a star in his field, and everyone had wondered if he'd be the one to make a breakthrough in quantum computing.

Xavier soon decided to start his investigation at Rousseau's research institute in Germany. There was something in this grieving physicist's eyes that triggered Xavier's mirror neurons. He would have to go to Munich.

+———

MUNICH, GERMANY
SEPTEMBER 11, 2041
10:02 LOCAL TIME

The Max Planck Institute didn't look as futuristic as Xavier had imagined. The gray-yellow building had clean Bauhaus contours, common on the streets of Munich. When he passed the bronze statue of Planck in the hall, Xavier lingered, surprised by its proximity to a statue of Saint Barbara. What did it mean, having this Catholic symbol here? Xavier could only imagine it was emblematic of the researchers' admiration for those individuals who adhere to their beliefs in the face of opposition.

A staff member led Xavier upstairs and down a long corridor to a small meeting room. Inside, Marc Rousseau sat waiting for him.

Xavier knew that since the deaths of his wife and son, Marc Rousseau had become a solitary person, still a figurehead, but no longer involved in the day-to-day running of the center.

How he spent his days, no one knew for sure.

"Dr. Rousseau, it's nice to meet you." Xavier smiled and sat down across from him, taking the measure of this man who'd gotten a double PhD in quantum information and condensed matter physics by age twenty-seven.

A low mumble emerged from Rousseau's unkempt beard. Even compared to the most slovenly PhD students, he appeared disheveled. His woolen shirt was stained and wrinkled, and his long, greasy hair was tied back carelessly. His eyes were bloodshot, but there was a cold light in them.

Quite the character, Xavier concluded.

"You have ten minutes." Marc Rousseau's voice was rough and faded.

"Well, I'm a representative of EC3, and my colleagues and I have encountered something we're having difficulty understanding. We'd like your professional opinion."

Xavier unfolded a flexible screen and pushed it across the table, watching Rousseau's face intently as he read Robin's email. Seconds and then minutes passed, but his expression remained wooden.

"What the hell are you getting at, showing me this crap?"

"It happened just two weeks ago. It's real."

"I need proof. You know European taxpayers have spent hundreds of billions of euros to build giant colliders, one after another, just to prevent science from sliding backward into metaphysical discussions of how many angels can dance on the head of a pin."

"The evidence is here. A private key of this P2PK address was cracked within ten minutes, and one hundred and fifty billion in bitcoin disappeared. The records are there, accurate to the millisecond."

"It's impossible," the doctor insisted, rubbing his eyes and studying the screen. "I could teach you the math in about a week, but I don't think you'd understand it. Unless . . ."

"Unless?"

"Unless the Americans have mastered new technologies we don't know about. Unless they've increased quantum computing power from a hundred thousand to a million qubits. But if that's the case, why would they go for an antique wallet? It'd be more like them to hold a press conference and crow about it to the world."

Xavier couldn't shake the feeling that Rousseau, now smirking, was hiding something from him. There had to be a way through his façade.

"Marc . . . can I call you Marc? Do you smoke?" Xavier knew Rousseau was a longtime smoker from the marks on his fingers.

Marc took a cigarette, lit it, inhaled, slowly exhaled a smoke ring, and relaxed.

"They say if Europe has any hope of competing with China and the U.S. for quantum supremacy, it lies with you. So, what is your research field?"

"They said that, did they?" Rousseau looked pleased. "Did you know that two sheets of graphene can become superconducting by rotating them at a certain angle and superimposing them?"

"The so-called magic angle?" Xavier had heard of this but couldn't quite wrap his head around it.

"Precisely. The same thing happens in the quantum field, but it's more complex, three-dimensional. Inspired by Professor Xiao-Gang Wen's work on quantum topological order forty years ago, we may be able to greatly increase computing power by adding limited qubits." Rousseau was silent for a moment. "Do you know why the ancient Egyptians built four-sided pyramids out of all the shapes they could have chosen?"

"I'm guessing it wasn't for beauty."

"Because they believed this shape would maximize and focus cosmic energy, and bring the mummies inside back to life."

Xavier shrugged. He was not interested in mysticism.

"They may have been right. To some extent, topology can indeed affect distribution of energy or information, and even improve conversion efficiency in ways humans can't imagine. We've done experiments and used AI to find the most effective quantum topology. It's a preliminary finding, nothing more, and far from ready for practical application."

It was like Rousseau knew in advance what Xavier meant to ask and answered directly. But this didn't dispel Xavier's suspicions. In fact, it put him on higher alert.

"I have a question I don't know if I should ask or not. It's about your family."

"My family? I don't see how that could possibly be germane to our discussion."

The conference room was silent. Xavier hadn't expected Rousseau's response to be so blunt. Feeling Rousseau's eyes on him, he rose from his chair.

Suddenly, a rapid buzzing came from them both, simultaneously.

Xavier tapped his interface, and there was the red "Breaking News" banner. He stared at it, his body going rigid.

Marc gazed out the window, motionless, gently exhaling smoke rings, as if he knew what was happening.

An angel had blown the trumpet of doom.

The attack had taken place two time zones away, in the Strait of Hormuz, an artery for one-third of the world's oil.

A pack of drones had descended from the sky like a swarm of black bees. They careened into the port, attacking with surgical precision the most critical links in the oil transportation system. Oil tanks exploded, pipeline pumps were damaged, supertankers capsized. The port was paralyzed as hellfire spread across the bay. The U.S. fleet stationed in the Persian Gulf hadn't had time to react. Now it, too, was in flames.

"It's beautiful, isn't it?"

Xavier looked up from the news, dazed, to see the physicist sitting as still as a statue. Marc turned slowly and said wistfully: "Like a grand fireworks display."

+———

NORTH SEA AIRSPACE, AMSTERDAM—THE HAGUE
SEPTEMBER 15, 2041
MIDNIGHT

Gasps rippled through the plane's passenger cabin, startling Robin awake. She drowsily put her face up against the porthole window, seeing flashing red lights down below like wounds in the distant night.

It was the Danish Straits, still reeling from the attack three days before.

The destruction had been unprecedented. Terrorist attacks had been staged at the seven major oil routes of the world. More than sixty million barrels of crude were transported from major production areas to the rest of the world every day, and most of it passed through a handful of narrow waterways: the Strait of Hormuz, the Strait of Malacca, the Suez Canal, the Danish Straits, the Bab el Mandeb Strait, the Turkish Straits, and the Panama Canal.

Choking these throats was like cutting off oxygen to a human body. The consequences would be swift: skyrocketing prices, market panic, inflation, traffic congestion. The collapse of distribution and service systems and then financial systems. No cars, no planes, no ships, no plastic, no alternative energy sources. Regional resource plunder. Rioting. Local wars. Total war.

No consumer goods or services existed independent of oil, apart from the agricultural economy. Investment in R&D for new energy technologies was too risky for many to take seriously. Various breakthroughs had come too late. They couldn't stave off the imminent full-scale crisis.

The International Energy Agency required member nations to keep in reserve at least a ninety-day oil supply, in case of emergency. The last time these supplies were used was after the Great Shift of the Pacific Rim Plate; before that, Hurricane Katrina, and before that, the Gulf War.

Humanity was not so lucky this time. The dream of a civilization built on oil was about to become a nightmare. An avalanche was poised to come crashing down.

Who was behind it? No organization had yet claimed responsibility.

So far, militaries had shot down a handful of the so-called Doomsday Drones. Their self-destruct systems always triggered before engineers could breach their defenses, leaving countless questions.

Where had the drones come from? How had they made it past air defense alert systems? What was their purpose?

No one knew.

That was why Robin was flying to The Hague.

Xavier had sent encrypted data to her. EC3 had captured a drone that failed to self-destruct, but they needed elite hackers. Some data suggested that the culprits might be the same forces that had whisked away Robin's bitcoin hoard. The time stamp indicating when the drone-controlling system was first activated was the same day that Robin had been robbed. Besides, Xavier needed Robin's hacker skills to aid in the legal imprisonment and interrogation of a suspect, Professor Marc Rousseau.

"We need you," the message read. "You're the first and only person who has fought this enemy."

Robin had trouble thinking of that encounter in the overturned ship as a "fight." After some hesitation, though, she had decided to come, with the promise of amnesty from EC3. Such delicate tasks could not be done remotely via robot.

Will and Lee were against it. They knew Xavier had been hunting their group for years. It could be a trap, they told her. But Robin had her reasons.

"Sometimes, in order to win, you first have to lose." It was a line Robin had once heard her grandmother say.

She knew Xavier better than anyone in the world—better even than did the man himself. She'd consolidated his past from all the flotsam scattered on the Internet sea. The data, so trivial and comprehensive, had been digested by AI and turned into a holographic model for calculating the emotions and behaviors of real human interaction. The algorithm had been designed for counterterrorism, but Robin had "borrowed" it to handle personal counterintelligence.

The more Robin understood, the harder it was to shake off a certain ambivalence. She felt like an insect trapped at the center of a spider's web, surprised to find itself sympathetic to the spider. She knew Xavier was trying to find his sister. She also knew Xavier was in for mostly bad news. Robin understood the situation of girls who'd been trafficked, a veritable hell on Earth. Maybe Xavier's ignorance was for his own good, Robin thought.

The Vinciguerra Group had evaded EC3 time and again via confrontation systems designed by Robin that hid, encrypted, and destroyed traces of crime. Be they human trafficking, online child abuse, data theft, or financial fraud, they vanished like water in water, shadowless, formless. These systems were a bargaining chip to protect her family from death threats. If she helped Xavier, it would be a public declaration of war against her employer, and it would violate the contractual spirit of the hacker world. She would spend the rest of her life running from her decision, perhaps only to turn up dead. She couldn't let the same fate befall her family.

A violent jolt to the fuselage announced their landing at Schiphol Airport, Amsterdam.

There's no time for regret now, Robin thought, emerging from the cabin into the darkness beyond.

WHEN THE MAN WAITING at the arrivals exit locked eyes with Robin, she saw his surprise. Maybe it was her gender that took him aback, or her youth or her good looks, or maybe all three.

Moments later, the unlikely pair were riding in silence in a bulletproof unmanned vehicle to The Hague.

Robin could tell Xavier had doubts he wished to voice. She appreciated his near-superhuman restraint. She turned her attention to the screen before her, the drone data EC3 had collated from various sources. Nothing she saw looked out of place: The weapons and power systems looked normal. The flight control system was operated by a built-in smart program that used high-performance depth-perceiving cameras to calculate optimal flight trajectories and resolve environments and targets. Its sophisticated anti-jamming encryption stood up to the latest generation of cognitive radio protocol cracking—or CRPC—tech. A swarm of drones could keep their data synced at high speed, in real time. They coordinated positions to avoid collisions.

A swarm was almost impossible to detect in advance, let alone to identify and combat.

The system reminded Robin of Mr. Blink, a legendary hacker who'd lived in seclusion for many years. He was rumored to have directed the incursion on NASA's control center in 2034, disrupting a rocket launch. But he had vanished from the scene years ago, giving up the outlaw life or perhaps—as many believed—dying.

"When you arrive at the hotel, get some rest," Xavier finally said, emotionless. "I'll pick you up at nine o'clock tomorrow morning."

"Let's go now." Robin didn't look up.

"What?"

"I'm not here on vacation. Every minute counts." Robin looked at Xavier coldly. He raised his eyebrows and instructed the car to change course.

Twenty minutes later, they arrived at a secret EC3 lab, Vulcan 7. The name was inspired by the race of rational beings in *Star Trek*, given the Vulcan-like demeanor of the lab workers, who prized logic and scorned speculation.

As Xavier and Robin entered the room, automatic lights clicked on to reveal a carbon-black mechanical creature lying on a titanium alloy workbench. It was held together with multicolored cables. The drone was so fragile and small it was difficult for Robin to associate it with the chaos reigning around the world.

Robin went to the central console, asking Xavier to call up the test logs. She scroll-gestured through the data at dizzying speed before stopping suddenly, like a conductor frozen in the final rest.

"Find something?" Xavier said, breaking the silence.

"It's dead. We need to bring it back to life."

"I don't understand."

"We won't have a chance of penetrating the anti-jamming system and accessing the internal rewrite program unless the drone is in mission mode. It'll be difficult even then, like aiming a playing card into the window of a Ferrari as it speeds by at five hundred kilometers per hour." Robin smiled slightly, her face pale.

Xavier sank into a chair. It would be a long night.

+——————

SCHEVENINGEN, NETHERLANDS
SEPTEMBER 16, 2041
14:31 LOCAL TIME

Only five kilometers from The Hague, Scheveningen was a beloved holiday destination for the Dutch. When the weather was
good, its pristine beaches would be dotted with windsurfers and
colorful kites.

No one would suspect that within these idyllic confines would
be an EC3 safe house. Inside was Marc Rousseau, looking more
haggard than ever.

"Sure you don't want to try it?" Xavier put a container of raw
herring on the table. The fishy odor was pungent. A toothpick
sporting a Dutch flag was stuck in the fish, to ensure tourists knew
it was a local specialty.

"I want a lawyer. This is illegal detention." Marc's voice was
hoarse, but still menacing.

"According to the EU's special witness protection clause, we
have the right to hold you like this." Xavier moved toward Marc
and lowered his voice. "There's a bounty on your head. Those
hackers want you gone, Marc. As long as that's the case, we're
going to keep you in this cute little house for your own good. Got
that?"

Furious, Marc lunged at Xavier. The smart restraint stopped
him in time. Within milliseconds, his clothes became shackles.
Marc crashed onto the table, arms outspread, his twisted face
pressed against the tabletop. Only his bloodshot eyes were free to
move.

"You couldn't trace it to me, am I right?" Marc exhaled, almost
laughing.

Xavier was playing a dangerous game, using Marc to attract
Robin, and Robin to contain Marc. "Protecting" Rousseau based on
faked assassination orders had worked, but the ruse couldn't go on
indefinitely. Evidence of Rousseau's involvement with the terrorist

attacks would draw the attention of every intelligence agency in the world. For now, Xavier would cling to his limited authority. But if Rousseau's involvement couldn't be proven, it would all come to nothing.

Xavier was gambling on intuition alone. He sensed hatred in Marc's eyes, the kind of hatred that drives people to do all sorts of crazy things. Xavier had also lost family. He understood that feeling. AI couldn't.

But the question was how Marc had done it. There had to be a greater force behind it all.

Xavier couldn't waste any more time. It was time to up the ante.

"Marc, you give me no choice. I'll have to use an extraordinary method to make you remember something you don't want to remember. If it makes you unhappy, I'm sorry. There are some things machines do better than people."

"What are you talking about?"

"You might have the best poker face in the world, but the AI interrogation system captures the subtlest micro-expressions and tone variations. It's a remarkable technology that I am loath to use. Not because it's still in the experimental stage, and not because several subjects have suffered irreversible brain damage from it, but because it requires filling out too many boring forms. But, like I said, you've given me no choice."

Xavier was not lying. This AI interrogation tech, BAD TRIP, used noninvasive neuro-electromagnetic interference to the limbic system to induce intensely painful experiences, both physical and mental—often including the playback of traumatic memories. Such playback was more realistic and immersive than the most artfully designed XR experiences. It was like a nightmare that amplified all emotional reactions and destroyed all reason.

BAD TRIP was known to leave test subjects with long-term psychological trauma, and its use was linked to suicide attempts. Some within EC3 wanted to abolish the method, but the rise of extreme terrorism in Europe had given the technology a reprieve.

A pair of technicians brought the equipment into the room. It looked like an octopus of metal and cable. They began to install it around Marc's head.

"Wait." Marc's bravado was beginning to crack.

Xavier, meanwhile, felt only disgusted with himself. What had become of him? And in the name of justice? But it was all to find his sister, he told himself. It was for Lucia.

"Wait!" Marc shouted. "If I tell you where the next wave of attacks will occur . . ."

Xavier raised his hand, and the technician paused. "My patience has its limits."

"Bring me paper and a pen! Untie me!"

A few minutes later, Xavier was holding a scribbled list of times and places, trying to connect to his encrypted channel with Robin. Marc shrieked in the background.

"I told you everything I know! Let me go!"

Xavier paused, and waved his hand. The technician had his orders. He continued securing the mechanical octopus onto Marc's head.

"You bastard! You'll regret it! You can't change anything." Marc's eyes widened. He struggled, the veins in his neck and forehead bulging, as the technician tightened a final strap on the machine.

"Sorry," Xavier said, barely audibly, leaving the safe house.

BAD TRIP started with a slight buzzing sound, like the compressor in an old refrigerator. A ray of green light flashed before Marc Rousseau's eyes. He tried with all his might to break free, but he froze. It felt like a skewer of ice had been plunged into his brain and begun to slowly drill down.

His eyes were full of tears, but he could no longer close them.

+———

BRUSSELS, BELGIUM
SEPTEMBER 17, 2041
7:51 LOCAL TIME

The twenty-fifth Global Science, Technology, and Innovation Conference (G-STIC) was held at the NEO II International Convention Center, next to the King Baudouin Stadium in the Atomium and Brussels Expo Park. Tech elites, investors, business leaders, political figures, and celebrities from all over the world had gathered to fill the hall.

On the third and final day of the conference, the atmosphere was tense. Security was always high at the G-STIC, but after the recent attacks on the global oil infrastructure, law enforcement was on especially high alert. Furthermore, Ray Singh, whose company, IndraCorp, was valued at more than a trillion U.S. dollars, was scheduled to give the concluding keynote. In recent years, he'd spared no effort to build a marine city, and he'd been attacked multiple times by extreme environmentalist organizations. The ATLAS Network's autonomous anti-riot vehicles roamed the grounds. SIU personnel in black combat armor were deployed at high-volume intersections, forming a kilometer perimeter that encompassed the central buildings. They looked skyward through enhanced eyepieces, watching for anything out of the ordinary.

Xavier and Robin sat in one of the cars, watching airspace-monitoring data. According to Marc's tip, a wave of attacks would commence in ten minutes.

The AI counterterrorism system, having analyzed previous attacks to find patterns, concluded that the probability of a strike here like those on the oil hubs was close to zero. This was no energy base or shipping hub. It didn't fit the profile of the other targets. Most of the world's military and law enforcement resources had been deployed in recent days to safeguard important energy infrastructure. That strategy was supported by data and logic. It hadn't been easy for Xavier to convince his superiors at EC3 and ATLAS Network to take his intel seriously.

He had bet his future on it.

Robin sipped her coffee and frowned. "I don't see anything apart from registered aircraft."

"No, Rousseau was telling the truth."

"But why would he foil his own plan like this?"

Xavier hadn't told Robin about BAD TRIP. He shook his head: "Maybe he's so arrogant, he wants to see us try to stop him."

Robin shrugged. "Looks like his arrogance was misguided."

A call came in. It was Dom, the operations commander in

charge of the grounds. His voice was uneasy: "Something's coming from the west, a flock of birds or something . . ."

Robin summoned up the interface parameters and tapped away. The screen changed to show a group of dense lights, flashing red, closing in fast.

"They look just like birds. They must have imitated the flight pattern of birds and evaded our detection system!"

"Shit! Reinforce the west side of the fire circle! Implement Alpha plan!" Xavier ordered Dom, then turned to Robin. "Are you ready? This isn't sitting in front of a screen writing code. We may die out there."

Robin offered a grim smile and put on her black helmet.

As the anti-riot car began moving forward, a heavy-duty tactical motorbike detached from the middle of the vehicle, with Xavier and Robin on it. The flock of ominous black birds filled the sky. They were too agile and quick for the slow-moving anti-riot vehicle. To keep up, Xavier and Robin had to rely on the bike, even though it meant exposing themselves.

Xavier gunned the engine. Robin seized the handrails as the bike launched out of the car, growling. They bounced twice on the ground, kicking up gravel, before peeling away in hot pursuit of the Doomsday swarm.

Robin raised her hands. The transmitters on her wrists deployed powerful short-range electromagnetic waves at the drones, but the drones were moving so fast that she knew the waves might not find their target. However remote the chance, Robin knew it was the only way to exploit the drones' comm protocol gaps, and give Robin and Xavier any chance of assuming control over them via reverse engineering.

Meanwhile, SIU personnel sprayed automatic fire from their positions. Strangely, the drones didn't retaliate. Even as lines of fire battered them into twisted shrapnel, the latecomers did not alter their strategy or route.

This scenic spot had suddenly become a battlefield. Smoke was billowing and explosions were everywhere.

"Hurry up!" Robin cried.

Xavier accelerated. The bike roared, leaping like a wild horse, then falling heavily.

"Two o'clock!" Even as Xavier shouted, the bike was already lunging toward a stray low-flying drone. It looked like an injured bird, listing, falling apart. Robin raised her fists and fired her EM waves again.

"Get closer!" she shouted.

Xavier cursed, navigating the vehicle around and over obstacles, up steps, in close pursuit of the drone. They had to catch up before it crashed, or their efforts would all be for nothing.

The interface in Robin's helmet flashed. Had the handshake signal succeeded? She quickly dispatched instructions, disguised as an innocuous inter-drone data exchange. Sure enough, the drone seemed to slow.

"Almost there, don't lose it!"

Xavier clutched the handlebars with sweaty hands. He felt like a stuntman in an outlandish Hollywood movie.

"Five seconds left!" Robin shouted. "Four, three . . . watch out!" She held her breath.

The drone hurtled into a hotel atrium. Below were six stories of exhibition space, the floors connected by escalators. The Belgians, ever conscious of ecological harmony, had planted trees and vegetation on each floor, creating terrace-like hanging gardens. Staring through the glass at this man-made paradise, Xavier faced a split-second choice: give up the chase or follow the drone by jumping into free fall.

"Fuck!" Xavier gasped. "Hold on!"

The bike crashed through the glass partition, into open space.

Robin's heart froze. She clung to Xavier as the bike plunged through the air.

Thinking fast, Xavier activated the compressed air jets lining the bike's frame, adjusting their trajectory. A tree caught his eye, three stories down and closing. He had almost no time to react. Holding on to Robin and leaping from the bike at the last second, he slammed into the canopy like a cannonball, Robin in his arms. His back took the brunt of it, though protected by his rear airbag, which had instantaneously filled with compressed air to buffer the impact.

They crashed through branches and landed heavily on a glass platform. Xavier groaned in pain as Robin climbed off him. "Are you okay?" Her concern surprised him.

"Alive," he managed. "The little bird . . . how is it?"

Robin cast about, and her apprehension gave way to hope. The black drone floated not far above their heads. Had she tamed it? Suddenly, a video transmission from the EC3 commander interrupted.

"Xavier! Robin! Where are you two? What are these things trying to accomplish here?"

The video cut to an exterior shot: Beyond the glass curtain wall of the NEO II Panorama Hotel, three surviving drones had withdrawn to form an equilateral triangle array that revolved around the building, as if scanning the hotel floor by floor.

Xavier glared at the machine floating above his head. "Can you stop them? Infect them?"

"I can try, but it'll take time," said Robin. "Why not just shoot them down?"

"There are still people in the hotel."

"What? I thought everyone had been evacuated!"

"There are some employees left, as well as G-STIC VIPs who didn't evacuate quickly enough."

"Shit!"

They found a shortcut to the armored vehicle, where Robin tapped away at her virtual keyboard. She felt a growing unease. Who or what were those drones looking for?

The tamed drone began to rise, heading for its three companions. Only within a certain range would the intra-swarm communication protocol activate, so that Robin's data-bait could take effect. She held her breath.

The Panorama Hotel rose eighteen floors above them. When the weather was good, from the top you could see the whole of Brussels and the glittering Senne. If at this moment a penthouse guest had looked out the window, he would have seen a strange black spot, like a stain on the glass that couldn't be wiped off. The spot would have drifted slowly from view, to be replaced by another from the opposite direction, then another.

This drone trio would be like three eyes watching you in turn, glinting with a cold, malicious light.

"Hurry up!" Xavier urged, eyeing their reprogrammed drone. But it clearly didn't have enough rising power. There were five floors to the top, and it had to ascend three more.

There was a faint burst of sound above, like a bubble had popped, and glass shards began to rain down outside the hotel. Suddenly, everything was in motion. The three drones had opened fire, pirouetting through the air and shooting from just outside the floor-to-ceiling windows of the presidential suite.

"Starting data packet transfer!" As they sheltered beneath the armored vehicle, Robin frowned, reading off the progress. "Executing!"

Xavier got to his feet, wincing, looking up at the battle unfolding on high. Suddenly, the drones froze, like four musical rests inscribed upon the blue sky.

Another update sounded through their headsets. Robin and Xavier exchanged glances. It was clear now that the three drones had not ceased fire because Robin had successfully infiltrated their code, but because they'd completed their mission.

Ray Singh, hiding in the presidential suite, had become the first name crossed off the Doomsday Blacklist.

+———

SCHEVENINGEN, NETHERLANDS
SEPTEMBER 16, 2041
15:00–21:00 LOCAL TIME

Marc didn't know how long he'd been cycling through BAD TRIP. It could have been minutes or decades. The worst part was not the physical sensations, but the agony wrought on his mind: the endless resurfacing of moments his brain had fought to suppress. Or rather, one truly devastating moment.

Five years earlier, Marc had taken his wife, Anna, and son, Luc, to California on vacation, to escape the cold German winter. They drove to the reconstructed town of Paradise to visit Marc's mentor, Paul Van de Graaff. They hadn't seen each other for many years, so the reunion was emotional. Luc was obsessed with the idea of hiking a part of the Pacific Crest Trail, so Anna decided to leave Marc and his friend immersed in their physics discussion and drive her son up the mountain into Plumas National Forest.

"We'll be back for dinner," Anna said, smiling as she left. "I hope we won't hear the word 'quantum' at the dinner table."

Marc remembered every word so clearly. It was the last sentence he ever heard from Anna.

After Anna left with Luc, Marc had explained his latest work and Paul had suggested a new transformation form to the quantum topological formula. A research dead end had opened up. Marc was so excited that he hadn't noticed time passing. It was late by the time he finally emerged from the engrossing academic discourse. He called his wife, but it failed to connect. It was then that a strange orange-red color appeared on the horizon, like a sunset but in the wrong direction.

Just then, his smartstream sounded piercing alarms, and so did Paul's. It was an emergency evacuation notice: a wildfire.

Marc followed Paul into his stick-shift Mustang. They called the police as Paul drove, weaving through the congested lanes of other people fleeing the fire. Marc was hoping to locate his wife's vehicle via satellite. The call was transferred to the AI emergency response service, and the sweet, synthesized voice said: "According to California data privacy protections, we cannot provide the vehicle location."

Marc hung up, frustrated. Paul hit the gas, accelerating to the car's limit.

They hadn't gone ten miles before they were stopped by a state police roadblock. The cops said it was extremely dangerous ahead and civilian vehicles weren't allowed. As Paul tried to calm Marc down, a firefighting convoy pulled up. They consented to take him with them, leaving Paul with the car.

"It's winter!" Marc exclaimed in disbelief.

"It's California." The firefighters smiled at his naïveté.

This land had once enjoyed a Mediterranean climate, complete with warm, rainy winters. But as global weather patterns became more extreme, winters grew hot and dry. Twenty-three years earlier, a super wildfire had destroyed the town of Paradise. Eleven thousand houses were turned to ash, eighty-five people died, and the fire consumed more than sixty-three thousand hectares. With additional strong winds, the risk of such mega wildfires was now constant.

The convoy drove over an orange-red iron bridge. Below, a clear, seemingly bottomless river wound into the distance. The slopes on both sides were lush.

Finally, Marc spotted his wife's Ford parked on the side of the road. It was empty. Had she and Luc ventured into the forest? After some pleading, two firefighters volunteered to accompany Marc on the search for his family—but he was running out of time.

The wind carried burning charcoal debris, which could ignite nearby vegetation at unimaginable speeds. Under extreme conditions, a mountain fire can travel 80 kilometers (50 miles) per hour. It can be difficult to escape in a vehicle, let alone on foot.

Marc and the two firefighters shouted for Anna and Luc. The three men spread out to cast a wider net as they moved through the woods. They could already see the fiery red sky before them, the forest in a golden halo. The air temperature was rising fast, and they could smell burning.

"We can't go any farther," one of the firefighters said, stopping with his partner. "Fire'll be here any minute now."

"Please," Marc said, practically begging. "They have to be nearby . . . help me!"

"I'm sorry," the other firefighter said, shaking his head.

Marc heard a faint sound, a bird call perhaps. He peered into the chaos of the forest before him, calling for his wife and child, shouting himself hoarse. He heard it again, a voice, clearer this time, the cry of a boy. Marc bolted in that direction, and the firemen came stumbling after him.

The wind shifted suddenly.

A wave of wicked heat almost knocked them to the ground. The entire forest was glowing. The red light surrounded them, like a monster opening its bloody jaws.

Marc spotted two indistinct figures, one lying prone and the other kneeling. They were at the base of a boulder. He was all but certain they were Anna and Luc. As he prepared to plunge toward them, the two firemen held him back. The trio fell to the ground as Marc clawed to get free. "What the fuck are you doing?" he snarled.

He'd barely gotten these words out when a dragon of flame swept past where he'd been standing. The dragon rode the wind, turning everything into red, sizzling coke. The two figures, not far away, were directly in its path.

And then they were gone without a trace.

IT TOOK SEVENTEEN DAYS to put out the fire.

At the funeral, all Marc had to mourn was a handful of scorched earth. Soon afterward he gave up on his work; he could focus on nothing but the wildfire that had taken his wife and son. As a rationalist, he could not accept it. It was so unfair; a person, institution, or system had to be responsible. Everyone blamed extreme weather, but that wasn't enough for Marc.

Anger and self-blame distorted his mind like slow-acting poison. He grew to hate his fellow humans. He came to believe that human arrogance and greed had killed his family. Anna and Luc were victims of a civilization that had chosen a path toward self-destruction. The epiphany from his mentor, shared just before the tragedy, was to become his weapon of vengeance.

In the quantum world, causality worked counter to human intuition. Cause and effect were intertwined.

Marc worked day and night. He began to communicate in extremist forums on the dark web, where illegal resources and sensitive information were freely traded. A scheme slowly took shape whereby he could have his revenge.

If Marc was to succeed in his plan, he would need to make a breakthrough in quantum computing power. He dedicated all his research to it. But on the dark web, computing power was a scarce resource, hunted by all, the most sought-after currency. He had to conceal the fruits of his research so as to keep it safe. He chose to fade from public view.

To most people, he was just a poor man indulging in grief, lost in his painful past, bound for self-destruction. They had no idea he was at the center of a shocking conspiracy.

BAD TRIP dragged Marc back to that devastating evening in California time and again, forcing him to experience the loss of his loved ones anew. Over and over he watched his wife and son turn to charcoal. If he hadn't been broken the first time, he was broken now.

When the technicians relieved him of the BAD TRIP hardware, night had fallen. He could finally close his eyes.

Marc Rousseau's only weakness was no more, and the slaughter on Earth had just begun.

+———

SILK ROAD XIII
ENCRYPTED CHAT ROOM [000137]
SEPTEMBER 17, 2041
20:51:34 COORDINATED UNIVERSAL TIME (UTC)

In the encrypted chat room, Robin used her usual avatar, a strange doll with dead fish eyes and a disgusted face created by the Japanese artist Yoshitomo Nara. Her partner Will was now an incarnation of Lone Sloane, the long-haired, red-eyed interstellar wanderer. Lee hadn't appeared yet, which was unusual. He was usually the most punctual of the three.

The virtual environment they chose was an eighteenth-century York dungeon, dark and cold. Dim candles flickered in crevices of the stone walls. From time to time, faint wails echoed from deep underground.

ROBIN: A fitting location.

WILL: Isn't it? People are biting the dust everywhere. This Doomsday Blacklist is getting out of control. AI predicts a total death toll of between twelve and fifteen hundred. All bigwigs.

ROBIN: So, have you found a pattern?

WILL: We fed the background info of the victims into machine cross-analysis and found no clear correlations. Nothing. These people basically cover every demographic, industry, age group . . . The only common thread is that they're all at the top levels in their respective fields, very influential. Maybe it's that simple?

ROBIN: I don't buy it. First the drones attacked oil hubs, then these elites. There must be some other link.

WILL: Could it be a diversionary tactic?

ROBIN: What do you mean?

WILL: To make the rabbit disappear, draw the audience's attention to the hat. The old magic trick.

ROBIN: Hmm . . .

WILL: You said yourself that it's too slow to hack in by spreading the digital virus through the drone comms network. You asked me if there's a way to break in via the hardware. Well, I've researched it. Turns out the problem has nothing to do with hardware. It has to do with epidemiology.

ROBIN: Say what?

WILL: When a drone swarm flies in concert, like they're programmed to do, comms frequency between individuals is actually quite high. Once a drone is infected and changes its behavior, comms between it and the swarm drops to almost nothing.

ROBIN: So, no matter how hard we try, we can't save many people.

WILL: The key is to find the source of the drones. Anyway, what's going on with Lee? Where is that joker?

A white fox had stolen into the room while they chatted. It transformed into a boy before their eyes, an avatar modeled on Lee himself.

WILL: You're finally here.

LEE: It took some effort and kung fu to lose a few stalking mice.

ROBIN: Xavier can't keep concealing Marc Rousseau's part in all this, unless he passes himself off as some kind of prophet that can foresee coming attacks. We only have a few hours to figure out how to break Marc before the powers-that-be complete hand-over procedures. I have to confront him. This may be our last chance.

WILL: You really have a thing for this Xavier, don't you? Don't forget he's always wanted to catch you and put you in prison.

ROBIN: How about you shut the fuck up. Lee?

LEE: Stop fighting, you two. I have news.

Lee gestured a screen into existence on the stone wall of the simulated dungeon. A cartoon started to play, showing decentralized yet organized crime operations on a global scale, in a blockchain and AI-based world. All transactions were encrypted. All manufacturing and transportation were automated. Criminals and crime were totally separated in time and space, so long as interlocking encryption tasks were set up. Weapons could be manufactured and deployed automatically. Drugs could be grown, harvested, purified, subcontracted to robots in uninhabited regions, transferred to the marketplace by unmanned vehicles, and delivered—all by drones. Buyers only had to access the dark web and click on what they desired, like ordering from a menu. Without human intermediaries, all the betrayals, leaks, and undercover agents of old gangster films no longer existed. Even if police got wind of a criminal enterprise, each stage in the process proceeded in a vacuum, allowing for efficient replacement and minimal losses.

LEE: In a world of automated terrorism, one person can destroy said world.

WILL: If he has enough money.

ROBIN: Well, let's focus on our pilfered Nakamoto wallet. Lee, talk me through it.

LEE: I've reviewed drone tech data on the Silk Road from the past five years. Although transactions are encrypted, publishing and browsing and discussions aren't. I used a semantic analysis program to group the discussion content according to relevance, and one group is rather suspicious. I'm talking automated drone assembly, swarm flight algorithms, encrypted anti-jamming systems, ultra-long-range energy modules, et cetera. Stack these technologies together and you've got a Doomsday Drone prototype. Most members of the group are anonymous users with encrypted IP addresses, but one IP is still exposed, even after a hundred concealment measures. Thanks to this IP, I figured out that this person has a keen interest in two other things. Any guesses?

WILL: Keep us in suspense and I'll strangle you!

LEE: Take it easy there, tiger. One is plutonium, the stuff flowing out of former Soviet nuke bases. The other thing is even more terrifying: how to build an intelligent model that can understand natural language and communicate like a real person, based on the social data of a dead person.

They were silent. Another inhuman wail echoed up from a distant chamber. The candle flames shivered in their sconces. It was like that moment in a horror movie just before the ghost arrives.

ROBIN: I'm not surprised he wants to build a nuke and destroy the world. But the other thing . . . that's interesting. He wants to create a ghost. Maybe this is our chance.

WILL: What do you mean?

ROBIN: Lee, you have two hours.

LEE: To make a ghost?

ROBIN: To make two.

+———

THALYS PLUS HIGH-SPEED TRAIN
BRUSSELS–SCHEVENINGEN, THE HAGUE
SEPTEMBER 18, 2041
00:32 LOCAL TIME

An exhausted Xavier put on his sleep-aid goggles. There was a faint rushing sound in his headphones, as if he were flying tens of thousands of feet high.

Black smoke rose in the distance. It changed shape like a murmuration of starlings, glinting in the sunlight. It was the Doomsday Drone swarm, released for flight from remote and hidden corners of the world, from unmanned factories disguised as hills. They lived on solar energy and haunted mountains and fields at night. Their programming drove them to imitate the formations and flight paths of birds, making them invisible to satellites.

As Xavier watched, the swarming drones grew in number and approached rapidly. Suddenly, they were threatening to engulf him. He couldn't escape. He was being subsumed by the swarm, becoming part of a new terrorist entity as it descended upon the Earth to carry out its kill plan.

Conference halls, luxurious penthouses, golf courses, cruise ships, limos, bank boardrooms: places redolent of money and status, razed by the grim reaper into level playing fields. Seeing faces distorted by fear, smart bullets piercing heads and chests, blood blossoming, Xavier realized that perhaps the cruelest thing in life was equal opportunity.

The violence became unbearable. Xavier turned away, as if he could escape that way. He saw a figure in the distance—it was his sister, looking just as she looked all those years ago, as if no time had passed.

Xavier tried to pass through the swarm, to take his sister's hand. But the frantic black birds collided with him, preventing him from taking another step forward. The machines' sharp edges cut his body. He bled black, sticky oil.

XAVIER CRIED OUT. He woke to see Robin, looking concerned.

"Nightmare?"

"Uh . . ." Xavier was dazed, unsure where he was.

"You said 'Lucia.' That's your sister?"

His heart breaking anew at the thought of his sister, Xavier turned to stare out the train window.

"I remember her," Robin said. "She had beautiful blue eyes."

"You've seen her?" He seized Robin's hand.

Robin pulled away. Of course she'd seen Lucia. The old photos Xavier stared at when he woke up in the middle of the night, the missing person videos posted everywhere . . . Those sapphire eyes were unforgettable. But Robin decided to lie. She wasn't sure why: sympathy, perhaps, or guilt, or maybe a conviction that in this terrible world, one should not be deprived of hope, even if it is misguided.

"When this mess is over, I will help you find Lucia."

She had Xavier's attention now. He trembled, but soon his will buckled and he couldn't help crying.

Robin wanted to comfort him, but her hands hovered helplessly in the air.

An hour later, they arrived at the safe house in Scheveningen.

Marc Rousseau seemed a completely different person from the man that Xavier had left with the BAD TRIP machine. He sat in the dark like a mad king, beard messy, eyes piercing. "Welcome," he said. "How many are dead?" Right to the point, and almost proud.

"Why do you care?" Xavier said.

"I don't. The algorithm does."

"The algorithm?" Robin glared at Marc. "Is that the algorithm for stealing wallets or the one for killing?"

Marc turned to Robin with an odd smile. "Sorry I had to expropriate your property, but it wasn't really yours, was it? Think of it like this: You bought an indulgence, a bond of atonement."

"You're the one who needs to fucking atone!" Xavier brought a fist down on the table.

"Yes, I suppose so. I need atonement, as do you two, and all those self-righteous accelerationists, all humankind. We all need atonement. And the time has come."

"Wait," Robin said. "You said 'accelerationists.' Is that your reason for killing these people?"

"Ha. Your precious anti-terrorism AI sees only quantified people and their various data tags: age, income, position, race, sexual orientation, company market value, consumer preferences, health status . . . it can't see beyond that. You people think technological change can solve all the world's problems, even if it brings bigger problems. You're always trying to solve them with brute force, never mind the carbon footprint. Human civilization is a car driving toward a cliff. Accelerationists keep stepping on the gas." Marc made an exaggerated gesture of explosion.

"So your plan is to punish humans by creating more explosions? What do you stand to gain from that?" Robin wanted to anger him, throw him off balance.

Marc's smile vanished. He leaned back, squinting, and said softly, "You'll see."

Xavier had gained no advantage. To get the information he wanted, he knew he'd have to attack Marc where it hurt. He would follow Robin's plan. "Marc, I'm sorry about Anna and Luc—"

"Don't," Marc said, glaring at him. "Don't even mention their names. I'm warning you."

"You can't blame all of humanity for an accident."

"Accident! Really?" Marc's tenuous self-control evaporated. He was shaking. "Fucking PG&E let their line age. That's what caused the wildfire. But no one will admit it—not the government, the company, the media, or even the goddamn public. Everyone blames nature, as if we're not part of nature! As if we've just been victims of a global climate anomaly for the past few decades. Like it's not our fault. It's idiotic!"

Xavier and Robin glanced at each other, then got up to leave.

"Marc, you need a little time to calm down," Xavier said. "We'll be back later to pick up where we left off."

And then Marc was alone again, a tyrant sobbing softly to himself.

He looked up in confusion as the lights flickered and then went out. Two faint blue flames appeared in the gloom. As they grew nearer, faces gradually became discernible in their cold glow. They were his dead wife and son.

"Anna? Luc?" Marc stared, agape. He didn't know if he felt terror or joy. "Is it really you? Am I hallucinating?"

"We're not quantum ghosts," Anna said, with the calm authority he remembered from life. "Of course it's us, Marc. And you haven't changed a bit."

"Dad," the boy called, timid, like he'd done something wrong. "I miss you so much."

"Luc . . ." Marc ached to take them in his arms but found himself tied to the chair. He cursed his confinement, tears streaming down his cheeks. "I miss you so much. If only I'd gone with you . . ."

"Don't blame yourself, Marc. It was meant to be. You'll need to move on one day."

"I'm fine, Anna. And we'll be reunited soon, very soon."

"Dad," Luc began, "why are you killing all those people?"

"They're destroying the planet. You loved nature and animals more than anything, right? I want to give Earth back to its original inhabitants. I'm doing it for you."

"But . . . will killing those people keep the planet from being destroyed?"

"Luc, listen. That's just phase one of the plan. When the last person on the list is dead, that's when the final stage is triggered."

"Please tell me, Dad. What happens next?"

Marc's expression changed. He looked more alert. Robin and Xavier, monitoring the conversation in the next room, had their hearts in their throats. Would these two holographic images, reconstructed from Anna's and Luc's residual network data, be convincing enough to fool Marc's tortured psyche? Maybe he'd already seen through the illusion. Maybe he was just playing along because he missed his wife and child, or because he wanted to lull his observers into a false sense of security.

"Luc," Marc said in the other room. "Remember the story I told you when you came to the institute?"

"The jig is up." Xavier's throat was dry. "He knows."

"Then we change strategy." Robin tapped at a keyboard, looking up the Max Planck Institute.

"I remember the statue of Planck," Luc said. "You said he created quantum theory. You said all the quantum technologies in the world today come from his radical ideas a hundred and forty years ago."

"Marc, maybe it's not the time for a seminar," Anna said.

"No, I'm talking about what was next to the statue: Saint Barbara. Her pagan father betrayed and killed her, because she wouldn't give up her belief in Christ. Planck and Saint Barbara are alike—these stories are about the power of belief. Only by committing ourselves unconditionally to our beliefs can we change the world and create the future."

"Dad . . . I don't understand."

"Anna, Luc, I love you. I love you very much, but it's time to say goodbye."

Marc closed his eyes as tears continued to slip from them. His voice was pained as he recited, "Fire color of gold from the sky seen on earth; heir struck from on high, marvelous deed done; great human murder, the nephew of the great one taken; deaths spectacular the proud one escaped."

"Marc," Anna said, "what is this? I want to talk with you a little bit more." She looked sad, holding Luc in her arms. The boy's imploring expression matched his mother's.

"Stop testing me, you demons!" Marc cried, voice trembling. "Get out of my sight! On the other side I'll see my real . . ."

He closed his mouth for the last time. Carefully, he ground one of his teeth. It took only ten microseconds for the neurotoxin contained within to reach his central nervous system. Marc's head tilted as his breathing slowed. He slumped in the restraint chair. Emergency personnel would have no chance of saving him now.

The ghosts of Anna and Luc drifted away into darkness.

Robin was horrified, but also puzzled. "What just happened?"

Xavier said in a low voice: "What did he mean . . . those verses?"

"The Centuries of Nostradamus. It seems the French have a tradition of playing prophet. Sounds to me like what's happening now." Robin recalled something Lee had told her: "Wait! Golden fire in the sky? Maybe that's what we're looking for. The last stage of the algorithm—"

Xavier interrupted her. A notification had come in showing that the Doomsday Drones had stopped attacking. They were in retreat.

The last name on the kill list had been crossed out: the author of the algorithm himself.

+———

THE HAGUE, PARIS, BAIKONUR, PLESETSK, SRIHARIKOTA, JIUQUAN, XICHANG, TANEGASHIMA, LOS ANGELES, CAPE CANAVERAL . . .
SEPTEMBER 18, 2041
03:14:51 COORDINATED UNIVERSAL TIME (UTC)

With the aid of the highest-level encrypted data channel at EC3 headquarters, Xavier was able to wake Eric Koontz, head of the European Space Agency based in Paris. Through Eric, Xavier circulated a warning to all major spacecraft launch bases around the world. It was just one sentence: "Stop all launches."

If Robin was right, completion of the drone kill program would automatically trigger the next level, like a video game. Piecing together Marc's dying words with the online clues uncovered by Lee, Robin believed she had identified the next threat. It was to be almost unfathomable in scope: An unknown number of nuclear bombs were disguised as ordinary space cargo, to be carried on upcoming commercial rockets and launched into space at any time.

"Why not just detonate on the ground?" Xavier said.

"Because Marc didn't buy enough raw material," Robin said. "He wasn't targeting specific regions or countries. He wanted to eradicate humankind. Radioactive dust from high-altitude explosions would be poison, spreading all over the world with atmospheric currents. No one would escape. Total Armageddon."

"In that case, why didn't he do it in the first place? Why all this Doomsday Blacklist nonsense?"

"You're right. Why kill a select few, *then* kill everyone?" Robin frowned.

Major launch sites around the world were responding, one after another. Reports of suspicious discoveries emerged. Eleven commercial launch projects were postponed, as unauthorized cargoes were detected. The sites were evenly scattered at various longitudes. Robin's intuition had been correct.

Two launch centers had not yet responded: the Kuru site in central French Guiana, South America, and the San Marco site, five kilometers off the coast of Kenya's Formosa Bay. In both cases, communication between staff and the outside world was cut off. Local military was on its way to both scenes.

Around the world, the silence was deafening.

ESA, NASA, CNSA: The space agencies of various countries were paralyzed with indecision. They passed the hot potato to the United Nations, where the secretary-general and his team raced the clock, urgently negotiating with heads of various nations and assembling an interdisciplinary advisory group to seek solutions.

Will and Lee were also hoping they could outrun this. They were trying to invade the central control systems of the two launch centers, which was how the hacker world solved problems.

Robin racked her brain. There had to be something she was missing. Marc couldn't have made such a redundant effort. He, or the algorithm, had a reason for every step taken.

"The Doomsday Drones didn't complete their mission," Xavier said suddenly. He was flipping through EC3's latest report.

"What?"

"They didn't kill all the people on their list. We rescued two hundred and seventy-four of them, but the next step was triggered anyway. Unless . . ." A terrible possibility dawned on him. He met

Robin's gaze. "Unless it was a padded list. Unless there were distraction targets alongside real ones!"

Robin quickly retrieved data on the last drone victim: Hikari Oshima, a leading information security scientist, one of twenty-three people in the world with a restart key for the DNS system.

Launched in 2010, DNS was a multinational cooperative project to ensure Internet security and domain name system integrity. Robin continued to study the names of the dead, finding yet more experts and scholars in fields related to network technology.

"He wasn't just targeting accelerationists," she murmured. "It was the Internet!"

"The Internet?"

"Marc was actually targeting anyone who could maintain network security, anyone with the knowledge and skills to restart the network."

"Restart . . . you mean he wants to shut down the entire network? How's that possible?"

Hundreds of millions of network servers existed on Earth, tens of billions of devices with network functionality—not to mention a Starlink composed of tens of thousands of communications satellites in space, and certain locked-down government and military data centers. The design was redundant in some ways, but that reinforcement prevented it from being completely shut down. Even if root servers were attacked and submarine fiber-optic lines cut, it was only a matter of time before the global network would be restored, as long as backup systems could take over.

"Maybe he just wants to pump the brakes on humanity."

Robin recalled Marc's words in the safe house: accelerationists hitting the gas, driving the car of human civilization over a cliff. If he really believed that, it all made sense. He didn't want to destroy the planet and terminate the human race. He just wanted to wind civilization back to its pre-digital state. He wanted to render humanity incapable of large-scale global cooperation. He wanted to reduce carbon emissions, end pollution, halt the ecological damage caused by petroleum energy. He wanted time for nature to recover.

Robin commanded the AI system to simulate the impact of two

nuclear bombs, detonated at different heights, on the global network over time. Two bright red spots bloomed over the eastern and western hemispheres of digital Earth. As she pressed play, the red light expanded like cancer. It would encompass the whole world in thirty minutes. The blue planet was fast becoming a flickering, ominous blood-red star.

"What is this?" Xavier said.

"High-altitude electromagnetic pulse, HEMP. Detonation in the middle stratosphere releases gamma rays and triggers Compton scattering. Upper-atmosphere atoms generate secondary ionization. Earth's magnetic field accelerates high-energy free electrons, stimulating more-intense electromagnetic pulses."

"And then what?"

"Power grids overload and collapse. Servers burn out, along with routers, switches, signal towers, and all electronic equipment."

"But we'd still have satellites."

"With no infrastructure on the ground to receive and process signals. If it were me, I might launch a communications protocol attack on the data link layer, on top of the physical attacks. So even if you could connect to the Internet, you wouldn't be able to complete ID verification or obtain any information."

Xavier stared at Robin like she was the terrorist. Wordlessly, he connected to the EC3 emergency info channel.

If Robin was right, hundreds of millions of people would die. Traffic control software, navigation appliances, and medical security systems would be paralyzed. Airplanes would crash. Vehicles would lose control. Ships would go down. The financial sector would nose-dive. The chain reaction would wipe out entire industries.

With the Internet gone and long-distance communications down, it would be difficult to coordinate logistics for food, medicine, fuel, and distribution of other necessities. Upheaval and panic would ensue. Local police and national guards would do their best to maintain order, but their reach would be severely restricted, since they could no longer broadcast orders or receive updates. They would have to rely on local decision-making.

In a few weeks, shortwave communications might be restored and some basic social order rebuilt. The rest of the lost network might be restored, too, in years or decades, dependent on professionals with relevant skills and knowledge. Large-scale human communication and collaboration, though, would be history.

In the meantime, the world would plunge into a long night.

WILL AND LEE FINALLY gained access to the central control systems of the Kuru and San Marco launch sites. They found overwriting that had switched the systems to automatic, unmanned launch mode. Personnel were locked out of the control centers, their movements restricted. The two rockets were being filled with fuel—the final stages of preparation before launch. Any signal interference might cause launch data errors: tilting, damage, rocket body explosion, all of it beyond a hacker's ability to resolve.

"There's only one option left," Robin said, looking at Xavier helplessly.

The Special Advisory Group for Planetary Emergency Services submitted their recommendations to the UN. They had been founded in 2025 and were composed of hundreds of experts across disciplines, aiming to address issues that required global collaboration, such as climate change and terrorist attacks. They called for military satellites in low-Earth orbit to shoot down the rockets with lasers before the weapons entered the stratosphere. The plan required a vote by representatives of every country. Following it would minimize global damage, but a high-altitude nuclear explosion would still cause hundreds of thousands of casualties on the ground. Regions nearer the explosions would doubtless suffer greater losses.

Not counting satellite altitude adjustment, the time allotted for target locking was less than sixty seconds.

The politicians had one minute to determine the fate of humanity. For them, it felt like an entire lifetime. For most of the planet's population, though, it was just another ordinary day. They were oblivious.

As the rockets' countdown to ignition began, the result of the vote came in: The faction in favor of shooting down the rockets had won by a narrow margin. The AI defense system calculated optimal kill windows, taking ground casualties and overall network damage into account. Even so, a global recession was inevitable, and subsequent collateral damage incalculable.

No one knew how future generations would come to view this momentous minute of decision-making.

5, 4, 3, 2, 1, ignition.

Two rockets rose into the sky atop raging flames. They were 257 seconds from the stratosphere.

Xavier looked at Robin in despair. He put a hand on her shoulder. "You've done everything you can. Now all we can do is pray."

Robin's thoughts kept flashing to the past. Since childhood, her training had turned her into a sophisticated machine. She had been raised to choose the best among many paths by relying on reason. She'd come to learn, however, that there were certain insurmountable defects in her cognitive framework. Such defects were called finite games. She made choices by looking at things through the lens of wins and losses. But life ought to be an infinite game, a pursuit of continuity, not a single great victory or loss.

224 seconds.

Maybe there was a third option, an alternative to shooting down the rockets and letting them end countless lives. The global Internet was about to be disastrously annihilated, and it was unclear when it would ever recover. There must be another way. But what?

Sometimes, in order to win, you first have to lose.

Grandma's words came back to her, seemingly at random, and then she understood.

"Get me through to whoever has final say!" Robin shouted. "Now!"

The secretary-general of the United Nations listened to the outlaw hacker's theory. And then, once the plan's feasibility had been confirmed by the experts of the Special Advisory Group, he gave Robin the green light.

176 seconds.

Robin's plan was for humans to cut off the power grid and submarine cable connections themselves. They would have to shut down root servers, signal transfer facilities, and all electronic equipment to minimize the impact of high-altitude EMPs and shorten the subsequent recovery time.

It would be shock therapy for the global Internet. It would realize the anarchist ideal countless hackers had dreamed of since the dawn of the Internet age.

To ensure the satellite lasers accurately targeted and destroyed the rockets at a specific altitude, the main communication network would have to be maintained until the very last moment. It could only be shut down once the rockets were hit, and once instructions were transmitted for cutting off the power grid and shutting down servers. In short, even if automation could handle most of the operations, the reaction time left for humans at the decisive moment would be no more than 750 milliseconds.

What Robin asked for was to hold the key to turn off everything firmly in her own hands. She was ready.

88 seconds.

With AI help, various nations were quickly divvied up into regions. Power grids and networks in remote areas were cut off first. Glowing continents in the eastern hemisphere swiftly dimmed. Darkness spread over the Earth.

31 seconds.

Robin's body tensed as she watched the rockets' trajectory on a monitor. The military satellites adjusted their positions. The laser weapons were locked on to their targets, waiting for the rockets to enter the specified range. With any luck, a thin high-energy cluster laser would knife through the vacuum, penetrate the atmosphere, and slice through the body of the hurtling rocket, cutting it in half with the efficiency of a scalpel. The rocket would explode. The wreckage would become a rain of fire falling to Earth.

Robin's thoughts were chaotic. Her forehead and palms broke out in a cold sweat. She had never experienced anything like this.

Something came to rest on her shoulder, warm and solid: Xavier's hand.

There was something complex in his eyes: concern, hope, admiration, maybe even a hint of tenderness.

"I believe in you," he said.

Robin was touched, but she didn't know how to respond. She nodded, tightening her lips, and returned her attention to the screen.

9, 8, 7 . . .

Robin's finger trembled, hanging over the button, ready to give the command that would transform life as they knew it.

3, 2, 1 . . .

It was as if two threads of a spider's web had streaked across the sky. The body of the first rocket split in two, then two became four. The white light of detonation filled the screen.

"Now."

Robin's finger descended.

Xavier looked out the window in horror. Nothing seemed to have changed, yet nothing would ever be the same.

The web connecting the world fell apart, and the rain began to fall.

+———

THE HAGUE, NETHERLANDS
SEPTEMBER 18, 2041
6:42 LOCAL TIME

Robin and Xavier stood on the empty beach, their tired faces illuminated by the morning light.

In the distant sky, flames bloomed like fireworks, or like a rain of fire, slowly expanding and falling to Earth.

Xavier glanced at his smartstream, but there was still no signal. This city should've been waking up by now. It was dead silent.

There was no electricity, no Internet, and no one knew how to restart the system. Half the people of Earth were beginning to wake up, and an unfamiliar world awaited them. Meanwhile, the other half of humanity had already fallen into chaos.

Many things had changed, but some had not. The force of gravity was the same, and the means of generating electricity. The sun was still rising and setting. There were still books, there was still knowledge, but it was scattered and isolated across space and minds. There were schools and teachers, like there had always been. As long as there were new generations, they would inherit old lore and invent new stories that would change civilization. These future humans would rebuild what their parents had made, bringing about a new and better world.

Xavier suddenly heard a child's laughter. It sounded like his sister's voice. He turned his head to look for Lucia, but there was nothing to see but the sea lapping at the beach. He knew it was time to let go.

"There are some things that can't be shut down forever," Xavier said. "They'll be back, but they will take time and patience."

"And faith," Robin added, looking at where the sea met the sky.

"Yes, and faith."

ANALYSIS

QUANTUM COMPUTERS, BITCOIN SECURITY, AUTONOMOUS WEAPONS AND EXISTENTIAL THREAT

Technology is inherently neutral—it's people who use it for purposes both good and evil. Disruptive technologies can become Prometheus's fire or Pandora's box, depending on the human using the technology. That's what "Quantum Genocide" is all about.

This story includes numerous technologies, but I will focus on two here. First, I'll describe quantum computing, which I believe has an 80-percent chance of working by 2041. And if that happens, it may have a greater impact on humanity than AI. It is truly general-purpose technology (like the steam engine, electricity, computing, and AI) that can help us dramatically improve science and understand nature. Quantum computing promises immense beneficial impact to humanity, as all general-purpose technologies have offered in the past. Quantum computers will be a great accelerator of AI, and quantum computing has the potential to revolutionize machine learning and solve problems that were once viewed as impossible. This story focuses on one negative usage, cracking Bitcoin encryption, which is likely to be one of the first major applications of quantum computing. But as we contemplate how to prevent a crime like the one in the story, we should not lose sight of the fact that quantum computing has many more great opportunities for good.

Autonomous weapons, like all technology, will also be used for good or evil. Autonomous weapons could save human soldiers' lives in an era in which wars are fought by machines. However, the threat of mass or targeted slaughter of humans by machines overwhelms any benefits. Autonomous weapons may inspire a new arms race that could spin out of control. They could also be used by terrorists to assassinate state leaders or anyone. I hope the atrocities in this story serve as a wake-up call to understand the severe consequences of this AI application.

QUANTUM COMPUTING

A quantum computer (or "QC," which is also used to refer to quantum computing in general) is a new computer architecture that uses quantum mechanics to perform certain kinds of computation much more efficiently than a classical computer can. Classical computers are based on "bits." A bit is like a switch—it could be either zero (if off) or one (if on). Every app, website, or photograph is made up of millions of these bits. Using binary bits makes classical computers easier to build and control, but also limits their potential for taking on really hard computer science problems.

Instead of bits, QCs use quantum bits, or qubits, which are typically subatomic particles such as electrons or photons. Qubits follow principles of quantum mechanics regarding how atomic and subatomic particles behave, which include unusual properties that give them super-processing capabilities. The first such property is *superposition,* or the capability for each qubit to be in multiple states at any given time. This allows multiple qubits in superposition to process a vast number of outcomes simultaneously. If you ask AI on a classical computer to figure out how to win in a game, it will try various moves and take them back in its "head" until it finds a winning path. But an AI built on a QC will try all moves extremely efficiently, holding uncertainty in its head, resulting in an exponential reduction of complexity.

The second property is *entanglement,* which means two qubits remain connected so that actions performed on one affect the other, even when separated by great distances. Thanks to entanglement, every qubit added to a quantum machine exponentially increases its computing power. To double a $100 million classical supercomputer, you'd have to spend another $100 million. To double your quantum computing, you just need to add one more qubit.

These amazing properties come at a cost. QC is very sensitive to small disturbances in the computer and its surroundings. Even slight vibrations, electrical interferences, temperature changes, or magnetic waves can cause superposition to decay or even disappear. To make a workable and scalable QC, researchers have to invent new technologies and build unprecedented vacuum chambers, superconductors, and supercooling refrigerators to minimize these losses in quantum coherence, or "decoherences," caused by environment.

Because of these challenges, it has taken a long time for scientists to increase the number of qubits in QC—from 2 in 1998 to 65 in 2020, which is still too few to do anything useful. However, even on a few dozen qubits, some computing tasks can be accomplished with QC over a million times faster than on classical computers. Google demonstrated "quantum supremacy" for the first time in 2019, proving basically that a 54-qubit QC can solve a problem (in this case, one that happened to be useless) in minutes that would take classical computers years. When will we have enough qubits to tackle real problems rather than useless problems? IBM's road map shows the number of qubits more than doubling every year for the next three years, with a 1,000-qubit processor due in 2023. Since 4,000 logical qubits should be large enough for some useful applications, including, for example, breaking Bitcoin encryption as depicted in the story, some optimists project that quantum computers will arrive in five to ten years.

However, the optimists may have overlooked some challenges. The IBM researchers acknowledge that control of errors caused by decoherence will get much worse the more qubits are added. To deal with this challenge, complex and fragile equipment must be built with new technologies and precision engineering. Also, decoherence errors will require each logical qubit to be represented by many physical qubits to provide stability, error correction, and fault tolerance. It is estimated that a QC will likely need a million or more physical qubits in order to deliver the performance of a 4,000 logical qubit QC. And even when a useful quantum computer is successfully demonstrated, mass production is another matter. Finally, quantum computers are programmed completely differently from classical computers, so new algorithms will need to be invented, and new software tools will need to be built.

Considering the issues in the previous paragraph, most experts believe it will take ten to thirty years to get a useful QC. Based on their expert opinion, I believe there is an 80-percent chance that by 2041 there will be a functional 4,000 logical qubit (and over a million physical qubits) quantum computer that can do what was described in "Quantum Genocide," at least as it relates to cracking the encryption used for today's bitcoins.

When such a multimillion-qubit QC really starts to work, one world-changing application will be drug discovery. Today's supercomputers can analyze only the most basic molecules. But the total number of molecules that could make a drug is exponentially greater than all the atoms in the observable universe. Tackling a problem of this scale requires quantum computers, which will operate using the same quantum properties as the molecules they're trying to simulate. QC can simultaneously simulate new compounds as new drugs, and model complex chemical reactions to it, to determine their efficacy.

As the famous physicist Richard Feynman said in 1980, "If you want to make a simulation of nature, you'd better make it quantum mechanical." QC will be able to model many complex natural phenomena that classical computers cannot fathom, even beyond drug discovery: for example, figuring how to counteract climate change, predicting pandemic risks, inventing new materials, exploring space, modeling our brains, and understanding quantum physics.

Finally, quantum computers' impact on AI won't just be a matter of making deep learning faster. Programming a QC involves giving it all potential solutions represented with qubits, and then scoring each potential solution in parallel. Then, the QC will attempt to find the best answer in very little time. This could potentially revolutionize machine learning and solve problems that were viewed as impossible before.

APPLICATION OF QUANTUM COMPUTING TO SECURITY

In "Quantum Genocide," the unhinged physicist, Marc Rousseau, uses a breakthrough in quantum computing to steal bitcoins. Bitcoin is by far the largest cryptocurrency that can be exchanged into other assets like gold and cash. But unlike gold, it has no inherent value. Unlike cash, it is not backed by any government or central bank. Bitcoins exist virtually on the Internet, with transactions guaranteed by computation that is unbreakable by classical computers. Bitcoins are also computationally guaranteed to be limited to no more than 21 million coins, which avoids oversupply and inflation. Bitcoins became particularly attractive after COVID-19, because more corporations and individuals are looking for safe assets impervious to inflation caused by central banks' quantitative easing. As an engineered safe-haven asset, bitcoins appreciated substantially. In January 2021, the total value of bitcoins exceeded $1 trillion.

Stealing bitcoins seems petty compared to the grand applications described earlier for QC, but it is actually a problem known to be solvable by a modest QC, and thus likely the first lucrative application of QC. While some of the quantum applications will take years to develop, breaking certain types of cryptography is relatively straightforward. All one has to do is implement the quantum algorithm in the seminal 1994 paper by MIT professor Peter Shor. If this algorithm is run on a QC with 4,000 qubits or more, it can break a class of cryptography algorithms under "asymmetric cryptography," with RSA as the most well-known of such algorithms. Some people credit this paper with igniting interest in quantum computers.

The RSA algorithm is used for Bitcoin and some other financial Internet transactions and digital signatures. The RSA algorithm, like all asymmetric cryptography algorithms, uses two keys, the public key and the private key. The two keys are very long sequences of characters that are mathematically related. The transformation from private to public is very simple, while the reverse is practically impossible to accomplish on classical computers. When you send bitcoins to me (say, for a purchase), you send them with a script of information that effectively serves as a publicly posted "deposit slip" (or transaction) that has my account (or Bitcoin wallet address) as the public key. While everyone can see this public key, only I have the private key that serves as the digital signature that can open the deposit slip. I complete the transaction by signing with my private key. This process is perfectly secure, as long as no one has my private key.

With quantum computing, all that is changed, because unlike classical computers, QC can quickly generate the private key from any public key for RSA or similar algorithms as used by bitcoins of today. So, the quantum computer simply accesses the public ledger (where all transactions are posted), takes each public key, uses the QC to generate a private-key digital signature, and takes all the bitcoins from the accounts that are not empty.

You might be wondering: Why would people post their wallet address and public key openly to the world? This was an early design flaw. Bitcoin experts later realized that was both unnecessary and dangerous. In 2010, essentially all new transactions shifted to a new format that includes the address but hides it, which is a lot safer (though not completely immune to attack). This new standard is called P2PKH. But there are still two million bitcoins stored in the old vulnerable format (called P2PK). And at the January 2021 price of $60,000 per bitcoin, that comes out to $120 billion worth of bitcoins. This is what thieves went after in "Quantum Genocide." If you have an old P2PK account, put this book down now and go secure your wallet!

Why don't people using the old P2PK scripts move that money to secure wallets? Well, they could, but most haven't. I can think of three explanations. First, many of the wallet owners have lost their private key, because the keys were too long to be remembered, and people didn't care as much when bitcoins were not that valuable a decade ago. Second, these bitcoin owners were unaware of this vulnerability. Third, about half of the two million belongs to the legendary Satoshi Nakamoto, the mysterious inventor of Bitcoin, who seems to have vanished. Thus, this story is referring to "Satoshi's treasure."

Why were all the transactions published on a public ledger? Because this was designed to keep bitcoins safe from any one company or individual. This public ledger is stored in a decentralized way on many computers, which makes it impossible for any one computer to modify or forge it. This was a brilliant design as long as no one could reverse engineer the private keys from the public keys on the ledger. This approach also made possible blockchains, which will have many valuable applications to keep information unalterable (such as deeds, contracts, and wills).

When a bitcoin heist happens, there is no way to report the crime or sue the perpetrators, because perpetrators cannot be easily determined. Bitcoins are not controlled by any government or company, and their transactions are not governed by banking laws. Anyone with the right private key can take the bitcoins in a wallet. There is no legal redress.

Why didn't Marc Rousseau go after banks? First, banks do not have a public ledger with public keys from which private keys can be computed. Second, banks have software to monitor for anomalies such as large suspicious transfers. Third, the movement of money between accounts can be traced, and prosecuted if laws are broken. Finally, banking transactions are protected by a different cryptography algorithm that will take a bit more effort to decode.

What could be done to "upgrade" our cryptography? Quantum-resistant algorithms exist. In fact, Professor Shor also showed that impregnable cryptography could be built on QC. A symmetric cryptography algorithm based on quantum mechanics is impregnable even if the intruders have a powerful quantum computer. The only way to infiltrate this cryptography is if the principles of quantum mechanics are found to be incorrect.

But quantum-resistant algorithms are very expensive computationally, so they are not being considered right now by most commercial and Bitcoin entities. Perhaps only after the inevitable quantum bitcoin heist happens will people wake up to revamp the algorithms. I hope it won't take that long!

WHAT ARE AUTONOMOUS WEAPONS?

Autonomous weaponry is the third revolution in warfare, following gunpowder and nuclear arms. Its evolution from land mines to guided missiles was just a prelude to AI-enabled true autonomy—the full engagement of killing: searching for, deciding to engage, and obliterating another human life, completely without human involvement.

An example of an autonomous weapon in use today is the Israeli Harpy drone, which is programmed to fly to a particular area, hunt for specific targets, and then destroy them using a high-explosive warhead, nicknamed "Fire and Forget."

But a far more provocative example is illustrated in the viral video "Slaughterbots," which shows a bird-size drone actively seek out a particular person, and when found, shoot a small amount of dynamite point-blank through the person's skull. These drones fly themselves and are too small and nimble to be easily caught, stopped, or destroyed.

A "Slaughterbot" such as the one that nearly killed the president of

Venezuela could be built today by an experienced hobbyist for less than $1,000. All the parts are available for purchase online, and all open-source technologies are available for download. In the near future, robots will be able to perform the same function, when costs come down. This is a testament to how AI and robotics are becoming accessible and inexpensive, a fact I've tried to underscore in this book. Imagine, a $1,000 political assassin! And this is not a far-fetched danger for the future, but a clear and present danger.

We have witnessed how quickly AI has advanced, and these advancements will accelerate the near-term future of autonomous weapons. Consider how quickly autonomous vehicles evolved from L1 to L3/L4 (as defined in chapter 6). The same will inevitably happen with autonomous weapons. Not only will these killer robots become more intelligent, more precise, more capable, faster, and cheaper, but they will also learn new capabilities such as how to form a swarm, with teamwork and redundancy, making their missions virtually unstoppable. A swarm of ten thousand drones that could wipe out half a city could theoretically cost as little as $10 million.

THE CASE FOR AND AGAINST AUTONOMOUS WEAPONS

There are benefits from autonomous weapons. First, autonomous weapons could save soldiers' lives if wars are fought by machines. Also, in the hands of a responsible military, they can be used to help soldiers target only combatants and avoid inadvertently killing friendly armed forces, children, and civilians (similar to how L2 and L3 autonomous vehicles can help save a driver from making mistakes). Also they can be used defensively against assassins and perpetrators.

But the liabilities far outweigh these benefits. The strongest such liability is moral—virtually all human ethical and religious systems view the taking of a human life as a contentious act requiring strong justification and scrutiny. UN secretary-general António Guterres stated: "The prospect of machines with the discretion and power to take human life is morally repugnant."

Autonomous weapons lower the cost to the killer. While there have been suicide bombers, giving one's life with certainty for a cause is still a high hurdle for anyone. But with autonomous assassins, there would be no lives to give up for killing.

Another major issue is having a clear line of accountability—knowing who is responsible in case of an error. This is well established for soldiers on the battlefield. But when the accountability is assigned to an autonomous weapon system, the accountability is unclear (similar to accountability ambiguity when an autonomous vehicle runs over a pedestrian). What makes this worse is that it may absolve aggressors for injustices or violations of international humanitarian law. And this lowers the threshold of war.

Another danger is that autonomous weapons can target individuals by using facial or gait recognition and by tracing phone or IoT signals. This enables not only assassination of one person, but a genocide of any group of people. In "Quantum Genocide," we saw the targeted killing of business elites and high-profile individuals.

Greater autonomy without a deep understanding of meta-issues will further boost the speed of war (and thus casualties), and potentially lead to disastrous escalations, including nuclear war. AI is limited by its lack of common sense and human ability to reason across domains. No matter how much you train an autonomous weapon system, the limitation on domain will keep it from fully understanding the consequences of its actions. That is why in the story, Xavier and EC3's anti-terrorism operations are still operated by humans, and not robots.

AUTONOMOUS WEAPONS: AN EXISTENTIAL THREAT?

From the Anglo-German naval arms race to the Soviet-American nuclear arms race, countries have coveted military supremacy and made it a national priority. This will surely be exacerbated with autonomous weapons, because there are many more ways to "win" (the smallest, fastest, stealthiest, most lethal, and so on). It could also cost less, lowering the barrier of entry. Smaller countries with powerful technologies, such as Israel, have already entered the race with some of the most advanced military robots, including ones as small as flies. With the near certainty that one's adversaries will build up autonomous weapons, ambitious countries will feel compelled to compete.

Where will this arms race take us? Berkeley professor Stuart Russell says: "The capabilities of autonomous weapons will be limited more by the laws of physics—for example, by constraints on range, speed, and payload—than by any deficiencies in the AI systems that control them . . . One can expect platforms . . . the agility and lethality of which will leave humans utterly defenseless." This multilateral arms race, if allowed to run its course, will eventually become a race toward oblivion.

Nuclear weapons are an existential threat, but they've been kept in check and even helped reduce conventional warfare, because of the deterrence theory. Deterrence means having nuclear weapons could deter a more powerful adversary, provided your nuclear weapons cannot be taken out by a surprise first attack. Because a nuclear war leads to MAD (mutually assured destruction), any country initiating a nuclear first strike likely faces reciprocity and thus self-destruction. But with autonomous weapons, the deterrence theory does not apply because a surprise first attack may be untraceable, and there is no threat of MAD. In the story, the difficulty of tracking down the Doomsday Drones is one example. While hacking the communication protocols of a Doomsday Drone could give clues, that would only work if a "live drone" could be captured.

As discussed earlier, autonomous weapon attacks can quickly trigger a response, and escalations can be very fast, potentially leading to nuclear war. The first attack may not even be triggered by a country, but by terrorists and non-state actors. This exacerbates the level of danger of autonomous weapons.

POSSIBLE SOLUTIONS FOR AUTONOMOUS WEAPONS

There have been several proposed solutions for avoiding this existential disaster. One is the human-in-the-loop approach, or making sure that every lethal decision is made by a human. But the prowess of autonomous weapons largely comes from the speed and precision gained from not having a human in the loop. This debilitating concession may be unacceptable to any country that wants to win the arms race. It is also hard to enforce, and easy to find loopholes.

A second proposed solution is a ban, which has been proposed by both the Campaign to Stop Killer Robots and a letter signed by three thousand people, including Elon Musk, the late Stephen Hawking, and thousands of AI experts. Similar efforts have been undertaken in the past by biologists, chemists, and physicists against biological, chemical, and nuclear weapons respectively. A ban will not be easy, but previous bans against blinding lasers, chemical weapons, and biological weapons appear to have been effective. The main roadblock today is that Russia, the United States, and the United Kingdom all oppose banning autonomous weapons, stating that it is too early. In 2021, the U.S. National Security Commission on AI, a body headed by former Alphabet chairman Eric Schmidt, recommended the United States reject calls for an autonomous weapons ban.

A third approach is to regulate autonomous weapons. This will likewise be complex because of the difficulty of crafting effective technical specifications without being too broad. What defines an autonomous weapon? How do you audit for violations? These are all difficult short-term obstacles. But since this book is about the long term, please indulge me to fantasize about a 2041 treaty—by then, could all countries agree that all future wars will *only* be fought with robots (or better yet, only in software), promising no human casualties, but delivering the classical spoils of war? Or perhaps a future in which wars are fought with humans and robots, but the robots are permitted to use only weapons that will disable robot combatants but are harmless to human soldiers (like laser tag)? These notions are clearly not practical today, but perhaps they will inspire something more feasible sooner.

I hope I've established that autonomous weapons are already a clear and present danger, and will become more intelligent, nimble, lethal, and accessible at an unprecedented speed. The deployment of autonomous weapons will be accelerated by an inevitable arms race that will lack the natural deterrence of nuclear weapons. Autonomous weapons are the AI application that most clearly and deeply conflicts with our morals and threatens humanity's continuity. We need to engage the experts and the decision-makers to weigh different solutions to prevent the proliferation of autonomous weapons and the termination of our species.

CHAPTER EIGHT

THE JOB SAVIOR

STORY TRANSLATED BY ANDY DUDAK

BEFORE I LET THAT STEAM DRILL BEAT ME DOWN,
I'LL DIE WITH THE HAMMER IN MY HAND.
                    —AMERICAN FOLK SONG "JOHN HENRY"

NOTE FROM KAI-FU: This story explores an issue that has caused much hand-wringing: What will happen to human jobs as the steady march of AI seeps into more industries, making human tasks redundant? As AI decimates routine jobs, a new industry arises—job reallocation firms, which are called in to retrain and reassign displaced workers. But what exactly are the new jobs—and will they satisfy many humans' desire to feel productive and useful? Who is most at risk, and how can humans flourish in the post-automation era? I'll give my thoughts on these questions in the commentary at the end of the chapter, describing how technologies like robotics and robotic process automation will continue to evolve to take over tasks for both white-collar and blue-collar workers.

IN THE DARKENED training room, Jennifer Greenwood and twelve other trainees gazed attentively at the imagery scrolling in midair before them. Accompanying the visuals was a male voice, narrating softly, like an oracle pronouncing divinations.

"Everything began to change in 2020," said the voice. "The epidemic led to social isolation and restricted travel. Business owners were forced to evolve, turning to robotics and AI to replace human personnel."

The floating scene shifted to an abandoned Times Square, a run-down mall, an empty Disneyland, images of barred factories and silent assembly lines. Next came images of crowds on the street wearing PPE, holding up placards protesting large-scale layoffs, and more-disturbing pictures of looting and riots.

The narration continued: "In 2024, the White House changed hands, and the new administration spearheaded a universal basic income program. UBI guaranteed each citizen a monthly stipend, paid for by taxing the ultrarich and the billionaire tycoons who'd made a fortune from companies powered by new technologies and data collection. Addressing the structural unemployment brought about by advancements in AI had become urgent."

A collage of headlines from the era appeared on the visual scroll above Jennifer and her fellow trainees, accompanied by dramatic sound bites from news programs. An animated graph showed the stock market fluctuating violently, followed by images of citizens reacting to a UBI deposit notification on their smartphones.

The voice continued: "Despite its initial popularity, UBI led to unintended consequences. Many ex-workers, left with little to do, became absorbed in VR games, online gambling, drugs, and alcohol. City centers once again became magnets for crime as major corporations and the wealthy abandoned the urban core. Too many people had been made redundant by the quick and aggressive development of AI, with little guidance from leaders on new pathways to employment. Cascading failures led to a high suicide rate. Some began calling for the repeal of UBI. By 2028, social media was consumed by the Great Debate over the pros and cons of UBI. The Senate and the House sank into protracted back-and-forth warfare over a proposed plan to abolish UBI.

"The UBI program was formally abolished in 2032. In its place, legislation was enacted that empowered a new field, occupational restoration, to meet the challenges of this historic moment. Job 'reallocators' help people train in new vocational skills and identify new job opportunities. The government used part of the tax revenue formerly earmarked for UBI to propel the job reallocation industry, in the hopes of resolving the social ills that had become pervasive features of modern life. It was around this time that I was inspired to found Synchia."

Jennifer knew the backstory to Synchia so well already that she could have given this speech herself. Now—in the era of occupation restoration—corporations were required to engage the services of a company like Synchia in order to enact layoffs.

Occupational restoration companies, in addition to being funded by government grants, negotiated package prices with employers. In general, this cost was lower than what employers would've spent on large-scale severance payouts. Companies like Synchia made up the difference by charging third-party hiring agencies for introductions to occupationally restored unemployed workers, similar to the costs borne by traditional "headhunters." The revenue went toward new training for those who weren't immediately re-employable.

Finding new jobs for people wasn't as easy as matching openings with job seekers. As the range of sustainable careers narrowed, job reallocators like Synchia conducted skill and personality tests on those affected by structural unemployment to create exhaustive vocational survey maps that included the latest economic figures to track shifts in society. Armed with data, the allocators provided job seekers with customized plans for new employment—with a heavy dose of empathy. That's where Michael's charm came in.

As the background lesson in job reallocation and Synchia came to an end, the holographic image of Michael Saviour appeared, magician-like, in the middle of the wall screen. He began to address Jennifer and the other trainees. It was Saviour's gentle voice that had been narrating the scroll. Not yet fifty by the look of him, Saviour was a tad plump, graying at the temples, sideburns carefully trimmed. He wore a tailored navy suit, attractive but not ostentatious.

It was Michael Saviour's reputation, not his looks, that had drawn Jennifer to seek out a job with Synchia. Jennifer had absorbed everything about him online, videos of him giving talks and discussion forums where others analyzed his company's techniques. Commenters marveled at how Saviour seemed to captivate the audience in every space he entered. Perhaps it was this subtle control of the room and its moods that made him the country's best job reallocator. Saviour's words were measured; his tone, his expression, his posture, all contributed to a sense of his professional trustworthiness. He knew how to make people feel better.

If I can just become Michael's assistant . . . The idea still seemed so ridiculous. Jennifer tried not to let herself fantasize about it. After all, this was just the first week of her internship at Synchia.

Don't be foolish. Jennifer forced herself to refocus on the task at hand.

Floating in midair, Michael was still talking, while gracefully waving his hands, like a conductor. The video window obeyed his gestures and fluctuated, growing, shrinking, and tilting in concert with Saviour's movements. "As you have seen, signs of change were already manifest before the global pandemic. The virus merely accelerated things. A large amount of previously offline economic activity went online as people maintained social distancing. The traditional service and manufacturing industries suffered heavy losses. Both were spheres where machines had an advantage."

As Michael gestured, an array of people from different professions appeared and vanished in the midair scroll: cashiers, truck drivers, seamstresses, factory workers, fruit pickers, telemarketers, well-dressed office workers, even doctors. The images kept coming, faster and faster, crowds like ghosts, vague and indistinct.

"Humanity's competitor was AI," Saviour continued, "which could learn and improve continuously, twenty-four/seven, without rest. Jobs that had been performed by humans just a month earlier were suddenly and ruthlessly overtaken by AI. This race had been under way for over twenty years, but suddenly it was apparent to everyone. And that's where we are today. There's no end in sight, not for the foreseeable future, anyway. Many employees are like turkeys on the farm, waiting nervously for Thanksgiving. And, of course, in this era of anxiety, certain extreme behaviors have become the new normal."

Images of protests and violent clashes once again filled the scroll. To Jennifer, these scenes were still shocking. The worst of the Great Debate had been hidden from children at the time.

"The government tried UBI. They tried shortening the workweek. History has shown that policies like UBI only prolong despair. They cannot solve the fundamental problem: that people, without the sense of achievement they gain from substantive work, feel lost and hopeless. Without a sense of self-worth, they turn to literal or figurative narcotics. And who can save them from this predicament?"

Michael paused, a slight smile on his face, looking around at the faces in the training room.

Jennifer couldn't help but feel Michael was looking right at her. She raised her hand.

"Ah! The young lady here in front. Please give us your name."

"Jennifer Greenwood from San Francisco."

"Excellent, Jennifer. And your answer?"

"Us, Mr. Saviour. We're just like your name. We're the saviors."

Everyone laughed, and Jennifer blushed. She wasn't trying to flatter.

"Thank you, Jennifer. There is an important lesson here. People are not machines," Michael intoned before the trainees. "We're more complicated, more adaptable, more driven by emotion, so there will be high demands on everyone here. You must become the best possible occupational restorationists. Your job is to save people—to restore not just their jobs, but their dignity."

AMID THE APPLAUSE THAT followed Michael's speech, the lights in the room brightened. Then, to Jennifer's surprise, Michael himself emerged from behind the screen in San Francisco. He had not been beaming in long-distance from Seattle headquarters after all—yet another of his magic tricks, it seemed.

Saviour's sudden appearance was exciting, but also made Jennifer a bit nervous. She wondered what had brought him here. She understood intuitively that, with few exceptions, when someone of Saviour's stature appeared in person, there was inevitably a major crisis needing resolution. Jennifer had heard rumblings about a big project that Synchia was up for: thousands of laid-off workers at Landmark, one of the country's largest construction firms. Could this be the reason Michael was here in the flesh?

As her mind raced with possibilities, Jennifer couldn't help thinking that Michael reminded her of her father, an insurance company worker who had loved instilling in his daughter a proactive, opportunity-seizing, "give it your best shot" attitude. At least, that was once her father's MO.

In that moment, Jennifer decided to seize the spirit of her father. She quickly wrote a short message on her smartphone, and after hesitating a moment, hit "Send."

THREE MONTHS LATER, in the plaza in front of the towering Landmark general headquarters, Jennifer forced her way through a crowd of protestors, looking out of place in a formal business suit.

Her impulsive email during her training week had gotten her a one-on-one coffee sit-down with Michael Saviour himself. He had admitted that her message had moved him with its lack of polite convention. In her message, she had alluded to her father's lost and defeated life and how this was her inspiration in joining Synchia. Michael told Jennifer that her pluck had impressed him. The twenty-minute coffee stretched to an hour. As it happened, the position of Michael's assistant was vacant. After their meeting, he'd invited her to try out for the role, pending, of course, HR's assessment. After all, plenty of young Synchia staffers wanted this chance. Luckily, Jennifer had not disappointed Michael, or herself.

Jennifer had joined the company amid a period of huge opportunity—and stress. Michael was determined to win the Landmark construction contract. With the traditional construction industry evolving toward digitalization, Landmark had opted for wide-scale replacement of traditional human roles with automation, prefabricated 3D-printed components, and AI designers. The result would be a major round of layoffs, primarily of blue-collar workers. To go through with the plan, the government required Landmark to retain the services of an occupational restoration company. In one fell swoop, thousands of workers would need to be steered toward new jobs. If Synchia won the bid, it would mark one of their largest single reallocation contracts to date. But there was competition.

Michael had gotten a tip that a mysterious new company was also vying for the Landmark contract. Who were they? Did they have some kind of an edge? Michael had ordered Jennifer to dig up intel. So far, she had come up empty-handed by searching public channels. A few days earlier, however, on her walk to the office, she had noticed some mysterious graffiti left on the plywood barrier to a Landmark construction site on her route. To her surprise, when she held her phone up to the image, it detected a QR code that took her to a secret online forum. As she scrolled through the forum, she quickly realized it was the online base for a group of

Landmark workers organizing against the layoffs—a kind of underground automation-resistance movement.

On the forum, anonymous workers had called for a day of protest. Now Jennifer found herself in the middle of a group of construction workers who were demonstrating and obstructing traffic in midtown San Francisco outside the Landmark Tower. And they didn't bring only signs. Heavy-duty cranes, concrete mixers, and wrecking ball trucks were arrayed like tanks preparing for a military parade. Workers in orange hard hats and reflective vests had brought along the tools of their trade, hoisting sledgehammers on their shoulders and lugging toolboxes, as well as protest signs. One placard read: "The machine oppressors eat The People!" Organizers with bullhorns led the crowd in a chant: "Robots fuck off!" It was deafening.

Police in riot gear formed a defensive line like an indestructible dam, keeping the protestors away from the headquarters' doors.

Jennifer realized it was a four-way tug-of-war. The government wanted stability, Landmark wanted lower costs, occupational restoration companies like Synchia wanted to win a profitable contract, and the most vulnerable party, the workers, wanted to keep their jobs, or at the very least have a sense that they were being taken care of, instead of just shoved aside in favor of machines. When three players had formed a seeming alliance under the table, the remaining party played the only card it could, a bluff, a blustery show of strength, in order to exert pressure on all parties.

There had been sporadic cases of anti-layoff demonstrations in recent years, but nothing of this magnitude. The workers were up in arms about Landmark's latest offer. Word was spreading that if Landmark didn't put forth a severance plan that satisfied the workers, the protest activity would escalate. As her eyes darted between the hulking construction vehicles and the police—armed to the teeth—Jennifer worried about the prospect of violence.

As she pushed through the crowd, Jennifer felt her phone buzz. She answered, cradling the tiny device deftly with one hand while clutching a black faux-leather handbag with the other.

"Jenny, where are you?" It was Michael. "We're in a race against time here and you're AWOL."

"I'm doing the job you taught me to do!" Jennifer shouted, straining her voice to be heard above the clamor of the protestors.

"It sounds like you're at a concert? A ball game? Or a party? But it's ten in the morning, so I know you can't be."

"Sorry, I'm conducting a little field investigation . . ." Jennifer trailed off. She had spotted what she was looking for amid the crowd of demonstrators: a man wearing a St. Louis Cardinals baseball cap. "Opposition research!"

"Christ! Don't tell me you're . . . look, get out of there right away and come back! It's dangerous!"

"I'll see you soon." Jennifer hung up and struggled toward the man in the Cardinals cap. "Hey, are you . . . SLC422?" It was the man's forum ID, perhaps his birthday or a lucky number, Jennifer assumed.

The man turned and gave a barely perceptible nod. "You're the one who messaged me on the forum—the reporter?" The man gazed at Jennifer's attire. "You don't seem much like a reporter."

"Doing our work requires a bit of disguise." She winked and fished out a notebook and pen. "I'm on your guys' side. Can't let those big corporations get away with this!"

"Good to hear. Thousands of jobs taken by robots. We're not planning to take this lying down!"

"In your post you said there's talk of a company bidding on the severance program—one that is saying it can guarantee one-hundred-percent occupational reassignment. Is that for real?"

"Yeah, it's for real. I got a buddy who works in HR. He tried it. He said the company was called something like Omega. Omega-Alliance. That's what I've heard."

"Wow. Sounds like it might be good if they get the deal."

"I don't know. These allocation companies generally give you shitty jobs, or they need you to relocate, get retrained and whatnot. I'm not thrilled with the idea of being allocated right out of my current life. We want something that is *good*—not just better than nothing." He raised a fist and gazed over the restless crowd around him.

"Okay, Mr. SLC422, you have my email. If you hear anything

else, contact me at any time. Good luck." As Jennifer made her exit, a vortex of bodies nearly swept her off her feet.

"Take care of yourself!"

The faint benediction was quickly overwhelmed by waves of protest noise.

THIRTY MINUTES LATER, Jennifer was back in Michael's office, the two of them gazing at each other in dismay.

"You say a hundred percent, but what does that mean?" Michael was incredulous.

"It means literally a hundred percent." Jennifer leaned in the doorway, unable to resist rolling her eyes.

"But it makes no fucking sense. You know how hard we work to get our percentage . . . the training, the relocations. Even if California doesn't need construction workers, maybe Pennsylvania does. Even if the States don't, maybe Europe does. We work our fingers to the bone to get our twenty-eight-point-six percent. A hundred? Ha! Just fucking shoot me already!"

In that moment, Michael seemed more like a disgruntled old man than a visionary, Jennifer thought.

"But that's what he said. He has no reason to lie. I mean, who would believe a lie like that?"

"But this . . . OmegaAlliance . . . that's a lousy name, incidentally, and I've never heard of them. This is your so-called intelligence from a morning spent sweating and getting jostled around?" Michael had tasked his AI assistant with checking out a possible competitor, but the assistant had come back empty-handed. Either the company hadn't been registered long, or something fishy was going on. *A code name? A shop sign? But why such pretense?* Michael was puzzled and annoyed.

ALONE IN HIS OFFICE, Saviour looked at his buzzing phone. The caller ID said "Allison Hale." His eyes widened. He and Allison, six years his junior, had once been an item, back in business school. They'd ended up in the same profession, becoming rivals of a sort. It had been quite a while since he'd heard from her, or about her.

"Well, hello, Allison. This is unexpected. You're in town? Lunch? Let me take a look . . . well, why not? I know a good place."

Michael hung up, intrigued—and a bit puzzled. Was this unexpected call somehow related to the arrival of the unexpected competitor? He felt compelled to find out.

LUNCH WAS AT A Cantonese-style restaurant called Three Treasures. Dim sum crowded the red tablecloth, like lotuses blooming in a midsummer pond. Michael arrived first. As he watched the host lead her to his table, he marveled that Allison didn't seem to have changed at all. *Maybe she's been doing the telomere restoration therapy,* he speculated.

"So, spill it," he said after they'd made small talk over seafood rolls and steamed buns. "I know you wouldn't seek me out if everything was fine."

"Ha! You don't think I'm capable of simply wanting to catch up?" Allison put down her chopsticks.

"It's just that I have another meeting in a little while. I'm sure you understand."

"Okay, Michael. You see right through me. Here's the deal. How long has it been since you founded Synchia? Five years? Eight?"

"So you came to poke and prod me, eh?" Michael eyed Allison from across the table. "This can't be a coincidence. The Landmark mess turns up a rival I've never heard of, and suddenly you show up and suggest lunch. Are you working for these Omega people now? Don't tell me you believe this hundred-percent conversion bullshit."

"Ninety-nine point seventy-three, to be exact. And yes, I believe that bullshit."

"Smells fishy to me. Is it legal? And where does it leave the workers?"

"Well, if you signed a few documents, you could be privy to all our trade secrets. There's a lot you don't know. What do you say?"

"So, you're offering me a job. Or trying to buy my company? And if I decline?"

"Then I'll have to suppose sad old Michael has been comfortable for too long and can no longer distinguish a rib eye from a bone in his mouth."

Michael stared at Allison. He recalled those years of daily heated debate in classrooms. She was intelligent and radical, the kind of MBA student who loved *Atlas Shrugged*. She'd always liked to attack Michael's worldview, suggesting it was a hypocritical sham—that his "community spirit" masked the profit motive, which she saw as the quintessence of human virtue.

These clashes had the effect of creating romantic sparks, but their fundamental differences had made the relationship fizzle out quickly.

"You haven't changed a bit," Michael said, laughing.

Allison shook her head. "Look, I'm offering you the chance to help way more people than you do now. Isn't that in keeping with your doctrine? Why not say yes? Think about it, and give me a call."

Michael watched Allison stride out of Three Treasures. No matter what it took, he had to find out what the hell was going on with OmegaAlliance. And in his plucky assistant he had just the person to do it.

"SO, YOUR NEW ASSISTANT is working out?"

"Quite capable. More capable than I'm feeling at the moment!"

Lying on a spacious and comfortable Le Corbusier LC4 recliner, necktie loosened, eyes closed, Michael focused on his breathing. His lunch with Allison had brought on an anxiety attack, and he had called his psychiatrist, Dr. Trisha X. J. Deng, in a panic. Deng's face, beneath a helmet of neat, silver hair, now appeared on the screen across from him.

"This attack had no direct triggering incident. Completely different from last time—from the Elsa incident."

Michael pursed his lips, replaying the scene in his mind.

Although he was the CEO, Michael liked to boast that he still met with Synchia clients himself. In his memory, a woman, Elsa, entered the consultation office at Synchia and sat down. Michael scanned her file. "Ms. Gonzales, may I call you Elsa? It seems our situation is not so good, but you're very lucky—"

"We've met, Mr. Saviour," she interrupted.

"Oh, really?" Michael looked up, examining her face, searching his memory in vain.

"You don't remember? I was sent to your office about five years ago, when I lost my job. I was a warehouse manager and you shifted me to a job as a theme park attendant at Adventure World because you felt I was patient and liked children. You told me this work would be stable, that it would get me to retirement, sir. Your office was a lot smaller back then. But now it's happened again." Elsa's tone was flat.

"Elsa, I believe your memory, I believe that happened, but this was an unforeseeable structural change. They were late to it, but theme parks and large-scale entertainment venues have begun using robotic attendants. It just costs less and is more efficient. And I think children might prefer the robots." Michael blinked, his expression apologetic.

"So . . . where will you send me this time?"

"The city zoo has open positions. I think that seems like an excellent match."

"So, lucky me, I'll get to shovel elephant shit every day. And how long do I have this time? Three years? One? Nine months?"

Elsa was trembling. Her voice grew louder. "Every parent wants to be a hero in their kids' eyes. But right now, I feel more like a cockroach. I scuttle from one corner to another, snatching whatever scraps they feed me to survive. I don't want my children to see me as a cockroach, Mr. Saviour."

Watching this mother on the brink of emotional collapse had triggered in Michael thoughts of his own. Michael's mother, Lucy, a talented bookkeeper, once knew the same shame. Back then the enemy wasn't AI. It was just regular, fast-calculating accounting software. Mom switched jobs several times, each position worse than the last, until the jobs ran out.

Michael closed his eyes, struggling to expel the visions of what followed: Mom unemployed, defeated, drowning her sorrows in alcohol. He remembered very clearly what he felt for her at that time. Grief, sympathy . . . and, though he couldn't help it, a strange sense of loathing.

"Michael?"

Dr. Deng's voice returned Michael to the recliner, the real world. He opened his eyes and regarded the psychiatrist on the screen, perplexed.

"You wanted to rescue Elsa," Deng said, "just like you wanted to rescue your mother. And just like you want to rescue every person who walks in the door of Synchia."

Michael considered her analysis. "We're just a pressure-reducing valve, I guess, alleviating societal problems, but giving these people false hope. We're constantly lowering their expectations, like a slow poison, gradually getting them to accept their fate of being banished and marginalized by technology. Do we really help people? Or are we accomplices?"

"Listen to me, Michael. You help restore a sense of dignity in these people."

"But where does it end? It's no secret that we can't keep up with AI. No matter how hard we try, it's like watering flowers in a desert. This Landmark case is just the beginning. The whole construction industry is headed for a massive earthquake." Michael pulled at his collar as though an imaginary noose was tightening around his neck.

Dr. Deng was about to say something, but an alarm sounded. She pushed a button in her tablet and an automatically generated diagnosis and treatment report popped into Michael's in-box.

"As your doctor, I can only say . . . same time next week. As your friend, I recommend you accept your imperfections."

Michael had already begun retying his necktie. No trace of inner turmoil remained on his face. He was reverting back to savior mode.

ON SATURDAY NIGHT AT the Silverline bar, you had to go hoarse to get the bartender's attention. The dive looked like it had been forgotten by time. Other than the TVs and digital cash register—hard cash hadn't been accepted in years by most Bay Area establishments—there was hardly a piece of technology in sight. Though the popularity of football had been declining in the United States for decades, over safety concerns, in this neighborhood watering hole, local guys, mostly middle-aged, mostly blue-collar, still packed the place on Saturdays for USC games. Jennifer's appearance triggered a few whistles. Even if technology had evolved, men hadn't.

She was a little wiser this time, ditching the pantsuit in favor of a USC sweatshirt and jeans.

She spotted SLC422, aka Matt Dawson, at the bar. He wasn't wearing his baseball cap this time, so his balding hairline was on full display. Somehow, Jennifer thought, it made him seem more downtrodden. Matt glanced over and beckoned when he saw Jennifer. She took the empty barstool next to him and motioned to the bartender to pour her a beer. The two of them sat awkwardly, sipping their beers.

Finally, it was Jennifer who broke the silence.

"You probably didn't invite me here just to drink beer. Are you here alone? Any spouse . . . or girlfriend?"

"My kids live with my ex-wife in Ohio." Matt took a swig, white foam lingering on his upper lip.

"Oh, I see." Jennifer took a drink, wondering about the story behind his words. "So, what specifically is your work?"

"Ten years doing scaffolding, fifteen years as a master plumber. Not to brag, but all I need is one glance at a blueprint and my hands know what to do. I'm not slower than those machines."

"I understand."

They fell back into their awkward silence, until Matt began speaking again: "When I was young, I always heard people say, 'The robots are going to take your job away.' The way they said it, it sounded like workers were the ones on the chopping block, and the managers would be safe. But I found out it wasn't so simple. The stuff that's hard for humans . . . it turns out that's the simple stuff for AI, stuff that runs on numbers, and analyzing documents, finding patterns. And work that seemed simple was AI's Achilles' heel, like taking care of other people, or figuring out how to fit a water pipe in a tight space. So I'd pretty much convinced myself I'd be lucky enough to work straight through to retirement. But I guess the robots got a little more sophisticated."

Jennifer sipped her beer while taking in Matt's words. "So, what's your plan now?"

"Dunno." Matt shrugged. "The protest got some attention, but nobody on the forum really has a plan about what to do next. I guess I'll see what comes. I heard a little bit more from the inside. My buddy in HR—he's in a position to know a bit of what's going on. What I'm hearing is there are two job reallocation companies bidding for this mess. One has fewer positions to offer, but maybe I could continue doing the work I know, though I'd have to move to a different city, or even another country."

"Sounds not so bad. And the other company?"

"The other one, yeah. It's the one I told you about, Omega-Alliance. I don't really get it. Supposedly everyone would get a new job after some simple training, but not in construction. Instead we'd work at home on computers, completing jobs through a VR portal, whatever that means. At first the pay would be low, but the contract is three years, and after that a bit higher. What do you think?"

"I don't know, Matt. It depends on what kind of life you want."

"Yeah. I've spent most of my life on constructions sites, banging on metal, and I'm pretty damn content with that sort of life. I dunno if I can stand wearing a headset every day and flailing my hands in the air. Sounds kind of ridiculous."

Jennifer suddenly had a thought. She lifted her beer mug and bumped it against Matt's.

"Cheers to a healthy perspective. But if it was me, maybe I'd demand OmegaAlliance provide a chance for a hands-on trial run. After all, this is new. Who knows what might happen? Better to try it out before signing a contract than regret it afterward, right?"

Matt looked at Jennifer, considering the idea. "Easier said than done. But . . . maybe we can go through the union, demand that a few delegates have a chance to try it out. It may already be too late, though. The union's deadline to respond to the latest offer is tomorrow."

"Well, promise me that if you make it happen, you'll save me some . . . source material?"

"Right. You're a reporter, aren't you, Jen?" Matt's eyes narrowed, and he grinned. "And if I get you this material, what do I get in return?"

Shit, please don't say it.

Jennifer took a deep breath. The implication made her furious— and a little bit sad.

"What do you want, Matt?"

"Hold on, Jen, you don't think . . . Jesus! I just want some company, a friend. These days of waiting for a stable job are hard to take."

Jennifer breathed a sigh of relief. She put a hand on Matt's shoulder, like a daughter reassuring her father.

"I get it, Matt. When you need some company, just give me a call."

The crowd once more erupted with applause. The game was over.

SUNDAY EVENING WAS STIFLING.

Michael sat on a bench in the park across from Synchia's office. The vantage afforded a wide view of the bay, the great suspended cables of the Bay Bridge, and, farther away still, the Pacific. Michael desperately needed the fresh air, but it wasn't clearing his head.

Inside information had it that Landmark management was leaning toward OmegaAlliance's bid, even though it was more expensive. It promised to find positions for more workers, eliminating the risks of a standoff between labor and management. And it was safer politically.

That afternoon several Synchia executives had unloaded on Michael, one after another, pressuring him to find a way to close the deal. The AI and robotics technology that was transforming the construction industry was just the beginning of a great change. If other companies saw Landmark succeed in ridding themselves of their human union workers, more would follow suit. And this was just one industry. Other fields and roles previously thought immune to automation were now in play. Hundreds of thousands of jobs would be involved. It had the potential to trigger a much bigger crisis and hurt Synchia's bottom line. And if Synchia lost the Landmark bid, news would spread. Synchia's reputation for a magic touch would go up in smoke.

"And then, Michael," one of his direct reports said bluntly, "your name will be remembered by people forever, not as a gold-standard job reallocator, but an out-and-out loser."

The glass façade of the Synchia building reflected the immense city sprawl glittering under the setting sun. Michael looked intently at its surface. It had been built by countless workers over two hundred years, brick by brick, tile by tile. It had gone through earthquakes, conflagrations, plague, pollution . . . but remained as relevant as ever, not collapsing, blossoming with vitality. Nowadays, every time he thought of these workers becoming redundant, discarded goods, a multitude of useless people, his heart grew heavy. He had given it his all, yet he was powerless. He unconsciously groped in his pocket, hoping for a cigarette to alleviate his worries, before remembering he'd quit several years ago. *Humans are so imperfect*, he mused.

"I thought I might find you here." It was Jennifer's clear, sharp voice behind him.

"Sit with me a bit. How long has it been since you've watched a sunset? I mean a real one, not a sim or game scene."

"It's been a while, I guess?" Jennifer sat on the bench, putting about a foot between herself and Michael, not knowing why.

"You've done your best, Jennifer. You know what moved me most in that cover letter that got you this job? Not intelligence, initiative, or determination. It was those stories about your parents. They weren't lies, were they?"

"Of course not!" Jennifer's face went bright red.

"Sorry, I didn't mean to offend. I just wanted to know. Sometimes people lie to reach their goals. Me, for instance. Every day I lie to the unemployed, tell them not to give up hope." Michael was silent for a moment, and then turned to Jennifer. "Tell me about your father. You said he was successively reemployed?"

"That's right. At first. About twelve years ago he endured that first layoff, when there was still no concept of an occupational restoration specialist. I was ten." Jennifer, gazing at the sea, sank into memory.

After his job had been taken by AI, Jennifer's father was transferred. He went from a position as a credit analyst who worked with back-end data, to a role as a customer-facing underwriter. The quantitative jobs were vanishing—they were the most easily replaced by AI. Other longtime staff at her father's company had opted for early retirement, getting by on UBI and social security. Some of the younger or more adventurous ones chose to walk a totally different path, retraining as social workers or nurse's aides, jobs that required empathy and strong social skills. But her father was an introvert. He was also stubborn—and he had an ego. He was not good with people. Being an underwriter was at the limits of his capabilities.

The company required him to use internal software to manage client data. A semi-intelligent assistant program, powered with machine learning, would pop up from time to time, helping him crunch numbers or tidy up, fill in forms, and generate notices— junior grunt work. This system, called RPA (robotic process automation), became the last straw for her father. Slowly, he noticed this helper getting smarter and doing more, sometimes even correcting his own small human errors.

When her father finally awoke to reality, it overwhelmed him. Every time the smart helper made a correction on a task, the AI noted the fix as a data point, helping it become ever smarter in the process, ever closer to replacing its human co-worker. A few years later, he was laid off for good. The underwriting process went completely online. Reports could be completed in seconds, with no need for inefficient human employees.

The experience had changed Jennifer's father. "He became a total stranger to me," she told Michael, "no longer the tender father

I'd known, but cynical, passing his days drinking and tinkering in the garage. He ignored my mom, and she left him. I resented him, thinking it was all because of his lack of initiative, that he'd brought it on himself by giving in to despair. I thought that right up until now, really."

Michael handed her a tissue.

"Thanks. Now that I've met all these people in the same shoes as my dad, I get it. Work isn't just a steady paycheck. It means dignity, self-worth. Maybe that's what defeated Dad, that sense of powerlessness, that he could no longer stand tall."

In that moment, Michael felt that exact sense of powerlessness. He vividly felt his age.

"This is why I so admire you and what you do, Mr. Saviour." Jennifer looked at Michael, her cheeks tearstained.

"Maybe I'll disappoint you, Jenny." Michael sighed deeply. "Tomorrow morning, Synchia will lose. I will become just another redundant employee in the useless masses."

Jennifer stared, stunned and wide-eyed, at this figure before her, a hero past his prime. Suddenly, her mobile buzzed. Michael saw her expression brighten.

"Mr. Saviour, maybe the game's not over just yet."

EARLIER THAT SAME DAY, under pressure from union negotiators, Landmark had made a request to OmegaAlliance for a hands-on trial: Take a sample of the workers made redundant by AI, and conduct an all-day, closed field trial for them to experience the VR reallocation jobs for themselves. With feedback from this trial run, Landmark—and the union negotiators—would be able to evaluate this new reemployment method against Synchia's more established methods.

Thanks to his HR buddy, Matt was among the guinea pigs. He signed an NDA, handed over his mobile, and was bused to a remote industrial park.

And he didn't forget his promise to Jennifer. Before leaving, Matt outfitted his Cardinals baseball cap with a miniature camera.

From its position in the bird's threaded eyeball, the camera could continuously upload data to the cloud. The button-size high-grade lithium oxide battery was good for a week.

"Don't forget, Jen," he told her before he left, "I'm breaking the law to help you out. Don't leak any of my personal data, no matter what. You found the data yourself, got it?"

"But Matt . . . the video's content will make it obvious who shot it."

"Oh . . . yeah. Well, shit."

In the end, Matt had agreed to trust Jennifer. Eight hours later, he was calling her while she sat on the park bench with Michael. Jennifer rushed to meet him at the bar.

WHEN JENNIFER ARRIVED, MATT seemed haggard, eyes red, like he hadn't slept well in days. He began talking as soon as she sat down. "I really couldn't live like that. I mean, it's ridiculous, those tasks they gave us, the design plans, the numbers . . . completely lacking common sense. I couldn't make heads or tails of it, any of it."

"Let me watch the video, Matt. Let me figure it out. What you need to do is go home and get a good night's sleep. Okay?"

Taking the liquid memory dot from inside the camera, Jennifer rushed back to the Synchia headquarters, where Michael was waiting for her in his office.

AI had edited the video down to its quintessence and key moments, a thirty-minute digest. They could pause it at any time, play it in slow-motion, and zoom in.

As the video played, Jennifer and Michael saw fifty workers assembled in a large hotel banquet hall. Each had a desk with a computer and VR outfit. A trainer from OmegaAlliance stood at the front of the room. The first hour was basic training, and then came the official work. But there was a twist. The system would score the work—offering rewards and penalties. Rewards came in the form of credit points equivalent to cash.

All the workers had to do was don VR goggles and gloves. There would be no getting dirty on this job. The screens were synced with the imagery in the goggles, only on-screen it was 2D, so staff could find problems and get the workers back on track. Matt placed his cap on the desk, the hidden camera aimed at the screen. Then he got to work.

In the video clips that followed, Jennifer and Michael watched as Matt rushed to follow on-screen instructions, learning to rotate his POV in a simulated space, to enlarge and reduce, to add components. The work interface seemed almost designed for the simpleminded, making use of bright colors, voices, and visual effects to guide this veteran skilled worker. The task: assembling a water heater system for a thousand-square-foot residence. It looked like he was playing a game.

Jennifer and Michael looked at each other. "What do you think?" Jennifer asked her boss.

Michael shook his head, brows knitted: "This sort of virtual workflow, humans and machines cooperating, was developed years ago in highly digitized fields like finance and banking. Why go there with the construction industry? There are AI tools that can do what Matt is doing, without need of Matt at all. I don't get it."

"Maybe . . . the costs are lower?" Jennifer was in brainstorming mode.

Michael pointed a finger and the video continued.

Following the headset's audio orientation that accompanied the scenes, Michael and Jennifer began to piece together the underlying logic. If OmegaAlliance was to be believed, these workers were providing preliminary component plans and assembly for "end-to-end integration" construction projects in developing countries. It was like what a firm called Katerra had tried a dozen or so years before. Only now there was low-lag-time, high-precision VR tech. Experienced workers could directly use their hand movements to remotely control robots and conduct fine operations, meeting the needs of high-end clients.

All sorts of work—carpentry, painting, masonry, concrete pouring—were transformed for a VR man-machine cooperative workflow. As the OmegaAlliance trainer explained, workers who had passed through simple system training could, based on their years of real experience, "edit" construction plans. "Editing" was the right word, because they wouldn't expend physical strength and needlessly touch real building materials. The system analyzed the work in real time, awarding credit points based on speed and quality feedback. At the front of the hotel ballroom, there was a leaderboard displaying the names of the workers with the most points.

It was clear to Jennifer and Michael that some of the workers in the trial were quicker learners. They observed Matt impatiently turning in all directions and chatting with fellow workers. Some looked like gamblers sitting before slot machines, hands in constant motion, expressions intoxicated. Noting his place on the leaderboard, one worker performed what seemed like a victory dance. Others looked glum.

"This is a video game, only the name of the game is 'work.'" Jennifer couldn't hide some disgust with the whole affair. The workers' reactions to the tech reminded her of her sad and defeated father.

Michael sat silently.

"That's it!" His reaction startled Jennifer from her reverie. He reached out and took Jennifer's hand. "Maybe this doesn't just *seem* like a video game. Maybe it actually *is* one. You need to help me find out."

"Find out what?"

"The architectural designs in this video . . . are they really being implemented somewhere?"

"You mean . . ." Jennifer suddenly understood. "Matt kept saying he felt something fishy going on."

"Jen, you were right."

"About what?"

"It is a game. And it's not over just yet." Michael had recovered his self-confident smile.

On the screen recap, the work simulation had ended, and the organizers were announcing performance results. Some of the workers seemed delighted by what they heard. But as the crowd dispersed, Jennifer couldn't help but notice Matt. He hadn't performed well. She watched his hand reach out for the baseball cap.

Suddenly, the cam went dark.

+———

THAT NIGHT AT THE TOP of the Mark bar, situated on the nineteenth floor of the Mark Hopkins Intercontinental Hotel, Allison looked out over the panoramic view of downtown San Francisco.

The AI had proposed a perfect match for Allison. The main colors of the décor, gold, red, and black, contributed to the bar's retro atmosphere, with its crimson carpet, pale yellow wallpaper, and dark wooden furniture.

As twenty-four-year-old Le Rêve Blanc de Blancs champagne rippled in her glass, Allison watched Michael finally walk through the door. Uncharacteristically, he wasn't wearing a tie, and his collar hung open.

"Sorry, a meeting held me up."

Allison smiled. She knew Michael's psychological warfare tactics all too well. They ate dinner in a kind of guarded pas de deux, feigning nonchalance and civility back and forth. Allison finally couldn't help taking the bait.

"So, where did you get that video?"

"You admit it's a scam?" Michael said.

After Jennifer and Michael had watched Matt's video, she had used pattern recognition software to analyze all the plans and data revealed by Matt's illicit recording: architectural drawings, sunlight angles, times, elevations, latitude and longitude, atmospheric conditions, and other such parameters to reverse calculate positions. She found no corresponding structures in the real world. It was, as Michael had declared, a fabrication. OmegaAlliance had provided the redundant construction workers with nothing more than an excessively beautiful sim game.

"So, what do you plan to do with your little discovery?" Allison took a sip of her champagne, as if unbothered by this turn of events. Her steely demeanor masked an inner fear. If Omega-Alliance's conduct was revealed to the masses, public opinion would be hard to control, and all the company's efforts might be for nothing.

"The public has the right to know the truth."

"So you'd choose to have more people lose their jobs to satisfy your morbid need to play the hero. Ah, Michael, it's been so many years and you haven't changed a bit."

"C'mon, Allison. Personal attacks aren't going to accomplish anything."

"Think about it . . . Synchia spends so much time and money on transfers and training. You're like a carriage racing a train. Face facts. You can't beat AI. Why not turn to a once-and-for-all method?"

"You really want to live in a world where all those people simply deceive themselves with fake work for the rest of their lives?"

"Michael, do you understand them at all? Do you understand normal human feelings? Look at the people around you, wearing fashionable new clothes, entering and exiting this noble, lofty place, professionals who bill by the hour. Who doesn't zone out during working hours, for entertainment, for small talk? You think they feel guilty when their paycheck lands in their bank accounts?"

"You are depriving them of their right to create real value for society."

"Maintaining stability from nine to five is work's greatest value."

"What you're doing is fraud!"

"Trust me, if you let people choose the red pill or the blue pill, they'd choose blue. Who wants to reckon with the fact that they're useless, reliant on AI's charity to pass their days? You're wrong, Michael. This isn't fraud. This is our chance to retain a shred of dignity for humanity."

The champagne bubbles gradually dissipated, dissolving into the air.

Michael shook his head and smiled.

"What are you laughing at?" Allison said, glaring.

"Sorry . . . I just realized that every time we argue, it ends in deadlock. As if this were a zero-sum game, as if only one of us can be left standing. It's just funny."

Allison laughed. The mood lightened. "I don't know why. Maybe deep down in my heart I've always had something to prove to you?"

"You want to hear some truth?"

"Please."

"You don't need to prove anything to me, because in my heart, you've always been perfect." Michael could no longer meet her gaze. "Maybe you don't believe me. Well, my plan was to go public, to force OmegaAlliance to abandon this business, but I've changed my mind."

"Why?"

"This may not be a zero-sum game after all."

"You mean . . . cooperate?"

"I've done the math. Let me guess. You take the unemployment benefits and training subsidies from the government, add in Landmark's compensation package. Then what? Pay it to the workers in installments? And then what? Your model solves some problems, but it creates others. Profits would be thin. Real work and economic activity aren't being generated. You can't rely on fees from companies laying off workers and government subsidies forever. So your plan is to sign contracts with the workers on an annual basis, robbing Peter to pay Paul. But what's your endgame? Back to UBI, more of the same. Am I right?"

Allison's expression conceded the point.

"You understand better than me why UBI will fail. It's too laissez-faire. But maybe there's another way."

"I'm listening."

"What if simulated work produced real value?"

Allison's brow furrowed: "But how? You know Landmark is expecting resolution now. They can't afford to let the standoff with the union escalate."

"Convince your bosses to come sit at the negotiating table. As for the union, I'll think of a way."

A YEAR LATER, JENNIFER met Michael once again in the park across from the Synchia building in San Francisco, only this time it was early morning. Everything—the city, the distant bridge and sea—was shrouded in reddish-gold mist.

She was carrying two cups of hot coffee, strolling along a narrow path and recalling the strange events of the previous day.

She'd had a videoconference with an unemployed single mother, Lucy.

Jennifer had graduated from Michael's assistant desk after the Great Compromise, as Michael came to refer to it. Through their strenuous mediation efforts, she and Michael had gotten Synchia and OmegaAlliance to jointly undertake the Landmark project, ensuring all the construction workers would be cared for. It was the dawn of a new method of occupational restoration, one that combined AI and human expertise.

Jennifer was now a junior reallocation specialist. She had dozens of online consultations per day. But this young woman, Lucy, had not been like the others.

Lucy had been a bartender, but her boss had sold the establishment to a restaurant chain, which transformed it into an automated dining hall. Lucy had lost her position, and she still had a hungry six-year-old boy at home.

But her choice of words, her unhurried cadence, her exquisitely refined makeup, and even her need to pause and consider questions before answering, as though her brain had a bit of a reaction lag, all made Jennifer feel she was not a real bartender.

But why would she lie?

As Jennifer tried to speed the conversation along, Lucy appeared unconcerned, like an actor with a script tossing out question after question, all of which touched on key points. How would Jennifer evaluate her skills? What training resources would be provided? Could Jennifer help her find a new post within the grace period? Unlike other redundant workers, Lucy betrayed no anxiety about her predicament, no uneasiness. And though she was a mother, she showed no concern for her son's needs as she considered job choices. This was unusual.

A spy from a competitor, perhaps, gathering intel? Jennifer was worried she was getting a taste of her own "field investigation" medicine.

"I'm sorry, Lucy, but I have another appointment. How about we call it a day? If a suitable opportunity comes up, we'll notify you."

Lucy finally seemed to understand. She nodded. Her farewell was thought-provoking: "Thanks. The information you provided was very helpful. I look forward to our next meeting."

Was she an automatic answering machine, or what? Jennifer's irritation with the encounter lasted into the next morning, until she spotted a familiar figure seated on a bench. The well-dressed middle-aged man was smiling to himself. Jennifer hurried over.

"Michael?" Not calling him "Mr. Saviour" took some getting used to. "I didn't dare believe it was really you. Are you here on business?"

Jennifer was overjoyed. After the Landmark deal, Michael had promoted her to partner—the youngest in Synchia's history—but she had passed on the opportunity to follow him to Seattle to work at the combined powerhouse of SynchiaAlliance, choosing instead to start as a reallocation specialist focused on serving small- and medium-size local businesses.

"Long time no see, Jenny. Looks like you're doing well."

"Thanks. Do you have a moment? I happen to have an extra coffee."

"Of course. Who can refuse a hot cup on a San Francisco morning?"

The two of them sat on the bench, separated by a foot or so. Jennifer held her cup in both hands, as if struggling to conquer some psychological barrier. She finally spoke.

"Michael, I know this is really stupid, but I still want to apologize to you, sincerely."

"For what?"

"You gave me such a good opportunity, but I refused. I hope you didn't see that as a betrayal."

"Oh, Jen, don't tell me you've been taking this to heart all along." Michael smiled bitterly and waved a hand. "Honestly, I was quite happy with your decision. Everyone said my assistants were cursed and would inevitably leave the company inside a year, so I'm just thrilled you broke the curse."

"Okay, since that's how you feel . . . how's it going with Allison?"

"She's great. We're great. At least now we are. Entering a relationship is a bit like starting a new job. You always need a breaking-in period, even if you've done that sort of work before. Tell me, is everything going well on your end?"

Jennifer smiled awkwardly. "At first I felt a lot of pressure. I was always dreaming about going into a crowded room to give a speech, and my opening remark was always, 'Thank you all for your long history of tremendous contributions to this company.' The standard opener for easing into layoffs and reallocation, as you know. But now it's much better, thanks to you."

"I should be thanking you, Jenny. If not for you and that worker . . . Matt, was it? You two helped tens of thousands of workers."

"I just hope my job can last a few more years. You know, they've begun trying out online consultation services with bots. Who's to say when we ourselves will become redundant?"

"Jenny . . ." Michael gazed at his onetime subordinate, hesitant to continue. "No matter how the world changes, I believe you'll do well, because you have a sincere heart."

Jennifer laughed. "Your scripted platitudes don't work on me, Michael. I've read the manual. Anyway, I have a morning meeting, so I won't disturb your view any longer. I hope it's not as long until we meet again."

Michael waved: "See you later, and good luck."

Jennifer vanished around the end of a glass curtain wall.

She's right, Michael thought to himself, alone on the bench. AI power was always expanding, intruding across the borders of tenaciously defended human professions. Maybe it was all just a matter of time.

Before this business trip, there'd been a policy meeting at headquarters, a vote on whether to introduce virtual workflow into SynchiaAlliance. If approved, it meant a portion of the junior reallocators' clients would be digital humans. Some were modeled on true humans, others entirely generated and driven by AI, simulating different sorts of redundant workers. The digital humans were so realistic it was hard for human staff to differentiate them from real people. They could continuously refine their natural language processing through deep interaction. In this process, the digital humans would evaluate human reallocator performance, selecting the best for promotion to management.

And the leftover people, perhaps they could only recapitulate the game that had already played out in other industries. It was absurd: Real humans helping the virtual redundant to seek virtual work opportunities.

They'd tried out a prototype secretly, enrolling it in a competitor's reallocation program. Michael had named the prototype after his mother, Lucy.

All the data indicated this innovation would improve organizational efficacy, screen for even more outstanding human talent, and optimize resource allocation streams. It seemed entry-level positions would become more about gaining practical skills than about generating actual value. The trend seemed inevitable. But, as before, there were voices of tradition laden with misgivings.

Management had raised its hands and voted three to three, a tie. Everyone had turned to the newest senior executive, Michael.

In an instant, he recalled that faraway afternoon, Mom's expression when she received news of the layoff. Who wins and who loses was perhaps insignificant in the face of history's powerful current.

Michael Saviour had tugged at his necktie knot, as if to give himself breathing room. Then he'd raised his hand and voted for the future.

AN ALYSIS

AI JOB DISPLACEMENT,
UNIVERSAL BASIC INCOME (UBI),
WHAT AI CANNOT DO,
3RS AS A SOLUTION TO DISPLACEMENT

Artificial intelligence can perform many tasks better than people can, at essentially zero cost. This simple fact is poised to generate tremendous economic value but also to cause unprecedented job displacement—a wave of disruption that will hit blue- and white-collar workers alike. In a lively narrative that imagines the frontlines of AI-related layoffs, "The Job Savior" describes a future in which AI will be doing everything from underwriting our loans to building our homes, and even hiring and firing us. This transformation of how work gets done will not only result in severe unemployment, but potentially ignite a host of social problems, including depression, suicide, substance abuse, widening inequality, and social unrest.

So where does that leave us? What can individuals, companies, and government do to mitigate potentially catastrophic consequences? What are the jobs that AI can and cannot displace? What is the future of work? Do we need a new social contract to redefine humans' fundamental expectations around employment? If long hours of work for economic output are no longer a necessary feature of human life, how will we spend our time?

The jobs most at risk of automation by AI tend to be routine and entry-level jobs. This trend will exacerbate existing challenges in society, as those who are poor become poorer. This complicated dynamic—AI's potential to create unprecedented efficiency as well as deep structural problems in society—can perhaps be best summed up by this question: When it comes to work, is AI ultimately a blessing or a curse?

"The Job Savior" imagines a future in which job reallocation companies emerge to retrain and counsel laid-off workers, placing them in other positions in far-flung places—or even training them to perform real work virtually. The story explores whether there will be enough jobs to go around—and whether, at some point, elites may begin to experiment with work simulations, with the aim of providing workers avenues to feel fulfilled even if they aren't performing "real" work.

Will we ever reach a point, at least in readers' lifetimes, where human work is so scarce that future work will turn into some kind of simulated game—the worst manifestation of an imagined firm like OmegaAlliance—just to pass the time? I don't think the horizon is quite so bleak. It is highly possible, however, that simulated virtual environments could offer support and training to human workers.

In order to assess the likelihood of that perhaps alarming vision, let's first take a deeper dive into how AI will continue to displace jobs.

HOW AI DISPLACES JOBS

AI's main advantage over humans lies in its ability to detect incredibly subtle patterns within large quantities of data. Take the example of loan underwriting. While a human underwriter will look at only a handful of measures when deciding whether to approve your insurance application (your net worth, income, home, job, and so on), an AI algorithm could take in thousands of variables—ranging from public records, your purchases, your healthcare records, and what apps and devices you use (with your consent)—in milliseconds, and come up with a far more accurate assessment of your application. Would you grant permission to such an AI underwriter to look at your personal data in exchange for favorable rates? Many people likely would, as we saw in "The Golden Elephant."

Such algorithms will displace routine white-collar work easily, just as software has steadily taken over routine white-collar tasks, such as bookkeeping and data entry. In "The Job Savior," we saw examples of affected white-collar workers ranging from bookkeepers to insurance underwriters. When combined with robotics, AI will also displace increasingly complex types of blue-collar work. As we saw in the story, by 2041, warehouse pickers—who perform routine tasks—will have long been displaced; many construction workers will have been displaced, as building practices shift toward prefabricated components built by robots that are easy to assemble en masse; and even plumbers will slowly be phased out, as human plumbers will only be needed primarily to service older buildings with complicated, retrofitted systems that require unique repairs. New buildings with standard prefabricated components will be serviced by robots.

How far will this job displacement go, and what industries will it hit hardest? In *AI Superpowers* I estimated that about 40 percent of our jobs could be accomplished mostly by AI and automation technologies by 2033. It's not going to happen overnight, of course. Jobs will be taken over by AI gradually, just as we saw with RPA (robotic process automation) and Jennifer's father, the underwriter, in "The Job Savior."

RPA is a "software robot" installed on workers' computers that can watch everything the workers do. Over time, by watching millions of people at work, RPA figures out how to do employees' routine and repetitive tasks. At some point, a company will decide it's better off letting the robot take over a given task entirely from the human. The number of employees on the company's payroll will shrink as the overall workload lightens.

Imagine a recruiting department of one hundred people. RPA might first be applied to conduct résumé screening and compare applicants against criteria in a job description. Let's say that there are twenty people doing that task, and that RPA can help these people assess applicants twice as efficiently. That's ten people who could be displaced. Then, as AI learns further from more data and experience, it might at a later junction replace nearly all twenty people. It's no stretch to see how RPA could take over email communication with candidates, interview setup, feedback coordination, hiring decisions, and even basic negotiation over job offers. Each of these tasks, when delegated to RPA, would displace more people.

AI could even conduct interview screening or first-round interviews, similar to the digital human Lucy's evaluation of Jennifer in "The Job Savior." That would save untold hours for HR departments and hiring managers. All of the above might whittle down the total number of employees needed in that recruiting department from one hundred to perhaps ten. And after recruiting is augmented by AI, then comes HR training, new employee orientation, and performance evaluation. And HR is just a single department. Once HR is AI infused, finance, legal, sales, marketing, and customer service will follow (or be transformed simultaneously). COVID-19 has accelerated the digitization of workflow by companies, which will make RPA and other technologies even easier to apply, thereby accelerating job displacements. While AI displacement is gradual, eventually it will also be total.

Optimists argue that productivity gains from new technology almost always produce economic benefits—that more growth and more prosperity always mean more jobs. But AI and automation differ from other technologies. As we've established in previous chapters, AI is an omni-use technology that will drive changes across hundreds of industries and millions of tasks simultaneously, both cognitive and physical. While most technologies were job creators and job destroyers at the same time—think about how the assembly line changed the automotive industry from artisans hand-assembling expensive cars to routine workers building many cars at much lower prices—the explicit goal of AI is to take over human tasks, thereby decimating jobs. The Industrial Revolution took over a century to spread beyond Europe and the United States, while AI is already being adopted all across the world.

JOB DISPLACEMENT CAUSES OTHER SERIOUS PROBLEMS

Soaring unemployment numbers are just a small part of the problem. A growing pool of unemployed workers will compete for an ever-shrinking number of jobs, driving down wages. Wealth inequality will go from bad to worse, as AI algorithms destroy millions of human jobs, while at the same time turning the tech titans who harness these new technologies into billionaires in record time. Many of the free market's self-correcting mechanisms espoused by Adam Smith (for example, the notion that high unemployment will drive down wages and in turn reduce prices to a point that eventually increases consumption and gets the economy back on track) will break down in an AI economy. Left unchecked, AI in the twenty-first century may bring about a new caste system, with a plutocratic AI elite at the top, followed by a relatively small subset of workers with complex jobs that involve wide-ranging skill sets and large amounts of strategy and planning, creatives (many of them low paid), and the largest contingent: the powerless struggling masses.

Even more problematic than the loss of jobs will be the loss of meaning. The work ethic born out of the Industrial Revolution has instilled in many of us the idea that careers should be at the center of how we derive meaning from our lives. In the coming years, people will watch algorithms and robots easily outmaneuver them at tasks they've spent a lifetime mastering. Young people who grew up dreaming of entering certain professions may have their hopes thwarted. This will lead to a crushing feeling of futility and obsolescence, paving the way for increased levels of substance abuse, depression, and suicide. (Spikes in suicides have already been traced to industries that have been heavily disrupted by technology, such as taxi driving.) Even worse, it will lead people to question their own worth and what it means to be human.

Recent history has shown us just how fragile our political institutions and social fabric can be in the face of disruptive changes, such as the pandemic. AI economics is arguably the largest disruptor, and if allowed to run its natural course, it will make today's sociopolitical tumult look like child's play.

It's a bleak picture. So what can we do about it?

UBI: A PANACEA?

The enormous challenges of AI and job displacement have breathed new life into an old idea called universal basic income (UBI), in which the government provides a stipend for each citizen regardless of need, employment status, or skill level. This stipend could be provided by taxing ultrarich individuals and/or companies. Ahead of the 2020 U.S. presidential election, candidate Andrew Yang promoted his variant of UBI, called "the Freedom Dividend," as the cornerstone of his campaign and a way to fight the automation tidal wave. A political novice, Yang gained far more traction than most pundits anticipated, and leads the poll in the 2021 New York City mayoral election, partly because of the appeal of UBI and partly because he articulated difficult truths about where the economy is heading—facts that other politicians have largely ignored, even though workers are beginning to feel the consequences.

I believe we must slow down the widening of wealth inequality, and UBI is a simple yet effective mechanism for just that. However, the unconditional distribution of UBI risks being too broad and wasteful. There are alternate proposals that add conditions or consider the needs of individuals, which would improve the efficacy of the UBI proposition and the public perception of it.

I've always liked the proverb that goes, "Give a man a fish and you feed him for a day; teach a man to fish and you feed him for a lifetime." That's what UBI should be striving for. In other words, UBI should help potentially endangered workers choose and train for appropriate new professions that are less likely to see short-term displacement. Left to their own devices, most displaced workers will not have the foresight to predict which professions might survive the AI revolution and therefore won't know how to best use UBI money to pivot their lives to better pathways. Unless training becomes a core part of UBI proposals, billions of people will suffer the same fate as the warehouse manager in "The Job Savior" who becomes a theme park attendant, only to lose her new job shortly afterward as the cycle repeats.

THE KEY QUESTION: WHAT CAN AI NOT DO?

In order to help guide people through AI displacement, we need to first understand what kinds of capabilities and tasks AI *cannot* do. We can then accelerate the creation of these displacement-resistant jobs, provide career counseling, and train more people for them, thereby nudging supply and demand toward an equilibrium.

These are the three capabilities where I see AI falling short, and that AI will likely still struggle to master even in 2041:

1. CREATIVITY: AI cannot create, conceptualize, or plan strategically. While AI is great at optimizing for a narrow objective, it is unable to choose its own goals or to think creatively. Nor can AI think across domains or apply common sense.

2. EMPATHY: AI cannot feel or interact with feelings like empathy and compassion. Therefore, AI cannot make

another person feel understood and cared for. Even if AI improves in this area, it will be extremely difficult to get the technology to a place where humans feel comfortable interacting with robots in situations that call for care and empathy, or what we might call "human-touch services."

3. **DEXTERITY**: AI and robotics cannot accomplish complex physical work that requires dexterity or precise hand-eye coordination. AI can't deal with unknown and unstructured spaces, especially ones that it hasn't observed.

What does all this mean for the future of jobs? Jobs that are asocial and routine, such as telemarketers or insurance adjusters, are likely to be taken over in their entirety. For jobs that are highly social but routine, humans and AI would work together, each contributing expertise. For example, in the future classroom, AI could take care of grading routine homework and exams, and even offering standardized lessons and individualized drills, while the human teacher would focus on being an empathetic mentor who teaches learning by doing, supervises group

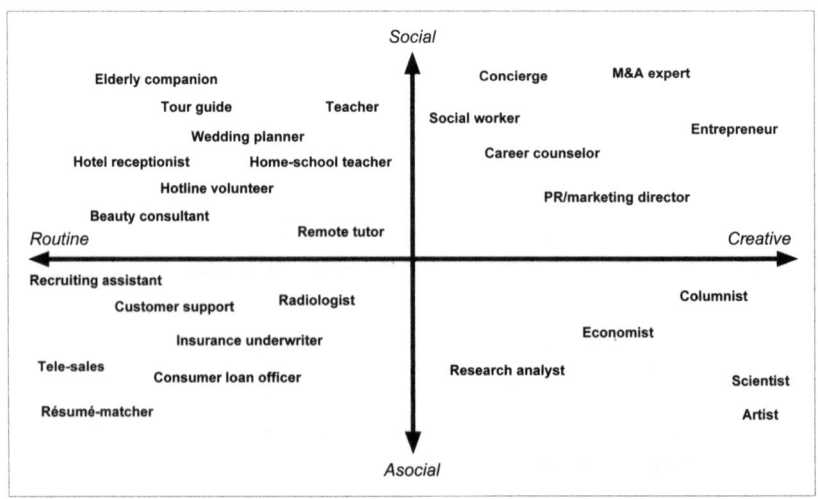

Cognitive jobs on a two-dimensional chart.
Upper right favors humans, and lower left favors AI.

projects that develop emotional intelligence, and provides personalized coaching.

For jobs that are creative but asocial, human creativity will be amplified by AI tools. For example, a scientist can use AI tools to accelerate the speed of drug discovery. Finally, the jobs that require both creativity and social skills, such as Michael's and Allison's strategy-heavy executive roles in "The Job Savior," are the ones where humans will shine. The figure opposite illustrates this for white-collar jobs.

The next figure shows a similar diagram for physical and blue-collar jobs. The y-axis represents social skills, and the x-axis represents the complexity of the physical job. Complexity is measured by the dexterity required and the need for navigating unknown environments. For example, caretaking that involves a task like helping an older person take a shower requires both social and dexterity skills, while quality inspection in an assembly line requires neither. Cleaning houses requires the ability to navigate unknown environments, while a bartender relies primarily on social skills, as a robot mixes better drinks than most human bartenders.

While it's clear that there are a lot of lines of work that AI will struggle to master—and thus would be safer for workers to pursue for their careers—these alone won't prevent a disaster for the legions of workers

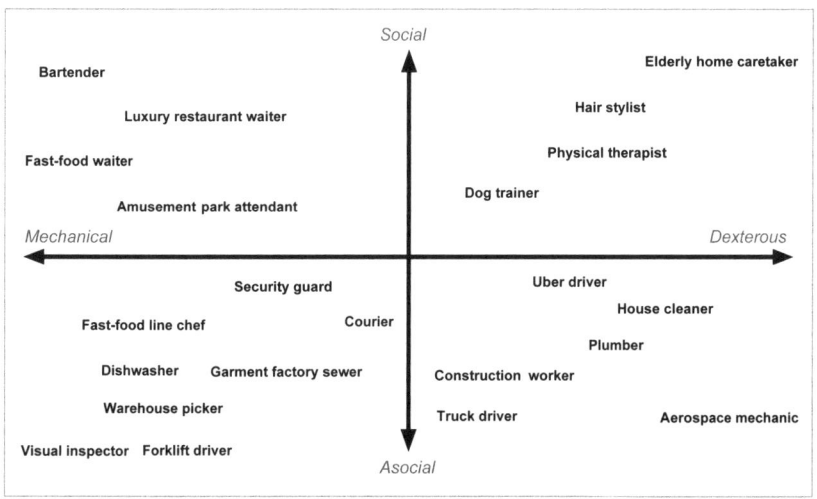

Physical jobs on a two-dimensional chart.
Upper right favors humans, and lower left favors AI.

displaced from roles that will be easier for AI. So what else can we do to help fulfill the basic human desire for a meaningful livelihood?

HOW CAN WE TRANSFORM THE HUMAN WORKFORCE?

To create more jobs and improve the readiness of workers for the transformation on the horizon, I propose the 3 Rs—relearn, recalibrate, and renaissance—as part of a gargantuan effort to deal with the central issue of our time: the AI economic revolution.

RELEARN

People in endangered jobs should be warned well in advance and encouraged to learn new skills. The good news is that, as discussed, there are skills that AI cannot master: strategy, creativity, empathy-based social skills, and dexterity. In addition, new AI tools will require human operators. We can help people acquire these new skills and prepare for this new world of work.

Vocational schools need to redesign their curricula to promote courses for such sustainable jobs. Governments could take the lead and provide incentives and subsidies for these courses, rather than blindly pursue broad-brush economic measures, such as universal basic income. Corporations could also provide programs like Amazon's Career Choice program, under which Amazon pays up to $48,000 for any employee to earn a degree in high-demand fields such as aircraft mechanics, computer-aided design, and nursing.

Pandemic or no pandemic, the importance and number of human-centric service jobs, such as nursing, will also increase as wealth and life span increase. The World Health Organization predicts we will fall short of the number of healthcare workers required to achieve the UN Sustainable Development Goal of "good health and well-being for all" by approximately eighteen million. Society has consistently devalued such vital human-centric service roles both in terms of how they are perceived and how much they are compensated, and we need to address this oversight. These jobs will form a bedrock for the new AI economy.

During his presidential campaign, Andrew Yang spoke often about rethinking what counts as paid work. Things that today we consider caregiving or volunteering may turn full-time in the future—roles like blood bank assistants, foster care providers, youth group mentors, homeschooling parents, or adults who leave the workforce to take care of a sick or elderly dependent family member. During this era of automation and upheaval, there will inevitably be a tremendous demand for volunteers to answer hotlines for displaced workers, helping them to deal with their concerns and challenges. These volunteers could potentially become eligible for extra payments and full-time work.

RECALIBRATE

In addition to relearning skills, we need to recalibrate what today's jobs look like with the help of AI, moving toward a human-AI symbiosis. The most prevalent and basic symbioses will be found in software AI tools. Software powers a human-PC interdependence, which has already revolutionized office work. Software AI tools can devise alternatives, optimize outcomes, or perform routine work for white-collar professionals in many fields. Specific AI tools will be customized for each profession and application—for example, AI-based molecule generation for pharmaceuticals, advertising planning for marketing, or fact-checking for journalism.

A deeper interdependence between AI optimizations and "human touch" will reinvent many jobs and create new ones. AI will take care of routine tasks in tandem with humans, who will carry out the ones that require warmth and compassion. For example, the future doctor will still be the primary point of contact trusted by the patient but will rely on AI diagnostic tools to determine the best treatment. This will redirect the doctor's role to that of a compassionate caregiver, giving them more time with their patients.

Just as the mobile Internet led to roles like the Uber driver, the coming of AI will create jobs we cannot even conceive of yet. Examples today include AI engineers, data scientists, data-labelers, and robot mechanics. But we don't yet know and cannot predict many of these new professions, just as in 2001 we couldn't have known about Uber drivers. We should watch for the emergence of these roles, make people aware of them, and provide training for them.

RENAISSANCE

Finally, with the right training and the right tools, we can expect an AI-led renaissance that will enable and celebrate creativity, compassion, and humanity. From the fourteenth to the seventeenth centuries, wealthy Italian cities and merchants funded the Renaissance, which saw a flowering of artistic and scientific output. We have reason to anticipate that AI will be the catalyst for a new renaissance centered around human expression and creativity. As with the Italian Renaissance, people will follow their passions, creativity, and talents once they have more freedom and time.

Painters, sculptors, and photographers will use AI tools to compose, experiment with, and refine artwork. Novelists, journalists, and poets will use new technologies for research and composition. Scientists will use AI tools to accelerate drug discovery. An AI renaissance will reinvent education, giving teachers AI tools to help each student find their own passions and talents. Education will encourage curiosity, critical thinking, and creativity. It will promote learning by doing and group activities that enhance students' emotional intelligence—and that bring them face-to-face with one another, and not just a screen.

As I imagine what this technology-induced Renaissance might entail, the words of second U.S. president John Adams seem prophetic: "I must study politics and war that my sons may have liberty to study mathematics and philosophy. My sons ought to study mathematics and philosophy, geography, natural history, naval architecture, navigation, commerce, and agriculture, in order to give their children a right to study painting, poetry, music, architecture, statuary, tapestry, and porcelain."

TOWARD AN AI ECONOMY AND A NEW SOCIAL CONTRACT

Turning some of the ideas above into reality would be an unprecedented undertaking for humanity.

The AI job-displacement tidal wave will eventually take away virtually all routine jobs, which tend to be entry-level jobs. But if no human takes an entry-level job, how will they learn, grow, and advance to more senior and less routine jobs? As automation becomes pervasive, we need to make sure there are still ways for people to enter all professions, to learn by doing, and to get promoted based on their capabilities. The blurring of "made-up job," "practical training," and "real job" are likely to emerge out of necessity, along with the use of VR technologies to implement this. This is what inspired the OmegaAlliance-Synchia partnership in "The Job Savior," in which entry-level positions became more about gaining practical training than about generating value.

One thing is clear: We will need to retrain a massive number of displaced workers. We need to raise an astronomical amount of money to fund this transition. We need to reinvent education to produce creative, social, and multidisciplinary graduates. We need to redefine the work ethic of society, entitlements for citizens, responsibilities for corporations, and the role of governments. In short, we need a new social contract.

Fortunately, we do not need to create this from scratch. Many elements already exist in different countries. Take, for example, the "gifted and talented" education programs in Korea, primary education in Scandinavia, university innovations (like massive open online courses, or MOOCs, and Minerva Schools) in the United States, the culture of craftsmanship in Switzerland, service excellence in Japan, the vibrant tradition of volunteering in Canada, caring for elders in China, and "gross national happiness" from Bhutan. We need to share our experiences and plot a way forward globally, where new technologies are balanced by new socioeconomic institutions.

Where will we find the courage and audacity to take on this gargantuan task? We are the generation that will inherit the unprecedented wealth from AI, so we must also bear the responsibility of rewriting the social contract and reorienting our economies to promote human flourishing. And if that is not enough, think about our posterity—AI will liberate us from routine work, give us an opportunity to follow our hearts, and push us into thinking more deeply about what really makes us human.

CHAPTER NINE

ISLE OF HAPPINESS

STORY TRANSLATED BY BENJAMIN ZHOU

BE NOT AFEARD. THE ISLE IS FULL OF NOISES,
SOUNDS, AND SWEET AIRS, THAT GIVE DELIGHT AND
 HURT NOT.
 . . . AND THEN IN DREAMING
THE CLOUDS METHOUGHT WOULD OPEN, AND SHOW RICHES
READY TO DROP UPON ME, THAT WHEN I WAKED,
I CRIED TO DREAM AGAIN.
                    —WILLIAM SHAKESPEARE, *THE TEMPEST*

NOTE FROM KAI-FU: AI can make us efficient and wealthy, but can AI make us happy? This story is about an enlightened Middle Eastern monarch who wants to use AI as the elixir for contentment. But what is happiness, and how is it measured? The monarch summons an eclectic group of guests to explore this fascinating problem . . . within themselves. In my commentary, I will discuss the problems of measuring satisfaction and happiness—and whether AI will be equipped to answer them. I will discuss the privacy issues raised by the idea that AI may know all our deepest and hidden desires, and analyze regulatory and technological mechanisms to protect our privacy.

UNDER THE VAST arc of the desert sky, a black all-terrain vehicle could be seen surfacing and plunging like a fin breaking the water's surface. Charging up two- and three-story-high dunes, its engine at full throttle, the four-wheeler coughed up a cloud of dust behind it.

Viktor Solokov's white-knuckled grip on his seat was all that kept him onboard. The turbulence blanched his face. "There's no way automated driving could handle this!" Khaled yelled over his shoulder, grinning. The Algerian driver's booming voice could barely be heard over the deafening roar of African electronic music and the paint-thick dust blasting the windshield.

On a satellite map, these dunes formed a vast zone, cutting across the Qatar peninsula from northwest to southeast. An eastbound traveler from the hinterland of the desert might encounter tourist-targeted retro camel caravans, abandoned oil refineries, mudflats, and mirage-like ultramodern cities—giving the impression that the young country's miraculous economy had spontaneously self-assembled from a random jumble of wasteland.

"By the way, why don't you just fly there? I mean, that would be a lot faster," said Khaled, pointing off in the distance to the coastline, and beyond it, Al Saeida. The artificial island on the Arabian Sea northeast of Doha's Lusail Marina was a place most people chose to visit by private jet or yacht.

Viktor shrugged: "I have nothing but time on my hands."

"You crazy Russkies!"

Khaled's passenger was nothing like the typical larger-than-life Russian tycoon, however. His frail body bundled in a black tracksuit, Viktor had a face that was youthful but somehow projected a depth of feeling unusual in a man his age. His most striking feature was his apparent indifference to the scenery and sites around him. He was like a shopper in a gourmet supermarket hell-bent on buying a single loaf of bread.

Until recently, Viktor had been one of the world's most celebrated entrepreneurs under forty. After building an e-sports platform in northeast Asia and selling broadcasting rights around the world, he pivoted the business to franchised crypto-backed gambling. In fewer than ten years, he'd accumulated a fortune that dwarfed the GDP of entire countries. But at the pinnacle of his success, Viktor shocked everyone by vanishing from public life. In the wake of his unexplained absence, conspiracy theories were rampant. Some speculated that the government planned to seize his business; others whispered that Viktor was suffering from a terminal illness.

The truth was that Viktor had simply lost interest. Suddenly, he couldn't bear being CEO, business genius, media darling, and success guru, all in one. Juggling all of these roles made him feel like a poorly made marionette: The more he acted, the closer he came to collapse. Retreating to his luxurious Black Sea estate, Viktor, in the months after his departure from the world stage, had hosted a revolving parade of friends and other assorted hangers-on, where they indulged themselves with alcohol, drugs, and escorts. But far from lifting his spirits, the party life was a disaster, culminating in a scandal when an acquaintance was found dead, having drowned in Viktor's pool while intoxicated. Viktor was remanded to a rehab center, where doctors diagnosed him with severe depression.

Viktor knew that the prescribed "treatments"—drug regimens,

support groups—were nothing but Band-Aids. He couldn't shake the voice in his head that was constantly mocking him, insisting that his once-idealistic dreams—bringing joy and fun to all through video games—had been waylaid by greed and reduced to mere obsessive moneygrubbing.

When Viktor returned from his stint in rehab, he found, sitting amid a pile of junk mail, a mysterious envelope marked "Al Saeida" in a lavish font and inscribed with the famous Latin phrase "Carpe Diem." The letter inside was an invitation—for Viktor to travel to Qatar for an "all-expenses-included visit to the world's most luxurious bastion of happiness."

He asked around—but none of his connected friends could provide any clarification, only that Al Saeida was the name of an artificial island recently constructed by Qatar. Maybe a little adventure was what he needed to snap out of his funk?

Viktor booked a flight.

KHALED GAVE HIS BUSINESS card to Viktor and dropped him off at the coast, where the Russian boarded a waiting yacht emblazoned with the same whimsical Al Saeida font as the invitation.

It was only a short ride to the island, which the setting sun had bathed in a radiant glow, like a pot of gold seen from afar. As he disembarked the boat, Viktor noted the island's squat buildings, which seemed to consciously evoke Qatari cultural symbols of domes, dunes, and pearls, and he was amazed by what seemed to be a translucent veil suspended above the island like a fine silk headscarf, wafting gently with the light breeze. Baffled, Viktor tried to locate the veil's endpoints, but he could see nothing. It was simply floating in the air.

The friendly synthetic voice of a robotic attendant greeted him at the dockside. The six-foot bot was clad in a gray robe that hung all the way to the ground, making it impossible to see how it moved.

"Hello, Mr. Solokov. I'm Qareen, your humble servant. I will attend to your every need here on Al Saeida."

"Hmm . . . I had assumed that Qatar's upper class preferred human servants."

"You're absolutely right! But here on Al Saeida, we've got our own unique style of smart service."

Frowning at this mechanized reply, Viktor mumbled, "So I guess I can't expect any grand welcome party."

"Of course you can! The festivities will ensue immediately following your acceptance of the terms of service."

"Terms?"

A glowing blue screen extended from Qareen's arm, displaying a dense chunk of text. Viktor scrolled through the terms, skimming through the keywords. It seemed as though whoever was behind the island getaway was requesting access to his most intimate personal data, from health records and financial accounts to Viktor's social media passwords and history, including all audio and video records. It constituted the totality of data that made up an individual's everyday life. A line of repeating crawl text read: ULTIMATE SERVICES FOR YOUR HAPPINESS.

"So how do you guarantee my data security?"

"Mr. Solokov, I can assure you that Al Saeida is equipped with the most advanced middleware technology. That means all your personal data will be fully encrypted and locked, so it can be accessed by AI only in service to a specific individual, which is you, with traceable services and content. If you feel that's still insufficient, we provide full transparency via an open-source algorithm notifying any client of hacking or malware installation.

"Well, what do I have to lose, I suppose. Where do I sign?"

Viktor couldn't tell if he was more curious about the mystifying island or about that crawling term HAPPINESS. Shrugging off his qualms, he clicked "I consent" on the screen, and then completed a quick verification of his iris, voiceprint, and handprint.

As Viktor lifted his palm from the screen after completing the verification sequence, a blue light burst forth from the robot's arm, seeming to flow to the ground and then expand outward, rippling out across every corner of the island.

Well, there goes my data. Viktor was shocked—but also intrigued. It seemed that a visitor's personal history activated the space. Viktor's sense of being an insect under a microscope triggered slight misgivings, but before he'd had time to consider them, Qareen had scooped up all his luggage.

"Mr. Solokov, let me show you the way *home*. All has been arranged."

When Viktor entered the dune-shaped holiday house, he understood why his robot servant had emphasized the word "home." The decorations, furniture, ornaments, and even the bear paw mounted on the wall over the fireplace were exactly the same as those of Viktor's villa in Moscow's Rublyovka district.

"How is this possible?" Viktor murmured, knowing there was no way for even 3D printing to conjure these objects so quickly. But he quickly realized the trick: None of the visible items were real. Every surface in the guesthouse was programmable. It was all an illusion.

Suddenly stung by a weird feeling, Viktor looked around. Through the window, he spotted a dark shadow that instantly vanished into thin air. Was someone spying on him? Before Viktor had a chance to ponder the strange incident further, Qareen announced that they should proceed to the guesthouse's cinema room for a surprise.

IN THE CINEMA ROOM, Viktor sat on a plush leather couch in front of a gigantic screen. As soon as he took his seat, a film began to play—what amounted to a biopic about none other than Viktor Solokov. The movie had been automatically compiled and edited by AI, retrieved from public channels and Viktor's private collections.

The life-stream revisited his unhappy childhood; his adolescence, full of competition and misplaced anger. There were images of every highlight of his adult life: awards, summits, IPOs, mergers and acquisitions, charity galas—all the trappings of Viktor's former life that he had turned his back on. Weary of the monotonous montage, Viktor half-closed his eyes. As he did so, he was unaware that all his physical metrics, including facial expression, body temperature, heart rate, blood pressure, bioelectrical signal, and levels of adrenaline, serotonin (5-hydroxytryptamine, aka 5-HTP), and dopamine, were being meticulously recorded—through the nondescript leather sofa he was sitting on.

For most of the viewing, Viktor's vital signs had been as stable as a Zen master's. Only once did his heart rate vary—when a family photo from Viktor's childhood flashed onto the screen. His gaze didn't linger on the face of his father or mother but hovered instead on the image of Margaret, his golden retriever.

When the film ended, a list of questions appeared on the screen. Qareen beckoned Viktor to read them.

"'Do I have warm feelings toward almost everyone?'" Viktor glared at Qareen, clearly offended. "What kind of stupid question is that? Do I have to do this?"

"Mr. Solokov, you must understand that happiness is a highly subjective feeling. The questions are merely intended to help us better understand your baseline state, so we can provide services tailored to your specifications. We'd like you to answer the questions as honestly as you can, using a six-point scale to represent the different degrees from 'strongly disagree' to 'strongly agree.'"

Viktor stared at the robot. "Idiot," he muttered. But he turned back to the screen and began to push the virtual buttons.

The questions seemed endless, and answering them tested what little patience he had left. But just as he was on the verge of quitting—the questions ended and the overhead lights clicked back on.

"So, what now?"

"Congratulations, Mr. Solokov. We can now proceed to meet your fellow guests. You will like them."

THE BANQUET WAS CONVENED in a restaurant the shape of a clamshell, partially open to the night air. Every detail was enchanting, from the flickering candlelight to the waiters' crisp sleeves to the intricate arabesque designs on the dinner service. The guests were treated to modern takes on classic Arabian recipes like *saloona*, Qatari-edition *warak enab*, and a platter of *majboos*. The fruits and vegetables, a waiter announced, had been flown in from southern Europe that same day.

Looking around, Viktor sized up his fellow guests. The banquet seated just thirteen in total—six, including himself, were evidently the guests invited to the island from around the world. He recognized some familiar faces: a film star, an encryption artist, a neurobiologist, a mountaineer, and a poet—all of them celebrities. The rest seemed to be Qatari. Although they were dressed plainly, their white *thawbs* couldn't conceal the fact that these particular locals were a cut above the common.

Princess Akilah was resplendent in a golden robe and fine jewelry. She struck a silver spoon on a wineglass, signaling for the guests' attention. She first apologized for the absence of her brother, the thirty-six-year-old crown prince Mahdi bin Hamad Al Thani, the chief designer of Al Saeida, expected to inherit the Qatari throne, who was feeling "unwell."

"My brother often says that no technology is good if it can't bring us happiness. We . . . he . . . Crown Prince Mahdi designed this island in hopes of finding man's ultimate path to happiness through the power of technology. And by coming to our island, you will enjoy the privilege of witnessing this unprecedented undertaking by the Qatari royal family."

Her crisp English accent was irresistible—and a little cold. Though he couldn't take his eyes off her, Viktor could tell that she was reciting a script.

"Your Highness, while I sincerely admire the crown prince's vision and determination, there remains an issue." It was the neurobiologist speaking. He cleared his throat before continuing. "We've already learnt that endogenous cannabinoid brings pleasure, dopamine controls the neural reward system, oxytocin enhances emotional connections, endorphins relieve pain, GABA fights anxiety, 5-HTP improves self-confidence, and adrenaline stimulates energy. But to date, we still haven't identified any neurotransmitter directly related to our sense of happiness."

The poet chimed in, raising his wineglass. "Whereas Emily Dickinson wrote 'How happy is the little Stone,' Raymond Carver wrote that happiness 'comes on unexpectedly.' So everyone has their own way of looking at it, right?"

"As far as I'm concerned, people are all trying to live happily, and they only vary in their skill and capacity for action. The best can even deceive themselves, and that's how people survive." The weary-looking actress sucked in a lungful of smoke from the hookah, held it a moment, then slowly exhaled.

After nodding patiently to everyone's opinions, Princess Akilah pounced. "This is exactly why you're here, because you all hold different attitudes toward happiness. But most importantly, you are all personally *not* happy."

"How dare you!" The poet leapt up, rattling the tableware. Red lights blinked in the eyes of several robots circling the table, and he quickly dropped back to his seat with an air of resentment.

"I've had enough of this madness!" the encryption artist erupted. "I thought I was coming here to discuss how to help ordinary people break through the trap of consumerism and seek a spiritual alternative. I never expected to be joining a billionaires' club."

Viktor couldn't hold his tongue anymore: "Relax, my friends. For mid-to-low-income people, wealth does bring a degree of happiness. But upon exceeding a certain critical level, it will deliver diminishing marginal benefit, and even have adverse effects. "

The princess nodded approvingly: "Daniel Kahneman puts the threshold at seventy-five thousand U.S. dollars."

"I personally doubt that number," responded Viktor.

"You're just speaking for the top one percent. All of you are. I refuse to be part of this farce, and I'm out of here!" The encryption artist threw aside his napkin and stood up.

There was a long silence. Everyone looked to Princess Akilah.

She smiled implacably, as if she was above any such momentary awkwardness. Rising gracefully and holding her wineglass aloft, she started to slowly circle the dining table.

"I assume that all of you read our terms when you arrived at the island. We've made it quite clear that except in case of serious injury or force majeure, no signatory is allowed to quit, and any early exit will be deemed a breach of contract. The extent of the penalty is synced to the total value of your assets but will be reduced as the experiment advances. In other words, should you quit now, you are choosing to make yourself penniless."

The encryption artist's face paled at the princess's remarks, his lips quivering. Like everyone else, he hadn't bothered to read the terms of service in any detail. A buzz broke out around the dining table even as the Qataris looked on, blank faced.

The mountaineer spoke up. "It's just like mountaineering. You'll fail unless you reach the summit. Even if you don't make it back alive, you would win the highest honor. So, Your Highness, when is this island adventure considered concluded?"

As Princess Akilah passed Viktor, she bent down to softly clink her wineglass with his.

"When the island deems that you've succeeded in finding happiness, it will be time for you to leave Al Saeida."

As Akilah passed by, it occurred to Viktor that the princess was the woman in black he'd spotted lurking outside his holiday house.

LIFE ON THE ISLAND proved to be more interesting than Viktor had expected, although his neighbors weren't as pleasant as Qareen had made them out to be.

Each time Viktor encountered one of the other guests, they would greet each other courteously and make small talk.

Viktor found these conversations rather boring—with one exception. Each time Viktor encountered Akilah on one of his strolls around the island, they fell into engrossing conversation. Through these exchanges, Viktor learned that Princess Akilah had earned a doctorate of psychology from the Institute of Psychiatry, Psychology and Neuroscience at King's College in London, majoring in the psychology of happiness.

"So this is why your brother chose you as his deputy?" Viktor asked during one of their encounters.

"Well, not exactly. Do you want the truth? Mahdi is not here," admitted the princess awkwardly. "He's remotely monitoring everything that happens on the island to avoid interfering with the experiment. He has learned from experience that people will always deviate from their typical behavioral patterns in the presence of someone like him."

"From experience? So we're not the first group of guests?"

"Right. We've experimented with some locals. My brother is a bit obsessive about this, and he wants the experiment to cover different cultures, classes, and races. For him, this is the Qatari royal family's contribution to mankind's pursuit of happiness, so he's anxious for every detail to be as perfect as possible."

"It doesn't sound like you're very confident about the project."

"Well, my brother and I have some minor differences of opinion." The princess paused briefly before adding, "Next week there'll be a show at the central theater. I hope you'll join me? We can talk more about it then. Meanwhile, don't forget to ask Qareen to tell you more about the island's middleware technology."

Smiling at the princess, Viktor raised his glass of whiskey and drained it.

OVER THE NEXT FEW DAYS, during Viktor's strolls around the island, Qareen, like a competent museum guide, detailed the development of middleware technology.

Over the past thirty years, countries had tried various means to restrict tech giants' swelling digital hegemony, by tightening governmental regulations, enforcing antitrust breakups, and promoting privacy protection laws to limited success. Then middleware gradually arose as another option.

"Middleware is the most promising way out," Qareen explained while guiding its master to the central theater. The robot servant's capacity to comprehend and generate natural language was so outstanding that Viktor would often forget that he was talking to a heap of silicon and iron.

"Why do you say that?"

"Look around you. All these buildings, devices, and services are instantaneously changing their parameters, just for you. Without middleware, we would have no way to capture your data—stored as it is on scattered platforms—to let AI use it to provide maximal satisfaction of your needs and desires."

For the past twenty years, a growing number of open-source communities and blockchain companies had been working to develop a middleware AI system that combined the benefits of distributed computing, open-source protocols, and federated learning. But to acquire sufficient data, a trustworthy mediating entity was needed. Through a strategy of "recentralization," Qatar's national AI plan had managed feats that remained beyond the reach of commercial platforms. Al Saeida's role was to connect data on all major platforms through such middleware, offering protection—and perfection—to its users.

And it was true: The island's AI vastly surpassed that of any product Viktor had ever experienced. In his previous business life, Viktor had required massive quantities of data to make even the smallest strategic business decisions, but here, everything had returned to its simplest origin—one's own feelings. The wallpaper in his rooms changed pattern to fit his mood, the running trails would guide him on different routes to avoid repetitive views, and the waiters would recommend dishes that suited his tastes while retaining an element of surprise. Notifications popped up on his smartstream that provided information on topics of interest—seemingly at the precise moment Viktor became curious about something. All this pleased Viktor enormously. He was so pleased that he had set aside his misgivings about the fact that, since his arrival on the island, his every sentence, facial expression, and gesture were captured by the ubiquitous cameras and sensors, relayed to AI for interpretation, and fed back to the environment.

"The island now knows me better than my shrink," Viktor said to Qareen. "It's comfortable, but it's not exactly fun. It might even be a little . . . boring?"

"That all depends on how the middleware's objective functions are set, which is the reason you're here. Ah, we've arrived at our destination!"

Viktor stared at the robot. They were outside the doors of the central theater.

QAREEN GUIDED VIKTOR TO a VIP box, where Princess Akilah was already seated, dressed for the occasion in a violet robe.

"Please be seated, Mr. Solokov."

"Just call me Viktor, Your Highness."

"Very well, Viktor. I hope you enjoy tonight's show." Akilah passed him a pair of XR glasses, crafted of metal and leather shaped to suggest the Arab falconry hood.

Viktor put them on. "Just the two of us?"

"Here in Al Saeida, everything is tailored just for you."

A troupe of performers in traditional Arabian garb took the stage, dancing *ardah* to the *al-ras* drumming. The performance of the night was "The Man Who Never Laughed Again," a classic tale from *The Thousand and One Nights*.

The story revolved around the son of a wealthy man. When his father dies, the son gives himself over to feasting and debauchery, till he squanders his inheritance and is reduced to working menial jobs.

One day, as the teenager sits against a wall, an old man, ugly but well-dressed, asks him if he would be willing to serve some old men at his residence in return for good wages. The boy agrees.

But the old man imposes one peculiar condition: *"If thou see us weep, that thou question us not of the cause of our weeping."*

Curious though this seems, the boy agrees. He follows the old man to a large house surrounded by elegant fountains and a verdant garden. Inside, he finds ten old men dressed in funeral clothes, weeping. The boy nearly asks the reason for their distress but remembers the condition and holds his tongue.

Through the XR glasses, Viktor was able to see the superimposed virtual backdrop changing as the plot developed on the stage. The performers' actions triggered various animation effects. Their Arabic lyrics were simultaneously translated into Russian captions that floated in midair, retaining the original Arabian flavor.

Viktor couldn't help but turn to Akilah: "This is amazing!"

The princess put a finger to her lips, signaling him to keep watching.

In the narrative, twelve years pass. The boy grows into a young man, and death claims the elders one by one, till there remains only the man who had hired him. Eventually, this one also falls sick, and as he lies at death's door, the young man goes to him and asks the cause of the old men's weeping and wailing.

"My son, I have vowed to Allah that I would acquaint none of His creatures with this, lest the hearer too be afflicted with what befell me and my comrades," the old man answers, pointing to a tightly locked door. *"If thou desire to be delivered from that into which we fell, open not yonder door. Open it and thou shalt learn the cause of that which thou hast seen us do; and knowing it, thou wilt repent, when repentance will avail thee not."*

With these words, the old man breathes his last. The young man buries him with the others and lives alone in the house till one day, as he sits pondering the last words of his dead master, his curiosity gets the better of him. He breaks the locks and opens the door.

```
Would you do the same if it were you?
```

Viktor noticed this virtual caption suddenly appear in the air above the stage, and just as quickly disappear. It was obviously not an official subtitle.

Viktor looked to the princess, flabbergasted. She wasn't speaking, but her throat vibrated as more virtual captions swam across the air.

```
It's me, Akilah, talking to you. It's the
only way that we can evade surveillance. Turn
around, look natural, and pick up your wine-
glass. There is a silicon film on the surface
of the liquid. Use your tongue to stick it to
the roof of your mouth and try speaking with-
out moving your lips. The film will convert
the electrical signals of your throat muscles
into text via an algorithm that can pick out
```

what you want to say—quite accurately, for
the most part.

Viktor followed the princess's instructions, finding it more difficult than she'd made it sound. At first his attempts amounted to
a meaningless word salad, but slowly, he got the hang of it; choosing common monosyllabic words improved the signal-text conversion accuracy.

The action continued onstage, where the young man is walking
through the door and down a bizarrely twisting passage before finally emerging onto the shore of a vast ocean. As he stands in awe
by the sea, a great eagle swoops down on him, seizing him in its
talons, flying away with him over the sea, and then casting him
down, dazed and bewildered, on an island. Days pass, and the
young man falls into despair, thinking that he will die on the de-
serted island. But one day, a vessel appears on the horizon.

Why are you doing this?

Long story short, Mahdi's algorithm can't
make you happy. He won't accept it, but I
know. Maximization of the objective functions
will just turn you into a hedonistic guinea
pig, always wanting more but never getting
anywhere.

Why don't you just tell him?

It's not that simple. Surely, you know the
challenges women still face in my country.
I know Mahdi too well. He'll never accept my
opinion of it.

Hearing this, Viktor recalled the cold performance of the princess at the banquet, which finally made perfect sense.

Viktor looked once again at the stage through his XR glasses. In the story, a ship built of ivory and ebony has been moored on the young man's island. In the ship are ten damsels of stunning beauty, who invite him onboard and sail to another land. They arrive to find the shore full of troops, each soldier magnificently arrayed and clad in full armor. The young man mounts a horse saddled with gold inlaid with precious stones, and rides to a palace under military escort. A king approaches, extending him a courteous welcome to the royal residence.

In the palace, the king invites the young man to sit down on a throne of gold, and removes his helmet to reveal his face. The king, the young man sees, is actually a beautiful and refined young lady. She tells him: *"I am the queen of this country. All the troops thou hast seen, whether on horse or foot, are women, for in our state, the men delve and sow and reap and occupy themselves with the tillage of the earth and other mechanical crafts and arts, whilst the women govern and fill the great offices of state and bear arms."* This speech astonishes the young man greatly.

What do you want me to do? And . . . why me?

When I was volunteering at London's Maudsley Hospital, I picked up a skill from doctors there—not for treatment, but for choosing patients. They always chose patients who showed a high degree of cooperation, those more likely to accept hints, and who had truly hit bottom. That way, the patients would see effects of their treatment quickly and create a positive spiral of recovery.

So that's why you chose me? That doesn't sound like a compliment.

Viktor, what you've said proves you're a unique person. You want to hop off the treadmill, and that resolution is critical to acquiring happiness.

But how can I do that?

A new algorithm. Mahdi chooses to let AI
keep satisfying your every sensual need and
desire. I choose to believe that happiness
is not so simple.

I'm all ears.

The show continued onstage: The queen orders the *vizieress,* a gray-haired old woman of venerable and majestic aspect, to fetch the qadi and the witnesses. She then turns to the young man. *"Art thou content to take me as your wife?"*

Astonished by the queen's bold proposition, the young man kneels to kiss the earth at her feet, saying, *"My lady, I am merely the least of thy servants."*

The queen points to the servants and soldiers, as well as the riches and treasures in front of them, saying: *"All these attendants will serve at thy pleasure, and all my treasures will be yours, save for . . ."* She gestures toward a closed door. *"Yonder door shalt thou not open, else wilt thou repent, when repentance shall avail thee not."*

Hardly has she stopped speaking when the *vizieress* reenters, followed by the qadi and the witnesses. They conduct the marriage ceremony and the queen orders a great wedding feast to entertain all her guests and troops.

A little history. Back in the 1970s, American
psychologist Philip Brickman performed an
experiment. He brought a group of lottery
winners together with a group of people
paralyzed by accidents, and used one-on-one
interviews to evaluate their level of
happiness. What do you think the results
were?

Not much difference?

Bingo! The lottery winners were no happier than the control group. While accident victims were less happy at the time of evaluation, their hopes for future happiness were no different from the controls.

How is that possible?

The brain measured its level of sensory stimulation against the level of stimulation it is already used to. The thrill of winning the lottery resulted in a significant upward shift in the winners' adaptation level, so they were less likely to find pleasure from the ups and downs of their everyday lives. And vice versa.

It's a fine theory. But what can be done about it?

You know Maslow's hierarchy of needs? Mahdi's algorithm probably works on people who haven't achieved the requirements at the bottom of Maslow's pyramid. But as people start to need love and belonging, esteem, and self-actualization, the algorithm won't be able to help. You're a perfect example.

I thought I was already standing on the capstone of the social pyramid.

To be honest, Viktor, our AI predicts that your suicide probability within the next two years is as high as 87.14 percent.

Silence enveloped Viktor, but something told him that the princess was telling the truth.

Onstage, the show continued, revealing the happy life of the young man and his queen over the next seven years, as the young man enters middle age. One day, however, the man thinks of the forbidden door, and says to himself: "There must be even more exquisite treasures hidden inside, otherwise, why would she forbid me to open the door?"

So he rises from his gold-and-gem-encrusted bed and breaks all the locks to open the forbidden door.

So your algorithm can help me? But how?

Only AI can know the unique psychological signature of each individual. We hope to discover more happiness-related biomarkers and add more diversified metrics to measure satisfaction—things like what makes people feel challenged, or gives them a sense of purpose, a more profound understanding of interpersonal relationships . . . But only if you agree to participate.

I'm not sure. This sounds risky.

Help me, and you'll help yourself. Time is running out. You don't know what awaits you.

The captions suddenly ceased, and the performers froze midgesture, as though someone had hit the pause button. Viktor realized that they, too, were robots.

"Here they come," murmured the princess. Her voice was nervous.

In the blink of an eye, the theater was lit up like daytime. As Viktor started to rise from his seat, the door burst open.

The intruders were his fellow guests, who looked to be in no mood for a show.

It seemed the encryption artist had succeeded in hacking his robot servant and putting him under his own control. Moreover, he had convinced the other guests on Al Saeida to reverse their roles and gain the upper hand over their hosts.

The mutinous robot stationed itself belligerently right beside the smashed door.

"We demand to be released from our contracts!" the encryption artist yelled at the princess.

"You're free to go anytime—as long as you pay the penalty," Akilah replied, impassively.

"We won't pay . . . anything at all. This wretched island . . . did nothing to make me happy!" the star actress slurred, sounding dazed. With AI's support, she had been building her alcohol tolerance.

The howling poet pulled at his hair above his red-rimmed eyes. "This island is like a monstrous Aladdin's lamp. All of our wishes can come true, but there's no inspiration or excitement. When everything is possible, nothing is interesting. I can't write anything at all, not even a dirty limerick!"

"The first time I ate white dessert truffle, it tasted like the food of the gods," said the mountaineer. "But at the second and third tasting, it was more and more insipid. I know it's not Terfeziaceae's problem, it's mine. Twenty years ago, when Qataris needed a drinking license to take a sip, that very sip alone would make you high. But now look at these drunkards." The mountaineer glanced at the star actress, not bothering to conceal his scorn.

Akilah and Viktor traded glances. The princess was right: Mahdi's algorithm was able to satisfy users' surface-level desires by spoiling them, but it couldn't provide sustainable happiness.

"You and your brother have made a bold attempt, using one black box to understand another. But you failed," said the neurobiologist, crestfallen. "We're still far from true happiness."

"Therefore we, as the victims of this failed experiment, deserve unconditional rescission," concluded the encryption artist.

"Plus compensation," added the muddled star actress.

Gripped by the sudden impulse to speak up, Viktor was held back by the princess, who was shaking her head.

"I'm awfully sorry that you haven't been able to attain happiness on Al Saeida. But as you've been aware all along, your data was imported into the middleware system in an encrypted form, and was automatically executed through smart contracts that no one is able to tamper with or destroy. That's how our system was designed to work."

"We demand to meet the real Big Brother. Why doesn't your brother appear?!" questioned the mountaineer.

"Mahdi has urgent business to attend to, so he has entrusted me—"

"This is an out-and-out scam. I will tell Al Jazeera and let them expose everything!" The poet raised his voice.

"Don't forget that you've also signed an NDA."

"It seems that we have to do this the hard way," said the encryption artist. "Jinn, seize Her Highness." At the artist's barked order, the robot turned to Akilah, taking clumsy strides toward her.

Viktor boldly inserted himself between the robot and the princess: "Hey! Everybody, calm down."

"What's got into you, Russian? Planning to marry into the royal family?"

"I just . . ." Viktor hesitated, not knowing how to explain.

"It's all right, Viktor. Al Saeida will protect me from all harm." Princess Akilah walked calmly to the robot, dwarfed by its looming bulk.

"As long as you cooperate, no harm will come to you," promised the encryption artist. "All right, let's go to the dock."

Escorted by the robot and followed by the others, the princess walked out of the theater. From afar, the group saw Doha's harbor, ablaze with high-power lamps, and the Museum of Islamic Art, afloat on the sea like a glowing iceberg. But opposite this extraordinary nightscape, a royal abduction was in progress.

Viktor racked his brain for ways to free Akilah. Before he could come up with a plan, he saw the princess's throat once again faintly vibrate, and almost instantly, caption lines appeared in his XR glasses.

```
At the count of three, you must drop to the
ground.
```

Viktor suddenly sensed something peculiar in the night sky, as if the constellations were changing their shapes and lowering. From a distance came a sound like hummingbirds.

As the captions counted from one to three, Viktor flattened himself. On covering his head with both hands, his eyes registered a flash of blue-and-white lightning. At the shockwave, everyone crumpled to the ground, all except the princess.

Helping Viktor to his feet, Akilah explained: "Don't worry. That was just an electric shock. They'll come to themselves in a few hours."

"How did you do that?"

"Fixed-wing UAV swarms."

The explanation reminded Viktor of the veil structure he'd seen when first setting foot on the island; now he understood why they'd seemed able to defy gravity.

"How do you plan on dealing with them?" Viktor motioned to the lifeless guests lying stunned on the ground.

"After daybreak, they will be transferred to Doha and tried according to local law. As for you . . . you may do what you will."

Viktor exhaled deeply. What had happened had made him re-evaluate Akilah's offer. He didn't want to become the guinea pig of a failed experiment, but he couldn't return to his old life, either. He had no choice.

"I fully accept what you propose."

THE ARCHITECTURE OF THE MIDDLEWARE SYSTEM allowed two sets of algorithms to operate in tandem, like two currents in the same ocean.

Viktor was still enjoying all the conveniences that Al Saeida afforded, though occasionally he would sense a lurking force trying to tease him, like a mischievous kid hiding around the corner: Annoying music would suddenly play; his smartstream would push a negative news story about Viktor's company; Qareen would become unexpectedly stupid and slow, and even bungle or contradict orders; and running trails would guide him into a boggy stretch of mud, to name only a few new annoyances.

He guessed it was the "challenge" feature that Akilah was experimenting with in the middleware AI.

As erratic as it had become, the AI had created many opportunities for Viktor to spend time with the princess. As they talked and bickered about how to improve the system, Viktor felt a kind of happiness. In his previous life, people in his retinue would behave with deep respect and humility toward him, or they would exude resentment. It had been a long time since he'd engaged in candid conversation.

A bond was forming between the two, sensed by AI—through those ubiquitous cameras and sensors, as well as the biosensor membrane on Viktor's skin—well before Viktor and the princess realized it. Micro-expressions and biomarkers never lie.

The new algorithm inspired Viktor to think about how he would apply the middleware system to his e-sports platform, as a way of upsetting centralized data monopoly and letting players experience anew the pure fun of video gaming. This would be a radical self-reinvention for the company, a great second act. But with his last public adventure having devolved into international scandal, Viktor was afraid. It was possible that such a revolution would bring everlasting infamy upon him and even put an end to his business empire altogether.

He shared this fear over drinks with Akilah, who shook her head, saying: "What you fear is not failure, but shame."

Viktor was dumbstruck. The princess had nailed it!

"Years of research have taught me one thing. The path toward self-actualization isn't always an upward journey. It's full of ups and downs."

"I don't quite get it."

"If you're overwhelmed by a sense of insecurity, you won't be able to find true love and a sense of belonging. Similarly, if overwhelmed by a fear of losing love, you can't attain true self-worth. Being on top of the mountain doesn't guarantee eternal happiness, because happiness is a dynamic process of constantly shedding low-level fears and conquering higher summits."

Viktor nodded his agreement. "What about you? What do you fear?"

The princess flashed her smile and peered into the distance. "I fear becoming the Akilah that Mahdi wants me to be. He loves me very much, but wants only that I live by his algorithms, like a fairy-tale princess, with no worries, only happiness. But I can't be like that. I want to bring true happiness to the world."

Helplessly shaking his head, Viktor raised his champagne glass and stopped Akilah from continuing.

"I don't think I can acquire happiness on this island, whether by AI's definition, or by my own."

They both fell silent. After a while, Akilah turned her head to address Viktor, as if she had just remembered something. "You still haven't seen the ending."

"What ending?"

"The ending of the show."

"Oh . . . 'The Man Who Never Laughed Again.' It sounds just like my own story." Viktor forced a smile. "So how does it end?"

"The man who married the queen breaks the promise he had made at the wedding. He opens the forbidden door, only to find the very bird that had brought him to the queen's island in the first place.

"The bird then seizes him in its talons, flies with him over the sea, and deposits him back on the distant shore from which it had first carried him off. Eventually, the man traces his way back to the house where he had dwelt with the old men. Looking at their tombs, the man finally understands that the same fate had befallen them, and that this was the cause of their weeping and mourning."

On hearing the ending, Viktor gazed into Akilah's eyes and tried to gather his thoughts.

"Such a sad story, isn't it?"

"Yes, indeed. People always make the same mistakes, and go back to square one," said Viktor with a sigh.

"Just like running on the treadmill."

"Maybe no one can really leave."

"You don't have faith that we can make you happy, do you, Viktor?" Akilah's eyes filled with concern and frustration.

Viktor shrugged and looked away, watching a distant sailboat smoothly gliding across the gulf.

The princess rose and departed without her normal courteous goodbye.

THE ADVENTURE ENDED JUST as suddenly as it had begun.

Viktor was informed by Qareen that he could leave Al Saeida that same night. He was booked on a red-eye flight to Moscow from Doha's Hamad International Airport. A speedboat would ferry him back to the mainland.

Princess Akilah didn't come to bid him farewell, but merely sent a recorded message through Qareen, leaving Viktor at a loss.

"I've done everything I can, and I hope you can understand." The princess looked very pale on the screen, as if she felt sorrow at the parting, and that cheered Viktor somewhat. "The others will be exonerated and granted their freedom, on the condition that they keep their mouths shut about everything that happened on the island."

The speedboat cut through the sea, trailing a long, white wake that pointed back to the island of happiness.

Viktor looked back at the cloud-like UAV swarms hovering darkly over Al Saeida and remembered Akilah's final remarks—it all was so unreal.

"I hope you acquire true happiness, Viktor. And hopefully Mahdi won't change his mind."

Change his mind? What did it mean? The thought disturbed Viktor.

HAMAD INTERNATIONAL AIRPORT PROVIDED all the amenities of first-rate terminal buildings. The departure lounge even featured a standard-size swimming pool and an indoor tropical garden studded with towering palm trees. Viktor was supposed to have plenty of time before boarding, but he headed to the counter to confirm his booking, just in case. His anxiety was eased by the Qatar Airways agent's soft smile—and then ratcheted up again when the search for his information took much longer than expected.

"Mr. Solokov, sorry to have kept you waiting. But the system indicates your ticket is now suspended, and you will need to contact your booking party."

Cursing under his breath, Viktor yanked out his smartstream to contact Akilah, but he had lost Internet connection. The screen was still displaying a news update that had appeared seconds ago: "Five foreign visitors have been sentenced for felony violations of local laws."

Had Mahdi changed his mind? His heart racing, Viktor was so frantically scanning his surroundings that he hadn't heard the agent's inquiry.

"Mr. Solokov, are you all right? I've contacted the airport staff, who will be here shortly to assist you."

Two charcoal-colored secubots, even more formidable than Qareen, were approaching rapidly. On seeing them, Viktor ignored the agent's blandishments and bolted from the terminal. Dodging oncoming traffic, he crossed several lanes to flag down a human-driven taxi.

"Good evening, sir. It's rare to have honest-to-goodness human passengers like you these days." The driver grinned at him. "Are you looking for fun somewhere? Legal or illegal, I'm your guy."

"Just drive, go!" Viktor roared. He fumbled for Khaled's business card. At that moment, he would believe only his own contacts.

As the engine roared to life, his smartstream flew out the taxi window and landed on the ground, flickering twice before going black.

IN SOUQ WAQIF'S LABYRINTH of alleys, the rough mud walls and exposed wooden beams seemed to transport Viktor back to ancient times, when Bedouin traders gathered at the market to sell jewelry, silverware, carpets, horses, and daily necessities. He was in no mood to relish the intoxicating night atmosphere, but the Russian got completely lost in the intermingled smells of hookah, bakhoor, honey, and dates, as well as the dizzying colored lights from mosaic lamps.

Separated from his smartstream, Viktor tried to rely on his own senses for directions. Anxious, he kept looking behind him, as if everyone turning a curious gaze on him could be one of Mahdi's underlings. He stumbled about the labyrinth of souvenir stalls before finally locating the old falconry shop illustrated on Khaled's business card. The Algerian driver who loved electronic music was also a part-time helper at the shop.

The shop's birds of prey were resting tranquilly on their individual nightstands, blind-folded but still a proud symbol of nomadic Arabian tradition. Each one was worth up to millions of Qatari riyals. As Viktor approached, the shop owner put his index finger to his lips to signal silence, and motioned for Viktor to wait while he called Khaled.

A few minutes later the Algerian driver with the booming voice appeared. Viktor explained what he needed. "So you want to cross the desert overnight? And enter UAE via its southwestern border? That doesn't sound like a good idea."

"The drive is only a bit over two hundred kilometers, nothing difficult for you at all. Someone will pick me up at the end, just like when I came."

"I'm not sure. It depends on . . ." Khaled rubbed his fingers in the universal gesture for money.

"You know us Russians," Viktor said, offering a disingenuous smile. "Money is no problem."

Khaled's all-terrain vehicle sped Viktor away from the bustling city of Doha. As they roared into the heart of the desert, a gigantic billboard loomed ahead, featuring an English-language slogan highlighted among all the Arabic text: THE FUTURE IS RESET. READY? Lost in thought, Viktor focused his eyes on the surrounding landscape as the lights of modern civilization diminished in the distance. The dunes extended beneath the moon like the swells of a tropical sea, while sand dust clawed at the windows of the speeding vehicle, creating a cocoon of white noise.

For once, Khaled's dashboard boom box sat silent, and the driver seemed jittery. "You know what? Those birds all have passports."

"What?"

"They're too precious to be smuggled out of Qatar."

"Oh."

Utterly drained, Viktor wanted only to drowse in rhythm with the bumpy ride. Just before he closed his eyes, though, the vehicle jolted to a stop as abruptly as if it had crashed into a brick wall.

"We're stuck." Khaled revved the engine over and over but couldn't reverse out. The tires, spinning uselessly, sprayed sand. "Could you help by getting out for a minute?"

The chill wind of the hostile desert night greeted Viktor as he climbed out of the car. He felt in his pockets for a smoke but found nothing but the taxi receipt. The swaying headlight beams illuminated particles hovering in the air, like streams of golden liquid.

"We're running out of time, Khaled," said Viktor, urging on the driver, still behind the wheel. "I don't want to be the first Russian ever to freeze to death in the desert."

"Sorry, Mr. Solokov."

"You needn't be. Just make it quick."

"Sorry," Khaled repeated, and the vehicle suddenly lowered to the ground. The smart tires deflated to increase ground grip, allowing the four-wheeler to exit the sand trap effortlessly. "I mean you no harm, but there's no way I can disobey them."

"What the hell—"

Viktor stood where he was, uncomprehending, as Khaled's vehicle made a sharp U-turn and accelerated into the night, back toward Doha. He ran after it for a few steps, but was blinded by the dust behind it and had to squat for a long round of coughing. When he reopened his eyes, the four-wheeler was nowhere in sight.

Viktor Solokov found himself alone in the featureless expanse of the desert. Cursing and roaring at first, he inhaled too much sand, and began to labor for breath. His voice gradually weakened and turned into sobbing. He tried to get his bearings back from the faint stars, like a Bedouin. He scanned the horizon for oases and scoured the surface for animal tracks. But soon he gave up and chose a path by vague recollection of the direction by which they'd entered the desert. The tire tracks had been obliterated by the sand, but he knew he had to keep going. He resolved to just keep putting one foot ahead of the other. By estimating the length of his aborted desert ride, he calculated that he should be no more than a few dozen kilometers from the Qatar-Saudi border, and convinced himself that it wasn't impossible—he could make it by covering the estimated distance before sunrise, well before the temperature would rise to 50 degrees Celsius and dehydrate him into unconsciousness.

Viktor lost all sense of how long he'd been trudging across the sands. He felt his throat burning, his eyes stinging with tears and dust, and his feet aching with every step. The dunes were all identical, and Viktor had to wonder if he was walking in circles. But he didn't dare to stop, even briefly.

THE DARK SKY DIMMED FURTHER, but Viktor was acutely aware that the sun would be lurking just below the horizon, waiting to deal his death blow. Scenes of his past played before his eyes, the classic precursor of death. Compared to the cruelty of ceasing to exist, all memories, even the most unpleasant ones, felt infinitely sweet.

Delirium was setting in. Viktor wanted to know what it all meant: how he had gone from a happiness-hunting trip to a lonely death in the wilderness.

A glimmer of light flashed over the horizon, but it was too late. Viktor Solokov stumbled, rolled down a dune, and lay in the sand, now warming with the morning sun. What was left of his will told him to stand up and move on, but his limbs wouldn't comply. He didn't want to die, not like this and not out here, but it seemed his time had come.

A familiar buzz descended in oscillating waves, calling back Viktor's consciousness from the brink. Was it a near-death hallucination? He struggled to turn his body to face the cloudless sky, where a spectacle hovered—a mirage? One moment the mysterious conveyance floated like a furled flying carpet, the next it looked like a boat without sails. Viktor's mouth worked silently, but he uttered no sound. *This must be the end.*

The flying carpet was in fact a passenger drone composed of a multitude of smaller fixed-wing unmanned aerial vehicles that were able to embody various configurations. Its landing whirled up a small dust storm, stinging Viktor's eyes shut. He felt only that he was lifted up and taken into a cool space, where an IV was inserted into his veins for the replenishment of water and electrolytes.

At long last, Viktor regained some vitality. He managed to open his eyes and was surprised to find himself looking into Princess Akilah's smiling face.

"Am I dead?"

"You're as healthy as a horse, Viktor. Just a little dehydrated."

"How . . . how did you find me?"

"Well, it was thanks to the sensors. In your clothes, shoes, body, and the desert. There's smart dust everywhere."

Viktor turned his head to look out the window, and saw the undulating desert glittering golden all around. Understanding began to dawn.

"So . . . this is also part of the algorithm?"

"Not completely. AI did help a little, but I'm the one that designed everything. I want to thank you."

"Why?"

"Your choice changed Mahdi's mind, not just about the algorithm, but about me as well. Would you care to join us? Your gaming platform can surely help optimize our algorithm."

Viktor hesitated before replying. "If my answer is no, will I be sentenced like the others?"

Taken aback, Akilah gave a brief silvery laugh.

"That was a piece of custom-fed fake news. All guests have returned to the island and are back to their usual tasks."

"Wait . . . so the guests I met were also . . . Oh, of course. I guess that explains their conversational skills. So, you really think this can help mankind achieve true happiness?"

"Look at yourself, Viktor, and tell me. How are you feeling now?" Akilah looked at Viktor with great tenderness and put her hand on his shoulder.

Viktor's mind went blank. The view outside the porthole had changed from desert to sea, and they were flying back to Al Saeida, the island of happiness. He burst into laughter, as if he had just understood the punch line of some cosmic joke. And as he laughed, tears ran down his face.

ANALYSIS

AI AND HAPPINESS,
GENERAL DATA PROTECTION REGULATION (GDPR),
PERSONAL DATA, PRIVACY COMPUTING USING
FEDERATED LEARNING AND
TRUSTED EXECUTION ENVIRONMENT (TEE)

In the earlier chapters, we covered short-term AI deep learning applications, such as optimizing financial metrics, classroom grades, health diagnostics, and the like. "Isle of Happiness" tackles a bigger question— and challenge: Can AI optimize our happiness? This is an incredibly complex and tough problem. The ambiguous outcome in this story suggests that AI's efforts to improve our happiness will still be a work in progress by 2041, with progress and early prototypes, but making no prediction about when, how, or even whether it will be solved.

Why is this problem so tough? I can think of four reasons.

First is the problem of definition. What is happiness exactly? There are countless theories of happiness, from Abraham Maslow's hierarchy of needs to Martin Seligman's positive psychology. Defining happiness will be even more complex by 2041, when society will have progressed, through AI technology, to a point where living standards for most if not all people are comfortable. Once people's basic needs are satisfied, what constitutes happiness? That definition may still be evolving around 2041.

The second challenge is the problem of measurement. Happiness is abstract, subjective, and individualistic. How can we quantify our happiness and continuously measure it? If we could measure it, how would AI guide our lives to be happy?

The third problem is data. To build powerful happiness-enabling AI, we will need extensive data, including the most personal forms of data. But where will this data be stored? General Data Protection Regulation

(GDPR) is a new standard gaining acceptance, and its goal is to have us each take our own data back under our control. Will GDPR accelerate or impede this grand quest for improving our happiness? What other approaches might be possible?

Finally, there is the question of safe storage. How can we find a trusted entity to store that data? History tells us that trust is possible only if that entity's interest is fully aligned with the users'. How would such an interest-aligned entity be found or created?

Now you can see why happiness-inducing AI is extremely hard! Let's dig into the four problems and possible solutions.

WHAT IS HAPPINESS IN THE ERA OF AI?

Setting aside AI for the moment, let's ask the most basic question: What does happiness mean anyway? In 1943, Abraham Maslow published his seminal paper "A Theory of Human Motivation," which described what is now known as "Maslow's hierarchy of needs." This theory is usually illustrated as a pyramid, shown below. This pyramid describes human needs from the most basic to the most advanced level. Each

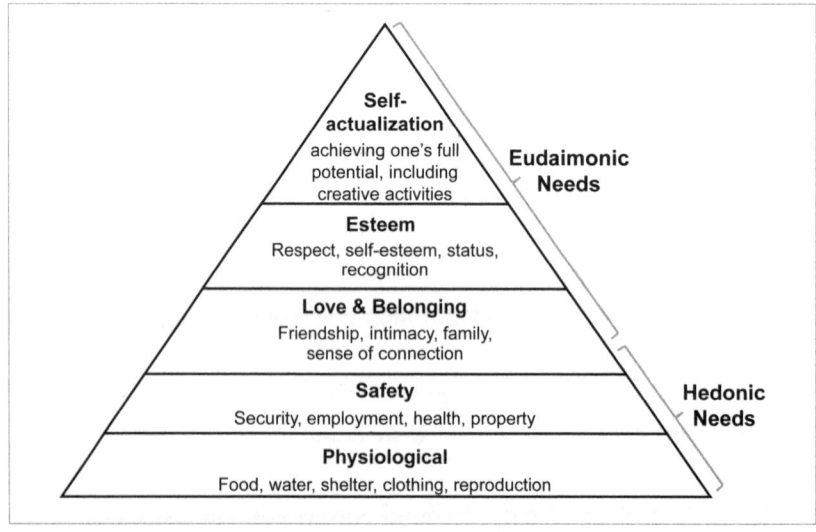

Maslow's hierarchy of needs—our happiness grows from the bottom up, as more basic needs are satisfied.

lower-level need must be fulfilled in order to move toward a higher-level need. The levels are "physiological," "safety," "belonging and love," "social needs" or "esteem," and "self-actualization."

Today, many people feel material wealth is the most significant component of happiness. Material wealth is mostly related to the bottom two layers of the pyramid—where sustenance or financial security are ensured by material wealth. Some people even associate material wealth with higher-level needs like power, esteem, and sense of accomplishment. But interestingly, research suggests that chasing material wealth cannot produce sustained feelings of happiness.

Psychologist Michael Eysenck introduced the term "hedonic treadmill" to describe our tendency to always readjust to a fixed level of happiness, despite monetary and possession gains (or losses). Studies have shown that people who come into sizable wealth (such as winning a lottery) are happy for a few months, but after that, their happiness usually drops down to their baseline level before coming into wealth. This is what doomed Crown Prince Mahdi's quixotic attempt to build his AI-enabled paradise in "Isle of Happiness"—his AI aimed to improve the guests' "hedonic happiness." When they first arrived on the island, the guests indulged in various pleasurable activities that produced short-term bursts of happy feelings, but over time, they were back on the hedonic treadmill, always treading, but never achieving lasting happiness.

In contrast with hedonic happiness (material wealth, pleasure, enjoyment, comfort), people who advance above the bottom two levels of the Maslow hierarchy pursue eudaimonic happiness (growth, meaning, authenticity, excellence). Maslow's hierarchy states that only after the levels of hedonic happiness are satisfied can people move up to eudaimonic happiness. In other words, once our material needs are satisfied, we will seek to belong, to love and be loved, to be respected, and to be self-actualized. This is why Princess Akilah wanted to replace Mahdi's hedonic AI with her eudaimonic AI—to help deliver to each person individualized happiness that is more experiential and purposeful, with love and authenticity.

It was in this context that this eclectic group of high-achieving visitors were invited to take part in the experiment, by becoming inhabitants of the island. Take Viktor. Before coming to the island, the successful entrepreneur was stuck on his hedonic treadmill. While he had achieved material wealth, success, and esteem, something was missing from his life. He was far from self-actualized and had sought refuge in mind-altering substances and other hedonistic escapes. These circumstances made him an ideal candidate for Akilah to recruit him to the island, where she would try to elevate him to eudaimonic happiness.

As the AI got to know Viktor during his stay on the island, he was given opportunities to build a relationship with Akilah. He was also put in situations that could satisfy his innate desire for adventure and offer boosts of self-esteem. He was offered the chance to seek self-actualization by using his game-designing skills to improve the Isle of Happiness. Viktor's goals were uniquely his, and AI tailored opportunities for him by understanding him and those goals. Whereas Viktor sought adventure, another person might have preferred just the opposite—serenity, for example—and for that AI would propose completely different experiences. By the end of the story, we knew Viktor would be happy, not because he possessed more material wealth, but because he was leading the life he wanted, growing his relationships with others and getting a chance to do important work that might help people. For him, happiness was not a binary state, but an ongoing pursuit.

Like the other stories in the book, "Isle of Happiness" is set in 2041. By then, societies will be richer, thanks to technological advancements, with AI taking over routine tasks, and robotics and 3D printing producing goods for next to nothing (this concept is known as plenitude; I explore it in more detail in chapter 10). If society is governed by good leaders, government will take care of all the people, assuring them material sufficiency. By 2041, in wealthier societies, people will find that their definition of happiness is evolving, as people advance from hedonic happiness to eudaimonic happiness.

HOW CAN AI MEASURE AND IMPROVE OUR HAPPINESS?

In order to build an AI to maximize happiness, we have to first learn how to measure it. I can envision three ways to do this, using technologies that are within reach today. The first one is extremely simple—we simply ask people. In the story, as the new inhabitants arrived on the island, they were required to answer a series of questions. Taking stock of people's happiness by asking questions is possibly the most reliable measure, but it cannot be done continuously, so there must be other measures as well.

The second way to measure happiness would consist of using the ever-advancing technologies of IoT devices (cameras, microphones, motion detection, temperature/humidity sensors, and so on) to capture user behavior, facial expressions, and voice, and then using "affective computing" techniques to recognize each user's emotions as determined from the IoT data. Observing people's faces, affective computing algorithms can detect both macro-expressions (usually within 0.5–4 seconds) as well as micro-expressions (0.03–0.1 seconds). These expressions reveal emotions. Micro-expressions are often detected when people try to conceal their emotions, and because they are extremely short-lived, humans usually miss them, while affective computing algorithms can recognize them accurately.

Other useful physical features to estimate an inhabitant's emotions include the hue of different parts of the face, which is caused by localized blood flow, and the pitch, loudness, tempo, emphasis, and stability of the voice. In addition, the trembling of the hand, dilation of the pupil, welling up of tears, patterns of blinking, humidity of skin (pre-sweating), and changes in body temperature are all useful features by which to estimate someone's state of mind.

With so many features, AI will be able to detect human emotions (happy, sad, disgusted, surprised, angered, or fearful) much more accurately than people can. This recognition can be further enhanced by watching multiple people throughout time. In the story, for example, AI observed that both Viktor and Princess Akilah were developing feelings for each other. This can lead AI to score them both higher for "belongingness and love needs" on the Maslow hierarchy. AI's ability to recognize human emotions already exceeds the average human, and this gap will grow much wider by 2041. Please note that this capacity does not imply that AI can convincingly show emotions or that it feels any emotions.

The third way to measure happiness is to continuously check levels of hormones that correlate with particular sensations and feelings. In this story, each inhabitant wears a transdermal biosensor membrane with a matrix of under-the-skin microneedles and an electrochemical sensor that continuously measures hormone levels as partial measures of happiness. For example, serotonin is correlated with well-being and confidence, dopamine with pleasure and motivation, oxytocin with love and trust, endorphins with bliss and relaxation, and adrenaline with energy.

Monitoring these features, the island's AI was able to note the activities, measures, and environments when an inhabitant was happy, and use these happy moments to train itself to recognize happiness. Then, the AI assistant Qareen could make recommendations or suggestions for activities or choices that would lead to more happiness (achievement, growth, or connection), or less unhappiness (sadness, frustration, or anger). Toward the end of the story, when Viktor was instructed to leave the island and go home, it was not because the experiment had ended, but because AI knew that by ending the game in this particular way, Viktor would opt to escape, because he loved adventure, and that experience would eventually bring him back to the island and make him happier.

In order to build a scientifically rigorous and robust happiness-optimizing engine, researchers will need to solve daunting challenges. First, what kind of happiness metrics can we use? We have some approximations above, but we know that our state of mind depends on unknown combinations of electrical (brain waves), architectural (brain structures), and chemical (hormone) components working in concert. The approximations above capture only some hormone levels, which are useful but surely incomplete. They do not tap into the electrical or architectural measures. Over time, we will need to read all three components and understand their interactions and causation for happiness in order to improve the training data quality for AI to learn.

Second, achieving higher levels of the Maslow hierarchy doesn't involve moments of instant gratification, but rather the long-term pursuit of meaning and purpose. AI learning across a long time span is challenging, because when a person's happiness goes up, the AI does not know whether it was a result of today's activities, or last week's, or last year's, or some combination thereof. This problem is akin to a challenge facing social media algorithms: How can Facebook train its newsfeed to help a user grow over the longer term, rather than simply entice immediate advertising clicks? When the person shows growth, how does the Facebook AI know which day's content or algorithms caused that growth? We will need to invent new AI algorithms to learn long-term stimulus-response amid a lot of noise.

By 2041, we will not gain a full understanding of what determines our state of mind, nor will we know how long-term eudaimonic happiness works. But by that time, AI's ability to read human emotions should be quite advanced, well beyond human capabilities, and there should be prototypes that try to improve human higher-level happiness.

DATA FOR AI: DECENTRALIZED VS. CENTRALIZED

Aggregation of data is a necessary step to building powerful AI. This is already happening today at giant Internet companies. Google knows everything you've ever searched, every place you've been to (through Android analytics and Google Maps, unless you turned off location history), every video you've watched, every email you've sent, everyone you've called on Google Voice, and every meeting you've scheduled in a Google calendar. Trained on this data, Google can deliver tailored services that are incredibly convenient for you. Google and Facebook have access to so much data, they can infer your home address, ethnicity, sexual orientation, and even what makes you angry. They can guess your innermost secrets, whether you cheated on your taxes, are an alcoholic, or had an extramarital affair. These inferences will have a fair amount of errors, but even the notion that these companies have the tools and your data to attempt to guess likely makes one uneasy.

These privacy concerns have led to discussions about government action. Countries ranging from the United States to China are looking at whether the power of the data has strengthened Internet companies into monopolies, and if so how to use antitrust laws to curb their power. Europe took action much earlier—the EU decided to put a stake into the ground on personal data by introducing GDPR (General Data Protection Regulation) terms, which the EU calls "the toughest privacy and security law in the world." Other countries are evaluating building their data laws with GDPR as a foundation. GDPR is a big deal, and it got off to a good start.

GDPR has the vision of ultimately giving data back to the individual, so as to help people control who gets to see and use their data, and even derive value from licensing their data. In the first few years of GDPR implementation, the law has achieved some successes. It has succeeded in educating the masses on the significant risks about personal data. And GDPR has required websites and apps the whole world over to rethink and refactor their applications to minimize malicious, erroneous, or neglectful abuses of user data. There are large fines for companies that violate GDPR.

But some details of GDPR are not practical, and in general GDPR is an impediment to AI. In its current form, GDPR stipulates that companies must be transparent to people about how their data will be used. Users' explicit consent for a specific purpose is needed in order for a company to start collecting that user's data (for example, giving your address to Facebook only for the purpose of facilitating e-commerce order delivery). Data must be protected from unauthorized use, leak, or theft. Automated decisions should be explainable, and escalation to human intervention should be available upon users' request.

I believe that GDPR's goals (transparency, accountability, and confidentiality) are all well-intentioned and even noble. However, the current implementation described above is unlikely to achieve these goals and may even be counterproductive in many ways. For example, it is difficult to limit the purpose of each piece of data collected, because AI is a sprawling exercise, and it is unfeasible to enumerate all purposes for each piece of data collected when the collection begins. For example, when Gmail saved all your emails, it was to help you to search and find any email. But later, when Gmail developed the new auto-completion feature, it needed to train on the old data. It is also impractical to expect that users will grasp each company's data-usage explanations every time they are asked whether to consent to data collection. (How many times have you encountered a complex pop-up on a site and just clicked "OK" without understanding or even reading the text of the pop-up?)

GDPR requires giving users the right to escalate to humans if a user is concerned with AI decision-making. But human escalation may cause havoc, as humans are not as good as AI in decision-making. Finally, GDPR's goal of data minimization and data retention requirements will seriously handicap the AI systems.

When considered independently, most people would want to take back ownership of their personal data using GDPR and other regulations. But this must be looked at in light of the fact that if all the data is ripped out, then most software and apps would become "stupid," if not entirely dysfunctional. In the story "Isle of Happiness," we suggest that rather than throwing out the baby (AI services) with the bathwater (data privacy concerns), another option when technologies mature would be a "trusted AI" to which we would give all our data to safeguard, hide, or give out. If that "trusted AI" knew everything that Google, Facebook, and Amazon knew about us, and much more, it would deliver capabilities well beyond today's Internet services. The many data swamps that have our data will be unified into a powerful data ocean. And when this "trusted AI" (let's call it the Isle) knows everything about us, we can have it respond to all data requests for us. So when Spotify wants to know our location, or when Facebook wants our address, the Isle will decide on our behalf if the benefits of the service is worth the risks of providing the data, based on what it knows about our values and preferences, and the trustworthiness of the company making the request. This will get rid of all the consent-seeking pop-up windows that confuse and annoy us. The Isle would become not only a powerful AI assistant, but also our protector of data, and our interface to all the apps. One could think of this arrangement as essentially a new social contract for data.

WHO CAN BE TRUSTED TO STORE ALL OUR DATA?

How can we be sure that the Isle can be trusted with all our data? If we are suspicious of Google and Facebook, the Isle is even more frightening, because it has so much more data than Google or Facebook. Also, the data guesses our state of mind and emotions, even when we try to hide them. How can this possibly work?

The fundamental issue is that when the interest of the AI owner diverges from the interest of the AI users, the users lose. We saw this in many earlier chapters ("The Golden Elephant," "Gods Behind the Masks," "Quantum Genocide," and "The Job Savior"), and we read it about Google and Facebook everywhere. The crux of the problem is that Facebook and Google's AI objective functions are necessarily business optimizing because they are publicly traded companies, which causes them to optimize goals that we as users have no interest in optimizing. And it is a nonstarter to ask Google and Facebook to use our objective function, for the simple reason that their profits will plummet. In order to find an AI owner we can trust, we need to find an entity not pressured to optimize commercial interests—one that will naturally embrace our interests without reservations.

What entities might have interests that align with ours? "Isle of Happiness" uses the perhaps fanciful example of a benevolent monarchy of a small wealthy country. Such a notion might seem out of place in the twenty-first century, but the monarch in the story was modeled after Frederick the Great of Prussia, who said: "My principal occupation is . . . to enlighten minds, cultivate morality, and to make people as happy as it suits human nature, and as the means at my disposal permit." An enlightened monarch like Frederick the Great believes his mandate to govern is contingent on improving the lives of the subjects. Thus, a benevolent monarch has strong trust from his or her subjects and possesses the courage to implement major changes. In the seventeenth and eighteenth centuries, enlightened monarchs were the key catalysts to usher in the Age of Enlightenment. So as we look for a catalyst of a powerful trusted data-aggregating AI, it is not so unreasonable to think of a benevolent monarchy as a starting point. I also predict that in the next twenty years, small countries governed by strong leaders with the support of the populace are most likely to make breakthrough decisions in technology adoption.

I can imagine other possibilities. What about a twenty-first-century digital commune consisting of people who share common values and are willing to contribute their data to help all members of the commune, based on a common understanding of how members' data will be used and protected? Academic projects looking at experimenting with this are currently under way, starting with volunteer university faculty, staff, and students. Another possibility is the development of a nonprofit AI, similar to Wikipedia or the open-source movement. Finally, someone could construct a distributed blockchain network that is not controlled or influenced by any single individual or entity (like Bitcoin). Storing personal data in a distributed network is a tougher problem than that of storing bitcoins but may not be unsolvable. Each of these types of entities is much more likely than a publicly listed company to align itself to the users' interest.

Over time, there may also be technology solutions that will allow us to have our cake (powerful AI) and eat it, too (with data protection even from the AI owner). There is an emerging field called "privacy computing" that is researching ideas in this area. For example, federated learning is an AI technique that trains AI across multiple decentralized devices or servers holding local data samples. It approximates centralized training, while disallowing the central AI owner to see the data. Another method known as homomorphic encryption encrypts the data in a way the AI owner cannot decrypt. AI is trained directly on the encrypted data. This doesn't work on deep learning yet, but future breakthroughs are possible. Finally, a TEE (trusted execution environment) reads encrypted and protected data, and decrypts the data for AI training on a chip in a way that guarantees that the decrypted data will not ever leave the chip. (One risk with TEE is that a chip company could put in a back door.) Each of these technologies still has bottlenecks or technical issues that prevent them from building powerful AI while fully protecting personal data. But over the next twenty years, with increasing scrutiny on data issues, I would anticipate significant progress in using privacy computing technologies to protect personal data. As the story suggests, privacy computing will likely not be totally pervasive by 2041, but these technologies will be mature enough to be applied to scenarios like in "Isle of Happiness."

For the skeptics out there, understand that the approaches proposed here are not a panacea, but possible paths that I believe we need to explore, along with GDPR and other methods. We humans have had so little experience with something as powerful as AI, and something as challenging as protecting so much data, that we must have an open mind about the solutions and balance thoughtful experimentation with preserving the status quo.

And if you still think giving our most valuable data to a third party is ludicrous, think about how most of us store our most valuable physical possessions with a secured third party, such as a bank safe deposit box. We also entrust our stocks to brokerage firms, and our bitcoins to the Internet. Why can't we do the same with our data? If we could give all of our data to a trusted entity that is interest aligned with us, then we could enjoy the most powerful AI to help us find lasting happiness, and we would no longer have to be puzzled with whether to consent to data uses by myriad applications, nor would we have to be worried about data theft and misuse. Whether this trusted entity was a benevolent monarchy, an open-source commune, or a distributed blockchain system, we could reap unprecedented benefits from this powerful AI, while remaining hopeful that new technology advances would continually make our data ever more secure.

CHAPTER TEN

DREAMING OF PLENITUDE

STORY TRANSLATED BY EMILY JIN

THOSE WHO LOSE DREAMING ARE LOST.
—AUSTRALIAN ABORIGINAL PROVERB

NOTE FROM KAI-FU: AI and other technologies will drive down the cost of almost all goods, most of which will be produced for next to nothing. For the first time in human history, developed countries could eradicate poverty and hunger. If this happens, would money be phased out? If so, what would take money's place to motivate people to live purpose-filled lives? Would any economic theory apply anymore? This story, set in Australia, explores a futuristic society that has introduced two currencies for a post-scarcity world: a card that provides for citizens' basic needs, and a new virtual currency for building reputation and respect through service to the community. In my commentary I will discuss how plenitude nullifies economic theories, and explore what might be next after plenitude: singularity.

STANDING IN THE foyer, Keira looked her new surroundings up and down. The home's entryway was spacious yet cozy, with precious staghorn coral specimens and Aboriginal art arranged atop a console table made from reclaimed wood.

She had been hovering next to her suitcases in the foyer for quite some time. As she waited for the home's owner to appear, Keira tiptoed around the adjoining rooms, paying particular attention to the pictures lining the walls. Most were mementos of a life spent on the water, featuring a dark-haired woman with a lively smile, laughing as she posed with various marine animals aboard a research vessel floating in the Coral Sea.

The woman, Keira knew, was a younger Joanna Campbell. A famed marine ecologist, Campbell had spent her entire adult life researching the preservation of coral reefs. Now seventy-one, with no children or other relatives, she had moved here, to the home Keira was now standing in—a unit within a smart retirement community located outside Brisbane.

Officially named Sunshine Village, the retirement facility was called AI Village by the locals. Each unit in the community had been designed by AI and assembled from prefabricated modules by robots. Informed by data collected from Brisbane's elderly population, AI had tailored every door, window, cabinet, appliance, and toilet to optimize residents' use of the space. Sensors measured the habits and physiological indicators of those living in Sunshine Village as the complex's AI offered personalized suggestions for its residents on a daily basis.

As Keira surveyed the walls of Joanna Campbell's unit, a piece of brightly colored Aboriginal art caught her eye: a classic Papunya painting teeming with dots of different colors in a psychedelic, dreamy swirl. She was mesmerized. The image reminded her of her home, Alice Springs, a small town located in central Australia, wedged in between the MacDonnell Ranges. Using her XR glasses, Keira scanned the painting for its information and saved it to a folder named "Home."

"Everyone who's visited loves this painting. Isn't it beautiful?"

Keira nearly jumped at the sound of the hoarse voice behind her.

It was Joanna Campbell herself, in an electric wheelchair. With silver hair and a frame made diminutive from the passing of years, she certainly looked different from the robust, vibrant woman in the pictures. Still, Keira noted, the woman's eyes were just as bright and sharp, scrutinizing her visitor.

"Yes, Ms. Campbell, I am Keira. I believe that the Sunshine Village Resident Services team informed you that I would be arriving today?"

"Well, no one told me that you would let yourself in, young lady. Or should I call you 'young girl'? I can never figure out how old you people really are."

Blushing, Keira scrambled to explain herself. "I'm so sorry! I rang your doorbell several times, but no one answered, so I entered with the password that the Resident Services team gave me."

"I still don't understand why they can't just send a robot over," muttered Joanna. "The last caregiver they sent couldn't stop staring at my paintings. I saw greed in his eyes, so I made sure he didn't last long. You wouldn't consider doing something foolish with one of my belongings, would you, child? What's your name again?"

"Keira," responded the girl timidly. "And of course not. My job is to help take care of you, Ms. Campbell."

"Ha! Guess this is what happens when you're old—you're left at the mercy of other people. How long will you be staying for?" Contempt laced the old woman's voice.

"The wristband matched me to this job. I guess I'll be staying . . ." Keira raised her left hand and showed Joanna her flexible smart wristband, glowing with colored lights. "Until Jukurrpa decides that my task is complete," she answered carefully.

"Please speak in plain English," Joanna huffed.

"Oh! *Jukurrpa* means 'dreaming' in the Warlpiri language. You know, the Aboriginal origin myth and all that. To be honest, it seems like the government is paying a bit of lip service by giving the program an Aboriginal name," said Keira, her tone unimpressed. Then she brightened. "I heard so much about you before coming over. You are amazing!"

The reality was that when Keira had met with the community's medical director at the Resident Services office, he'd warned Keira that Joanna Campbell would be tough to deal with. All of her previous caregivers had quit because they couldn't stand her temper.

"Oh, yes, 'the dreaming project.' Now I remember. A funny name. They've told me about it many times but my memory isn't what it used to be," Joanna went on, ignoring Keira's compliment. "How much are they paying you to babysit me, again?"

"Well, Project Jukurrpa pays me in Moola, not cash."

"More young-people things I don't understand," said Joanna, cutting her off. "I suppose you don't celebrate Australia Day either?"

"Um . . ." Keira smiled awkwardly. "Due to the problematic history of January twenty-sixth, we voted to reschedule the national holiday ten years ago. Now Australia Day is May eighth—sounds like *mate*, doesn't it?"

"Ludicrous," Joanna said, waving a hand dismissively. She turned her wheelchair around and headed for the living room.

Keira stood, dazed, until Joanna's voice rang from the front of the house. "Kala! Come help me find my glasses. I can't read anything without them."

"Coming!" shouted Keira. She took a deep breath and followed Joanna into the room.

OVER THE PAST YEAR, Joanna's smart home had determined that she was exhibiting early symptoms of Alzheimer's disease. First there was the frenzy with which she had taken to opening and closing the refrigerator door, and the growing delay in locating misplaced items, like her keys. Names and faces had begun to elude her. With wisdom gleaned from the accumulated health data of millions of Australians, the signs were unmistakable to the Sunshine Village AI.

However advanced, the smart home itself could not compensate for the rate at which her mind was deteriorating. Joanna's doctor had advised that human companionship could help alleviate the symptoms. The Sunshine Village Resident Services team had requested a companion for Joanna from Jukurrpa—or, rather, a string of companions, of which Keira was the most recent iteration.

Keira was far from unique in taking on work as a caregiver. In 2041, Australians aged sixty-five and over made up 35 percent of the entire population. At the same time, the growth of AI and the corresponding job automation meant that the unemployment rate had also skyrocketed. Now the country's job reallocation program struggled to keep unemployment to its current 12 percent of the population.

The age group hit hardest by the employment upheaval were those under twenty-five. Most vulnerable of all, thanks in part to their long history of entrenched disadvantage, were young people within the Aboriginal population, whose members had fallen well below Australian averages in terms of education, employment, social mobility, and life expectancy.

Even as many of its residents struggled, Australia could hardly count itself as underdeveloped or lacking in innovation. Its abundance of natural resources and its "AI prioritization" national development strategy had turned Australia into a global leader in new energy, materials science, and health technology. The government relentlessly advocated for renewable energies like solar and wind, which, together with low-cost, high-capacity lithium-ion battery arrays, had driven the cost of energy down close to zero. The country had also succeeded in eliminating greenhouse gas emissions altogether, making Australia one of the first countries in the world to achieve carbon-neutral status. Aided by advancements in genomics and precision medicine, Australia's life expectancy was now 87.2 years.

These advances—combined with the country's stable financial system, awe-inspiring natural environment, and comprehensive welfare system—had attracted millions of immigrants, most of them wealthy people looking to retire in Australia.

Still, for all that the country's leaders had done to address big problems and turn Australia into a magnet for the global elite, the nation's persistent inequalities had incited the anger of its young people. In their eyes, Australia had become wealthy while failing them—and failing to bring about economic and social justice to marginalized groups. In the early 2030s, young people in Brisbane and other cities around the country—feeling overlooked and robbed of a prosperous future—had taken to the streets in mass outpourings of frustration. A wave of violence, crime, and conflict rippled out from these initially peaceful protests, and the turmoil had spread throughout the country.

In 2036, in response to the social unrest, the Australian government had launched Project Jukurrpa and declared that "Australia would take good care of her people." The project, spearheaded by ISA (Innovation and Science Australia), consisted of two parts. First was the introduction of the BLC, or Basic Life Card, which guaranteed that every citizen who opted in would receive a monthly allowance to cover the cost of food, shelter, utilities, transportation, health, and even basic entertainment and clothing. All thanks to the abundance of wealth and nearly free clean energy generated by the technological revolution.

The second part of the Jukurrpa program was the establishment of a virtual credit and reward system called Moola. The system rewarded citizens for voluntary community work, such as caring for children and keeping public spaces pristine. Participants' smart wristbands collected speech data from the volunteer work and quantified it with the help of AI. The score was predicated on variables including difficulty, contribution to community and culture, degree of innovation, and self-improvement, as well as the most important factor: the satisfaction of the person or community served. The data enabled the wristband to calculate the Moola earned by a participant in real time. Moola scores were reflected on participants' wristbands, with high scorers' bands beaming with an array of bright colors.

With Moola, the government had intended to establish honorable service, a sense of connection and belonging, rather than monetary wealth, as a true measure of an individual's value. In reality, Moola had more practical benefits, too, operating as a kind of credit score that supplanted other forms of currency. For instance, when evaluating candidates for a job opening, employers could choose to prioritize applicants with a higher Moola score. Those who earned the country's highest Moola scores were even entered into a competition for a chance to become a reserve member of the Mars base.

But the program didn't always function as its designers—and the country's leaders—intended. Despite the government's lofty ambitions, many young people treated Moola as just one more indicator of social status, looking to the colors on the wristband as simply another label to boast about, a symbolic replacement of wealth. Some young people even tried to game the system by bribing service recipients and conducting fake conversations and phony interactions to improve their Moola scores in the shortest time span possible.

The data also showed that among the groups enrolled in the program, the Moola growth rate for the Aboriginal population was significantly slower than the overall average. Project Jukurrpa, from its very first day, had come under public scrutiny regarding its potential to exacerbate racism. Because the Moola score depended on other community members' affirming the successful completion of participants' Moola-earning tasks, would Aboriginal and other nonwhite participants encounter bias and thus have a harder time building up credit?

The government defended Project Jukurrpa in the face of these criticisms. Dr. William Swartz, Jr., a spokesperson for the ISA, gave a press conference calling the project a forward-thinking social investment. "A society without love, belonging, justice, and respect will no doubt collapse. The core of Project Jukurrpa is about rebuilding trust in the younger generation. We believe that every person can achieve their dreams in this land of plenitude, regardless of their race and ethnicity."

Project Jukurrpa's first target: the unemployed population below the age of twenty-five, where Aboriginals made up a whopping 35 percent of the demographic. This far exceeded their proportion of the entire Australian population, which was a meager 5 percent.

Keira Namatjira, aged twenty-one, was one of the Aboriginals who signed up.

IT DIDN'T TAKE KEIRA long to grow accustomed to life at Sunshine Village. In Joanna, she may have been assigned a cranky charge, but other residents welcomed the Aboriginal girl with long, curly dark hair into their community, and many came to adore her. In addition to caring for Joanna, Keira frequently performed small acts of service for others in the community who weren't eligible for a full-time caregiver. When she assisted them by making deliveries, hanging laundry, or walking dogs, residents showered Keira with positive feedback and never hesitated to click "Confirm service" on her wristband. It would then flash varicolored lights and hum a melody, notifying her that new Moola had arrived.

Keira's day-to-day work at Joanna's place involved less instant gratification. In addition to helping Joanna with her daily routine, Keira was also responsible for conducting a comprehensive checkup of the old woman's cognitive functions according to the Resident Services medical guidelines. Joanna's truculence ensured Keira had her work cut out for her.

"Ms. Campbell, can you tell me about the article you read just now?" asked Keira one day as they sat together at Joanna's kitchen table.

"It's about endangered marine life. Why do you ask? Do your schools no longer offer reading comprehension?" Joanna glared at Keira from behind her reading glasses.

"Ms. Campbell, do you remember where you put your pill box?"

"You think you can baffle me? I put it . . . wait." Joanna fumbled through her pockets, then shouted with glee as she pulled the box out of her pocket, like a child who had discovered a hidden treat. "Ha, I knew it! In my pocket!"

"Ms. Campbell, do you remember what we had for lunch yesterday?"

Joanna gave Keira a look and frowned. "Soup, egg custard, salad, and fruit. Oh, right, there was also filet mignon. They told me that the meat was lab grown and no animals were harmed in the process. That's why I agreed to try it. It tasted exactly like the beef I remember. So don't take me for a fool, Ms. Koala."

Keira grimaced; still, by now, she had grown accustomed to the older woman's ways—and felt compassion for her declining cognitive abilities, even when they manifested in rude remarks. "Actually, yesterday you said you weren't hungry, so we skipped lunch. Also, my name is Keira, K-E-I-R-A."

Hearing this, Joanna didn't fire back in her usual way. She fell silent, a stunned look on her face. After a few minutes, she let out a sigh.

"I don't know what's happening to me. The doctor said my symptoms are not so severe, and I only have to wait," she murmured. All of a sudden, she raised her head again, and a glimmer of hope kindled in her eyes. "Do you know when I can get the procedure?"

Keira knew Joanna was referring to genomic precision therapy for early stages of Alzheimer's disease. However, even with Australia's comprehensive healthcare system, certain high-end medical therapies were hard to come by, given the sheer number of people demanding treatment. For genomic precision therapy, it would take months—maybe years—on the waiting list. Keira worried that when the time came for Joanna to receive treatment, the older woman's symptoms would have advanced to the point that the therapy would no longer have any effect.

"Soon, in a few weeks," reassured Keira, knowing Joanna wouldn't remember this conversation. "I'll be sure to remind you when the day comes."

"It's strange. I can't even remember what I had for lunch yesterday, but memories of my younger days are just as vivid as ever."

"Tell me what you remember," said Keira. Half-crouching and pressing her palms to Joanna's knees, she looked into her eyes encouragingly.

"I remember . . ." Joanna's gaze drifted over to the sunlit world outside her window and grew unfocused as her thoughts spread their wings, took off into the wind, and embarked on a voyage to another space-time.

1992. JOANNA WAS IN the prime of her youth, her skin tanned from long hours in the scorching sun and her hair bleached a shade lighter from the ocean. She would spend months at a time at sea on a research ship, studying climate change and water pollution in the ailing Great Barrier Reef ecosystem. The Coral Sea, an aquatic kingdom of 4,791,000 square kilometers located in the Pacific Ocean northeast of Queensland, was home to hundreds of millions of marine creatures. However, it had been dying a slow death as a result of rising temperatures, unsustainable fishing, and outbreaks of coral-eating crown-of-thorns starfish. To counteract the destruction of the Great Barrier Reef, Joanna was ready to do anything.

2004. After ending her marriage, Joanna gave her full attention over to her beloved ocean—her constant companion and what had come between her and her spouse. In June that year, after the Australian government refused to recognize same-sex marriage, a group of activists planted rainbow flags on one of the uninhabited Coral Sea Islands southeast of the Great Barrier Reef, declaring the place an independent haven, in an act of protest. All by herself, Joanna journeyed out to the group, hoping she could persuade them to vacate the islands on account of their vulnerable ecosystem. However, when she told the protesters that the third global bleaching event, a result of ocean warming, would destroy 40 percent of the Great Barrier Reef, she was rebuffed with cries of *Don't you care about diversity?*

2023. Joanna was no longer fighting the battle alone. Leading a team of scientists, she was researching technology that might improve the Great Barrier Reef's resilience to climate change. Joanna, now silver-haired, carefully studied the innovations that were being churned out by a new generation of marine science innovators. They were using underwater robots to plant coral larvae in designated areas pinpointed by AI algorithms, and relied on sensors to monitor growth; they covered the ocean surface of the Great Barrier Reef with an environmentally friendly film made from biomaterials in order to reduce the intensity of the sunlight hitting the reef. Joanna was also excited by a proposal to genetically engineer zooxanthellae, a microorganism that played a pivotal role in many symbiotic marine relationships. Ocean warming, along with acidification, was impacting the health of the zooxanthellae, triggering in turn coral bleaching and the death of anthozoan coral polyps. Invertebrates and fish that had built their lives on corals would either leave or perish. As a result, the ecosystem would collapse.

"IF WE COULD IMPROVE the resilience and adaptivity of the zooxanthellae," said Joanna to Keira, who was listening with rapt attention, "the corals would return to their original state, and regain their color, and the anthozoan coral polyps would get the nutrients that they need. We really thought it could save the Great Barrier Reef."

Joanna was a different person when she talked about her work. Her gaze was no longer dim; her memory was sharp and refreshed. As she spoke, she radiated vitality, as beautiful as a blooming coral bush.

"But you did it! Now everyone calls you 'the savior of the Great Barrier Reef'!" exclaimed Keira. "I can't even begin to imagine the difficulties you've been through . . ."

"Let me put it this way—the greatest difficulty doesn't come from the outside, but rather from within yourself."

"I don't understand."

"It takes faith and courage, my child, to dedicate your entire life to a goal that appears impossible, especially when everyone else around you is busy making money, establishing a family, and rearing children," said Joanna with a smile. Her tone softened. "Now it's my turn to ask questions. Is earning Moola your only motivation for coming here?"

Keira could feel her cheeks burning. For a woman who often seemed to forget her name, it felt like Joanna had seen straight through her. Keira had struggled to find a stable job in an XR company, and signing up for Project Jukurrpa and coming to Sunshine Village had been her best option.

"Yes and no," said Keira. "It might have been my motivation at first, but now I'm beginning to feel that gaining the respect of others makes me happier than anything else."

"Well said, K . . . child. I will confirm your service to your wristband thingy, as long as you promise to do one thing for me," said Joanna, winking.

"I'll promise you anything!" Keira said hastily.

"You don't have to shout. My brain might be messed up, but my ears are not. I'll tell you more tomorrow. Good night, now!"

The old lady wheeled herself toward the bedroom. Once again, Keira was left standing, stupefied, with her eyes fixed on the photos of tropical fish that lined Joanna's kitchen.

JOANNA'S WISH WAS FOR KEIRA to take her to the ocean.

Before everything was wiped away from her memory, Joanna hoped to once again gaze at the Coral Sea that she had given so much to save—and which had given her life so much purpose.

Keira was torn. As much as she would have loved to take Joanna to a beach in Brisbane, arranging daytrips was outside the guidelines of her service. And despite their lucid conversation the previous day, Joanna's health was deteriorating. Keira was worried that Joanna's body couldn't handle the rigors of traveling, and she herself couldn't afford the possible consequences.

In the hope that Joanna would forget her wish, Keira came up with all kinds of excuses: bad weather, traffic jams, holidays. Joanna, though, was as stubborn as a child, and pestered Keira every single day.

"I heard there's a community party today, and they're having food, drinks, and a live band. Everyone's going! Don't you want to go?" suggested Keira, trying to distract Joanna.

"No," she replied instantly.

"Come on, Joanna," pleaded Keira. A week earlier, Joanna had asked Keira to stop addressing her as "Ms. Campbell," because apparently that was how people addressed real estate agents.

"You promised you would take me to the ocean! *You lied!*"

"No, I didn't promise that."

"Don't you want your confirmation anymore? The Moola score you care so much about?"

"Shhh . . . the AI system would deduct points from my score if it heard this conversation," whispered Keira. She took off her XR glasses and rubbed her eyes, sore from staring at the glasses' image projections. Recently, in addition to her duties with Joanna, Keira had begun volunteering on the side as a product developer of augmented reality experiences for an AR company named DingoTech. She hoped that, with the experience she gained, she could one day land a real AR job.

"Why are you always wearing glasses? As far as I understand, you're far too young to need reading glasses," grumbled Joanna, curious, as she reached for Keira's XR glasses. The moment she put them on, she cried out in surprise. "Wow! Everything is glowing!"

"Wait, let me adjust them for you," said Keira, fine-tuning the XR glasses' focus parameters to accommodate Joanna's sight. Now Joanna no longer saw fuzzy blobs of light, but varicolored dots with sharpened edges superimposed on her vision, a filter in the style of a Papunya dot painting. The AR algorithm would alter how the dots' effects were rendered in real time based on the surrounding environment and the user's head posture, turning reality into a kind of dot painting that metamorphosed every second with new patterns and colors, undulating and rippling like the ocean surface on a windy day.

Incredulous, Joanna exclaimed, "It's beautiful! Did you make this?"

"Yes," Keira said bashfully. "I've always dreamed of becoming an artist, but it would be next to impossible for someone like me. This is the next best thing."

"I don't think so," said Joanna, her face scrunched up in disdain. "Young people! Always looking for excuses—"

"No!" Keira blurted out, for the first time cutting the old woman off. She could feel a surge of emotions rising. "This isn't an excuse. I'm talking about the difficulties of navigating life as an Arrernte."

"I don't believe I've heard of your people before," said Joanna.

"My people have lived on this continent for thirty thousand years but look at what's happening to us now!" Keira's voice was loud, almost a shout. In that moment, she didn't care what her smart wristband heard. Keira took a deep breath. "Our language has almost disappeared. We are driven to settlements and assimilated into big cities after our homes are snatched from us. And for us young people—yes, *young people looking for excuses*—ensuring our next meal can depend on either becoming a criminal or this *goddamn Moola*!"

"Hey, watch your language!"

"I used to hope that Project Jukurrpa would usher in a new age of equality," Keira continued, "but I was wrong. Like everything else about our system, Project Jukurrpa favors certain people. People who already excel at earning Moola and waving their glowing wristbands—people who know how to please, deceive, or intimidate other people—will only have more opportunities to win Moola and, duh, *society's respect.* That's how the world works. No matter how hardworking or talented I am, people like you will always look down on people like me."

"I didn't—I didn't mean to—" stuttered Joanna, clearly taken aback by the response of her usually even-tempered caregiver.

"Ms. Campbell, please understand that not everyone in this

world is as lucky as you are. Not everyone can pursue their dream. But you are right about one thing: Everyone should have the courage to try. So, you've inspired me. I'm telling you right now: I quit."

With that, Keira left the living room and strode toward her bedroom, walking so fast that she completely forgot she had left her XR glasses behind.

THAT NIGHT, KEIRA had a nightmare.

A yowie, covered in long golden hair, emerged from under the bed and pounced on her. She wanted to run, but her feet were unmovable, her body utterly frozen; she wanted to scream, but no voice came out of her gaping mouth. She could do nothing but stare with horror as the ape-like monster's jaws closed around her.

She woke with a start, drenched in sweat. Day had dawned; the sky was a light shade of blue. A little unsteady from the nightmare, she stepped into the kitchen for water, and her eyes landed immediately on the front door. It was wide open.

"Joanna?" Keira called. No response. She walked into Joanna's bedroom and found the bed empty.

After searching the entire house, she found a scribbled note near the door, in the spot where Joanna usually put her keys:

K: I'm going to see the ocean.
I'll return your glasses when I get back.
J.

Keira cursed under her breath as she threw on clothes and made a dash for the security desk in the Sunshine Village Resident Services center.

ACCORDING TO THE SURVEILLANCE FOOTAGE, Joanna had left home about an hour ago on her electric wheelchair.

"Not to worry. The biosensor membrane on every elder person can help us track her whereabouts," said Nguyen, a staff member at the Resident Services center. Still sleepy-eyed, he pulled up the real-time tracking system on the computer, then paused. The blinking GPS icon for Joanna indicated that she was, in fact, at home. Nguyen's eyes widened, now alert and awake. "Wait, did she take the membrane off?"

"We need to get everyone to help track her down right away," said Keira, now frantic with worry.

"How far can she possibly get on that wheelchair?" Nguyen tried to reason with Keira, who wasn't having it.

"Now!"

Keira knew that for people with Alzheimer's disease, the greatest threats came from behavioral disorders caused by the deterioration of various cognitive functions: becoming distracted while taking the stairs and missing a step, forgetting their destination and pausing in the middle of a busy road to remember, and injuring themselves when using sharp objects. She was terrified that Joanna, out in public on her own, would end up in an accident. *If I hadn't been so hot-tempered yesterday, maybe Joanna wouldn't have left,* thought Keira bitterly.

Nguyen launched the emergency procedures, which sent human staff and drones alike out on a search. The alert also went to the Brisbane police, who could access surveillance footage of the surrounding area.

Amid the frantic scene, Keira had fallen silent. A shadowy half-thought hung in the back of her mind. She knew it was important—but she couldn't remember what it was.

The note. *I'll return your glasses when I get back.*

"Glasses!" Keira pulled out her smartstream. If Joanna was wearing her XR glasses, Keira could access the glasses' live vision field remotely and deduce Joanna's location.

A river of flashing multicolored dots appeared on the screen. Sure enough, Joanna was wearing Keira's glasses, and she hadn't switched off the AR experience demo that Keira had made. The frame was still. Slowly, dots of light flowed down the winding river, changing colors as they bobbed up and down.

"There are several rivers that look like that around here," said Nguyen, craning his neck to look at Keira's screen. "Can you connect to audio as well?"

The glasses' auditory sensors picked up various sounds of nature: the rippling and bubbling of the river, the chirping of birds, the rustling of tree leaves, a gentle swish of the morning breeze. It was overlaid by a rhythm of inhaling and exhaling, which presumably came from Joanna. After a while, they heard a rumble coming from the right, lasting for about three seconds before disappearing again.

"She's at Breakfast Creek!" Nguyen cried. "That's the train. There's a bridge there that crosses the creek!"

"Take me there now!" Excited, Keira grabbed Nguyen's hand. "Quick, tell everyone to meet us at the creek to search for Joanna!"

KEIRA TROTTED ALONG THE RIVERBANK, peering through the lush vegetation for any sign of Joanna. The singing birds and buzzing bees annoyed her, and sweat beaded on her forehead and dripped off the tip of her nose. Keira kept comparing the feed from her smartstream to the scene in front of her eyes. Finally, she saw a flash of long silver hair under a pine tree.

When she approached with the rescue team, she found Joanna sitting in her wheelchair in silence. On her inner wrist was a square of lighter-colored skin, where the biosensor membrane used to be. The woman appeared to be lost in a trance. Tears streamed down her face, staining the XR glasses' lenses with a foggy veil. Keira stepped up and pulled her into a tight embrace.

"Keira, you're here," murmured Joanna. *This is the first time she's gotten my name right,* thought Keira. "Your glasses brought me back. Now I remember. I am one of you."

"Huh?" Keira, her anxious heartbeat still loud in her ears, was taken aback by the old woman's words. She heard a camera flash.

"I am the stolen generation," whispered Joanna as the staff carried her into the ambulance.

KEIRA PUSHED JOANNA'S WHEELCHAIR along the pedestrian path by the sand. It was a beautiful day on Noosa Main Beach. Beachgoers laughed, children played in the sand, and surfers paddled in the sunshine. Joanna set her gaze to the northeast, where azure water stretched infinitely into the horizon.

"Do you see the Great Barrier Reef?" asked Keira, even though she already knew the answer.

"Well, I know she's there. I can feel her." Joanna grinned. "Thanks to you. The government should give you more Moola than you're getting. It's funny to think that this time next week, I'll be getting my precision treatment. I'd never imagined that I'd make it to the end of the wait list."

"I'm so happy for you. I'm sure you'll recover in no time." Keira laughed. "There's one question I never got to ask you, though."

"Shoot."

"That day at Breakfast Creek, you said that you were 'the stolen generation.' I didn't know what that meant, so I did some research. And I found that, starting in 1909, the Australian government separated up to one hundred thousand Aboriginal children from their parents, placing them in the care of either white families or in official shelters for assimilation. This policy ended in 1969, along with those shelters, which left many children homeless. You were born after 1969, though. How can you be one of them?"

A look of melancholy appeared on Joanna's face. "My adoptive parents were very kind people. They registered me with a later birth date, thinking that would protect me from the ugly truth. I was taken from my biological parents immediately after birth, and then raised by the church for the next few years, before I was eventually adopted. I was lucky to be assigned to such loving stepparents."

"So how did you find out? I mean, it's been so many years since it happened, and I'm sure a lot of those records were destroyed," said Keira, unable to hide her curiosity.

"I always knew I didn't look like my siblings. I could tell that I was different, from the way people at school treated me. But I didn't want to ask my parents the question. They gave me as much love as they gave the other children. So I buried the question and didn't think of it again until the genome sequencing report."

"Genome sequencing for the Alzheimer's treatment?" asked Keira.

Joanna nodded and pointed north of the Pacific Ocean. "The report indicated an 85-percent probability that I was a descendant of Torres Strait Islanders. When I found out, my whole life seemed to collapse. I didn't know who I was, or who my real parents were. What did it mean? I didn't understand."

"So you chose to forget?"

"I'm afraid that forgetfulness chose me, my child. My illness gave me the perfect excuse to deny the truth . . . until your artwork led me to it."

"My *what*?" Keira couldn't believe her ears.

"When I put on your glasses, I saw a magnificent world unfold before my eyes. Just like dreaming, my experience wasn't static or linear, but rather crossing space-time, spanning from the past to the present, even seeping into the future. I could feel something ancient rising from my heart and rushing through my veins, reconnecting me to this piece of land. It told me that I shouldn't run away from my pain, and I mustn't forget who I am. Being honest with myself was the only way that I could heal myself."

Keira, touched, stared speechlessly at Joanna.

"I need to thank you." Joanna grabbed Keira's hands and brought them to her chest. "There are not many of us left from the stolen generation. Many people have died carrying the weight of pain and confusion, just like what I had before. The government issued an official apology thirty-three years ago and began to declassify the history, but that's not enough to make up for what they took from us."

Keira could feel the sea breeze gently kissing her long curls. Never would she have dared dream that her creation could help another in such a way. The salty smell reminded her of the days and nights she had spent with Joanna.

"You know, I'm the one who needs to thank you," said Keira, her tone solemn.

"Why? Because I always piss you off?" countered Joanna.

"Well, for that, too." Keira brushed away a few strands of hair tickling her eyes and grinned. "You made me think about things that had never occurred to me. My hopes and dreams, Project Jukurrpa . . ."

"I'm listening."

"In my opinion, Project Jukurrpa has cheapened the bonds between people in our community and served to widen the inequality gap even more. Most people don't use it the way it was intended. It no longer serves to motivate people to live up to their potential. I've been thinking about what you said, and I started a discussion in the VRock community a few weeks ago. Since then, tens of thousands of people have joined in. What started as an Internet debate has now become a movement called 'dream4future' and the media can't stop talking about it. The conversation struck a chord with people—their dissatisfaction with how Project Jukurrpa was working. Now, parliament has proposed a law to revise Project Jukurrpa."

"Wow! What would the new version of the project look like?"

"BLC gave people basic life necessities and security, and that's not going to go away. But everyone, especially young people like me, should have the right to choose freely *how* they would like to live—and no one's dreams should be snatched away. When someone strives for self-discovery and actualization, just like you, she should be granted the chance. Project Jukurrpa should be providing everyone with equal opportunities to explore who they want to become and help them fully realize their potential. Whether it's developing leadership skills, uncovering the mysteries of Mars, restoring Aboriginal languages with AI, building environmentally friendly cities, you name it. Every step of an individual's road to self-realization, every effort and achievement made should be seen, recognized, and encouraged. That's the only way we can bring back hope. Otherwise, we are facing a new kind of stolen generation."

"Listen to you! Keira, you are amazing!" Joanna, enlivened by Keira's speech, clapped excitedly. All of a sudden, her hands halted in midair. "Does this mean you're going to leave me?"

"I'm sorry, Joanna, but yes. I'm here to say my goodbyes today," said Keira, leaning down to hug Joanna. "The picture of us taken by Breakfast Creek was posted everywhere in VRock. After all the media attention, Dr. Swartz from ISA invited me to join their project team. Together, we'll try to find a way to make these goals quantifiable and train a smarter AI to build a more equal, more inspiring Project Jukurrpa. I've always wanted a real job—I thought it was going to be in AR, but the chance to work to shape the possibilities of what young people can achieve is beyond my wildest dreams."

"I'm truly happy for you," said the older woman. Hesitating, she dropped her eyes, as if embarrassed. "But before you leave, there's something that I need to tell you."

"What is it?"

"I was so reluctant to confirm your service because I was afraid that you would leave me behind once you received the Moola," whispered Joanna, her voice trembling. "I didn't want you to leave."

"Oh, Joanna . . ." Tears were welling up in Keira's eyes.

"Don't cry, child. Don't cry." Joanna wiped the corners of her eyes and smiled at Keira. "You brought me to the ocean, and now it's my turn to make good on my promise to you."

The crisp, melodious tones of Moola cash-in dispersed in the sea breeze. With Keira pushing Joanna's wheelchair, the duo continued their long journey down the beach. Together they watched the waves ebb and flow, molding the shape of the shoreline, inch by inch. Just like they did a billion years ago. Just like they would in the future.

ANALYSIS

PLENITUDE, NEW ECONOMIC MODELS, THE FUTURE OF MONEY, SINGULARITY

We humans have long fantasized about the day when we no longer have to work, and everything is free. "Dreaming of Plenitude" depicts a 2041 future in which energy revolution, material revolution, AI, and automation take us halfway there.

While AI and other technologies are bringing about the fourth industrial revolution, a clean energy revolution is under way—one that will address the crisis of climate change while dramatically reducing the cost of powering the world. We are approaching a confluence of improved solar, wind, and battery technologies with the capacity to rebuild the world's energy infrastructure by 2041.

As the cost of energy plummets, it will also bring down the cost of water, raw materials, manufacturing, computation, logistics, and anything that has a major energy component. At the same time, production will shift from using limited or toxic materials (oil, minerals, some chemicals) to nature's plentiful and inexpensive building blocks (photons, molecules, silicon). Finally, from what we read in chapters 1 through 9, AI and automation will dramatically reduce the final cost component of production: routine human labor.

As the cost of energy, materials, and production fall at historic speed, we can look forward to plenitude. "Plenitude" is the word I have chosen to denote a new phase of human life, in which all people are entitled to a comfortable life, as goods prices approach free, and work becomes optional. Others have called it "abundance" or "post-scarcity."

But in "Dreaming of Plenitude," a society that at first seems like it might possess all the ingredients of a utopian paradise, in which everyone's basic needs are met (thus leaving people free to pursue higher purposes with their lives), is revealed to bring about as many problems as it solves. In particular, the story reveals the possible danger of how young people might react in a world where they have lost the traditional anchor of a sustaining career around which to build a prosperous life. The reason there will be so many bumps on the road to utopia is that economic models were designed for scarcity, not plenitude. When almost everything is free, what is the purpose of money? Without money, what happens to people who grew accustomed to working to make money as their meaning of life? What happens to economic institutions and corporations?

In this section, I will describe the energy revolution and the materials revolution, and how they will provide the fuel and the raw materials for AI-enabled automated production, making plenitude inevitable. I will explore how plenitude invalidates existing economic models and institutions, including money. Then I will talk about how money might evolve and explain the design of the new currencies in "Dreaming of Plenitude." Next, I will explain why we chose to end the book with plenitude, rather than singularity, which some futurists believe will be the defining technological moment of the 2040s.

Finally, I will leave you with a few overarching thoughts to consider about the future of humankind as we conclude both this chapter—and our journey through 2041.

THE RENEWABLE ENERGY REVOLUTION: SOLAR + WIND + BATTERIES

In addition to AI, we are on the cusp of another important technological revolution—renewable energy. Together, solar photovoltaic, wind power, and lithium-ion battery storage technologies will create the capability of replacing most if not all of our energy infrastructure with renewable clean energy.

By 2041, much of the developed world and some developing countries will be primarily powered by solar and wind. The cost of solar energy dropped 82 percent from 2010 to 2020, while the cost of wind energy dropped 46 percent. Solar and onshore wind are now the cheapest sources of electricity. In addition, lithium-ion battery storage cost has dropped 87 percent from 2010 to 2020. It will drop further thanks to the massive production of batteries for electrical vehicles. This rapid drop in the price of battery storage will make it possible to store the solar/wind energy from sunny and windy days for future use. Think tank RethinkX estimates that with a $2 trillion investment through 2030, the cost of energy in the United States will drop to 3 cents per kilowatt-hour, less than one-quarter of today's cost. By 2041, it should be even lower, as the prices of these three components continue to descend.

What happens on days when a given area's battery energy storage is full—will any generated energy left unused be wasted? RethinkX predicts that these circumstances will create a new class of energy called "super power" at essentially zero cost, usually during the sunniest or most windy days. With intelligent scheduling, this "super power" can be used for non-time-sensitive applications such as charging batteries of idle cars, water desalination and treatment, waste recycling, metal refining, carbon removal, blockchain consensus algorithms, AI drug discovery, and manufacturing activities whose costs are energy-driven.

Such a system would not only dramatically decrease energy cost, but also power new applications and inventions that were previously too expensive to pursue. As the cost of energy plummets, the cost of water, materials, manufacturing, computation, and anything that has a major energy component will drop, too.

The solar + wind + batteries approach to new energy will also be 100-percent clean energy. Switching to this form of energy can eliminate more than 50 percent of all greenhouse gas emissions, which is by far the largest culprit of climate change.

The above projection is contingent on continued technology improvements and substantial capital spending on the part of nations around the world to build an infrastructure for renewable energy, which means more-progressive countries will benefit from it sooner. This is why we set "Dreaming in Plenitude" in Australia, where renewable energy has grown ten times faster than the world average.

THE MATERIALS REVOLUTION: TOWARD INFINITE SUPPLY

We are experiencing what Peter Diamandis calls "dematerialization," or an age in which many physical products are made obsolete as their capabilities are absorbed by software and platform products like mobile phones. Recent examples include radios, cameras, maps and stand-alone GPS systems, camcorders, and encyclopedias. As dematerialization occurs at increasing speed, previously expensive products become effectively free.

In chapter 4, "Contactless Love," we discussed the power of synthetic biology in drug discovery and gene therapy (such as CRISPR), which will reduce healthcare costs, improve treatment efficacy, and increase human longevity. In addition, synthetic biology has shown promise in redesigning organisms for useful purposes by engineering into them new abilities. Synthetic biology will revolutionize the food industry. Meat can be grown in the lab using animal-sourced starter cells, with the same protein and fat profiles, and the same taste. This transformative technology will create "real" meat without harming animals or the planet. Future foods will not be limited to what we have tasted in the past. Working at a molecular level, scientists will be able to mimic existing foods as well as create entirely new food products, uploading them to databases to be produced in quantity for very low cost, just like software or commodity hardware.

Most vegetables and fruits can be produced on vertical farms that are really automated factories, and the cost will plummet with economies of scale. Eventually the primary cost of these farms will be electricity, water, and fertilizer. We know from the previous section that electricity and water will become almost free. Synthetic biology can engineer bacteria to supply the nitrogen that plants need to thrive, thereby ending toxic chemical fertilizers.

Synthetic biology can also be used to make rubber, cosmetics, fragrances, fashion, fabrics, plastic, and "green" chemicals. It can dissolve plastics and clean pollutants from the environment. Synthetic biology will revolutionize many industries by making them more sustainable while dramatically lowering the overall cost.

In June 2011, President Obama announced the Materials Genome Initiative, a nationwide effort to use open-source methods and AI to double the pace of innovation in materials science. In the past ten years, this effort has created an enormous database that has endowed scientists with the ability to build materials one atom at a time. These include materials that seem straight out of science fiction, such as artificial muscles and nanomaterials that make everything much lighter.

With these building blocks shifting away from limited and toxic compounds to plentiful materials from nature (photons for energy, molecules for synthetic biology, atoms for materials, bits/qubits for information, silicon for semiconductors), we will be one step closer to the dream of plenitude.

THE PRODUCTION REVOLUTION: AI AND AUTOMATION

As we have established in previous chapters, robots and AI will take over the manufacturing, delivery, design, and marketing of most goods. Autonomous vehicles will take us anywhere anytime at minimal cost and save us money by our not having to buy cars ("The Holy Driver"). AI service robots will do chores at home better than the best housekeeper ("Contactless Love"). AI will take over all routine jobs and tasks, white-collar and blue-collar alike ("The Job Savior"). AI works 24/7, does not get sick, does not complain, and does not need to be paid. AI will reduce the cost of most manufactured goods to a small increment over the cost of materials.

AI will also provide excellent service for many routine white-collar jobs. AI assistants will guide our lives better than the very best human assistants ("The Golden Elephant" and "Isle of Happiness"). AI teachers will teach engaging classes tailored to each student ("Twin Sparrows"). AI doctors will diagnose and cure patients better than human doctors ("Contactless Love"). AI entertainment will be realistic and immersive, yet virtual and thus nearly free ("My Haunting Idol").

Robotics will become self-replicating, self-repairing, and even partially self-designing. 3D printers—which will look more and more like the replicator in *Star Trek*—will enable sophisticated or customized goods (like dentures and prosthetics) to be produced for minimal cost.

Houses and apartment buildings will be designed by AI and use pre-fabricated modules that are put together like Lego blocks by robots, thus dramatically reducing housing costs. Just-in-time autonomous public transportation, including robo-buses, robo-taxis, and robo-scooters, will take us anywhere without our having to wait.

So, with nearly free energy, inexpensive materials, and AI-automated production, we will usher in the age of plenitude.

PLENITUDE: A TECHNOLOGY-MEDIATED INEVITABILITY

"Post-scarcity" describes a world where nothing is scarce, and everything is free. In "Dreaming of Plenitude," we encounter a future world in which countries are moving toward post-scarcity, although at different paces. In the last story, Australia, a highly developed country, is wealthy enough to give everyone basic necessities and comfortable living (through the BLC card). Poorer countries, we can infer from the story, would reach a state of plenitude sometime later.

Because the timetable will vary for different countries, I prefer the term "plenitude" rather than "post-scarcity." Also, strict post-scarcity will never be achieved. For example, no matter how much technology improves, there will never be more than twenty paintings by Leonardo da Vinci. Also, premium versions of goods and services—those offering unique human value (like a motivating private tutor) or complex or rare components and technology (like the earliest functioning quantum computers)—will still be scarce. But these high-end offerings will be the exception rather than the rule, just as most of us drink nearly free tap or filtered water, while pollutant-free water from under the Mount Fuji volcano is still scarce.

The age of plenitude will arrive when most things are no longer scarce, can be produced for next to nothing, and—most important—are made available freely or cheaply to all people. These "almost free" things will begin with necessities like food, water, clothing, shelter, and energy. Over time, plenitude is a process in which more and more goods and services are provided to more and more people, as technologies advance and costs come down to grant new free "indulgences" every year. I expect that plenitude would start with these necessities and gradually expand to provide a comfortable and gracious lifestyle for all, encompassing a provision for transportation, clothing, communication, healthcare, information, education, and entertainment. The citizens in "Dreaming of Plenitude" enjoy all of these benefits for free.

If you are skeptical about plenitude, consider that even today it's already here in certain segments of the economy. We can consume all the music and movies we want, on any device at any time, for about twenty dollars a month. We can select from a large collection of ebooks and audio books at a nominal cost. We can read or watch news for free. We can buy and sell stocks for no commission. We can search and access valuable information online that was once artificially made scarce and expensive.

You might argue that the examples above all relate to digital products, because they have virtually no marginal cost of manufacturing and shipping. But what about "real" things like food and shelter? In 2020, the United States discarded $218 billion worth of food, while the cost to eliminate hunger in the United States has been estimated at just $25 billion per year. In the United States, there are more than five times as many unoccupied houses as there are homeless people. So we already have theoretical plenitude in 2021 for food and shelter in the United States. Imagine trying to explain this accomplishment and imbalance to humans of five hundred years ago. As William Gibson said, "The future is already here—it is just not very evenly distributed."

ECONOMIC MODELS FOR SCARCITY AND POST-SCARCITY

For millennia, human economic systems have evolved under one fundamental premise—scarcity. Scarcity exists when human wants for goods and services exceed the limited supply for them. Scarcity has been the cause of wars, mass migration, capital markets, and every aspect of how civilization has developed. Scarcity is an assumption made by all theories of economics.

Economics is a social science concerned with the production, distribution, and consumption of goods and services. The field is concerned with how individuals, businesses, governments, and nations make choices about how to allocate resources. A fundamental assumption of economics is that society's wants are unlimited, but all resources are limited. Economic models are theories about how to produce, distribute, and consume these scarce resources efficiently.

The father of modern economics, Adam Smith, theorized that by giving everyone the freedom to produce, exchange, and consume based on their self-interest, economies will naturally balance and continue to grow. Karl Marx argued that the increasing power of capital will invalidate Adam Smith's theory, as it gives inordinate power to those who control capital, leading to inequality and exploitation of the working class. John Maynard Keynes shared the concern that "natural balance" will take too long but advocated using monetary policies to modulate the economy to increase demand and decrease unemployment. One thing all three theories have in common is the underlying assumption of scarcity.

In the future, if the scarcity assumption is nullified, so would these economic models. When there is no scarcity, then all mechanisms such as selling, buying, and exchanging will no longer be needed. Money will arguably also no longer be needed. What, then, is the economic model?

Science fiction has often made prescient predictions about the future. As far as plenitude is concerned, Star Trek provides a fascinating future vision. In his book Trekonomics, Manu Saadia describes the Star Trek economic model, which is encapsulated in Captain Picard's famous declaration that "people are no longer obsessed with the accumulation of things. We've eliminated hunger, want, the need for possessions." In Star Trek: The Next Generation, set in the twenty-fourth century, the replicator can make anything, thus eliminating the need to work and trade. Without these needs, money and labor become redundant. Employment becomes optional and voluntary, and social status and respect become the new currency, as more people move up the Maslow hierarchy and live for self-actualization. Among these people are the Enterprise crew, who achieve self-actualization through exploration of new worlds and pursuit of knowledge.

I believe that in the very long term, an economy similar to the one outlined in Trekonomics is achievable. It will be built on a new social contract that grants increasing basic services for a comfortable life, while redefining concepts like work, money, and purpose, as well as the role of corporations and institutions. The new system should be designed in a way that reaches the same kind of balance found in Adam Smith's theory: If people pursue their self-interest, then a virtuous cycle forms and everyone is better off.

Star Trek paints a fascinating end point that took three hundred years to reach but does not address how to get there. "Dreaming of Plenitude" describes a plausible waypoint for such an evolution, centered on one concept: money.

MONEY IN THE AGE OF PLENITUDE

In his book *21 Lessons for the Twenty-First Century* Yuval Noah Harari wrote, "Human society was developed based on thousands of years of 'stories'—stories that we tell ourselves. We are the only mammals that can cooperate with numerous strangers because only we can invent fictional stories, spread them around, and convince millions of others to believe in them." Professor Harari also said: "Money in fact is the most successful story ever invented and told by humans, because it is the only story everybody believes." Money has been a key part of human society since 5,000 B.C. If money is demolished because everything is becoming free, it will bring down many key pillars of our society along with it.

Money is a store of value, unit of account, and medium of exchange. But more important, we have been taught to accumulate money for centuries, in our pursuit of safety and survival. Money has become a status symbol that gives us respect as well as vanity. Our desire for money is often insatiable, leading to greed, but also providing a sense of purpose. In other words, money has become a key ingredient in the entire Maslow's hierarchy, and its emotive impact has become deeply engrained in us after thousands of years of storytelling. Money is not something that can be eliminated overnight; a very long-term gradual plan is needed.

In "Dreaming of Plenitude," this gradual plan takes the form of Australia's Project Jukurrpa program. Jukurrpa tries to reinvent money gradually, while also addressing changes brought about by plenitude and automation-induced job disruption, by giving citizens basic necessities and helping them through reskilling. It has three components: the BLC, Moola, and, later, a revised program brought on by the citizen-led dream4future movement.

The first is the Basic Life Card (BLC), which you can think of as essentially universal basic services. Unlike universal basic income (UBI), the BLC gives holders credits that can be exchanged for services that fulfill basic needs as well as allow for a comfortable life. Unlike a simple UBI, BLC credits can be used only for food, water, shelter, energy, transportation, clothing, communication, healthcare, information, and entertainment. This exchange limitation is important because unemployment is known to be correlated with alcohol and opioid use.

The BLC would provide for everyone the physiological and safety levels of the Maslow hierarchy, whether one has a job or not. Also, education and retraining are provided completely free, along with personalized assistance, because for people who want to continue to work, retraining is critical to avoid AI job displacement, as discussed in "The Job Savior."

The second component of the Jukurrpa plan, Moola, is a new "currency" intended to help some people move to the next level on the Maslow hierarchy—love and belonging, as exemplified in caring, friendship, warmth, camaraderie, trust, and connection. Unlike money and BLC, love and belonging cannot be spent. With money, the more you spend, the less you have. But with love and belonging, the more you give, the more you have. Your Moola grows as your wristband listens and senses the emotional well-being of the people around you. Are you helping and caring about others? Are you strengthening your community and forming beneficial relationships?

Powering Moola is an AI algorithm that scores people's empathy and compassion, as demonstrated in their community service work and other interactions, and follows the premise that the more you spend, the more you have. To protect each user's privacy, the wristband uses privacy computing technologies like federated learning and TEE to guarantee that private data is never transmitted outside the wristband and is permanently deleted once used. The wristband in "Dreaming of Plenitude" also suggests ways to further increase one's Moola with volunteer activities such as elder care, which is how Keira and Joanna meet each other.

The notion of a nonmonetary currency like Moola speaks to anticipated job displacement issues—as automation takes over routine tasks, the largest number of safe "jobs" will be service jobs that require human-to-human connection. Moola's AI algorithm guides people to opportunities that help them apply their empathy and compassion, making them more eligible for good opportunities in the service sector.

One of the design flaws of Moola was that when people accumulated enough Moola, they were rewarded with brilliant colors on their wristband. The program's designers wanted to encourage people to accumulate more Moola and lead purposeful lives filled with love and belonging, while becoming more empathetic and compassionate, which helps them fit the more necessary service jobs. But the designers overlooked people's need for vanity via accumulation. The greed for more Moola caused some people to game the system, by coaxing, threatening, and colluding to get nice things said near the wristband to get more Moola. The depiction of Project Jukurrpa in the story is an unchartered territory. To be successful over time, a country introducing such a program would need to frequently uncover the kind of design flaws suggested by the narrative, and its administrators would need to address them, with the goal of iterating toward a successful program over time.

The third and last component of Project Jukurrpa is revealed at the end of the story, as Keira explains to Joanna how she was part of a young people–led online movement, dream4future, that convinced Australia's leaders to rethink how the program could encourage people to realize their dreams and potential. Joanna's purpose-filled life as the savior of the Great Barrier Reef inspires Keira, just as Keira's dot-painting-like artistic work for the augmented reality maker inspires Joanna. This mutual inspiration leads Keira and Joanna to a shared belief that people must be encouraged to keep dreaming big in the age of plenitude, whether it is to restore the Aboriginal culture, to explore Mars, or to build sustainable cities. The dream4future movement inspires a third component of Jukurrpa, and the journey on which Keira chooses to embark. This last component, although it is complex and not fully defined as the story ends, clearly corresponds to the top of the Maslow hierarchy: self-actualization.

The dream4future enhancement to Project Jukurrpa would need to involve upgrading the AI algorithm from listening for empathy to also evaluating all higher layers of the Maslow hierarchy, with positive reinforcement beyond simple accumulation or addiction. Just as in "Isle of Happiness," in which AI learns to measure people's happiness, scientists would learn to build AI to recognize people's sense of respect, accomplishment, and self-actualization. Perhaps they will find that these virtues are what happiness is all about, thus connecting the two stories into one thematic whole.

While the specific ideas about the evolution of money in "Dreaming of Plenitude" are speculative, I hope I have convinced you that with plenitude, we need to design an inclusive new world that simultaneously provides for a young retiree to live comfortably, an industrious worker to gain new skills, a hobbyist to pursue her passion, a compassionate caregiver to spread his love, an overachiever to win respect, and a dreamer to change the world.

In a world transitioning into plenitude, we cannot simply assume everyone falls into the "useless class," nor that everyone will strive for self-actualization. As economic models move beyond scarcity and money, they should be reinvented to try to elevate human needs, encompassing love and belonging, esteem, and self-actualization. Abraham Maslow has said: "One's only failure is failing to live up to one's own possibilities." It is hoped that our future economic model will be both inclusive and aspirational, helping to elevate as many people as possible up the Maslow hierarchy.

CHALLENGES FOR PLENITUDE

I have painted a grand road map to plenitude. But this road is full of obstacles and even death traps.

First, reaching plenitude requires nothing short of a total financial overhaul. All financial institutions such as countries' central banks and stock markets will need to be reinvented or replaced. The disappearance of scarcity will cause deflation, leading to a collapse of prices and eventually markets. As we have seen in the two major financial crises that have already occurred in the twenty-first century, our financial system is fragile. To prevent a catastrophic financial crisis, the scope of the work and the depth of the issues required are monumental, including handling the deflation that will result from collapsing prices, distributing free goods and services, and transitioning from one economic model to another.

The second systemic problem is that corporations will refuse to accept the end of scarcity. Historically, whenever goods become cheap to make, the preferred choice of giant corporations has been not to altruistically lower their prices, but rather to create artificial scarcity, in order to perpetuate their profit. This has been happening for centuries. The discovery of plentiful diamonds did not drive down prices, as monopolist De Beers released limited quantities per year to create an artificial scarcity, while the industry used advertising to brainwash us that diamonds equaled love. The fashion industry indoctrinates us to believe that old designs are outdated and even embarrassing, so that we buy more clothing than we can wear, while they destroy unsold inventory. An average American bought sixty-eight pieces of clothing in 2017, while that same year Burberry alone destroyed $40 million of merchandise. It costs Microsoft virtually nothing to make another copy of Windows, but the company charges $139 to $309 for different editions of the operating system. The $139 version is basically the same as the $309 version, but with some capabilities turned off, thereby creating artificial scarcity for the $309 product.

Finally, the transition to plenitude requires a successful societal overhaul. All the changes described in this book will lead to unprecedented disruptions, which include disgruntled workers displaced by AI, governments dealing with transition to the age of plenitude, wealthy people watching their net worth plummet, and companies that refuse to reduce prices when things become non-scarce. If the disruptions result in social unrest, class polarization, or even revolutions, then all bets are off for our future.

In summary, a successful transition to plenitude would require an improbable shift for corporations to prioritize social responsibility over profit, an unlikely cooperation of nations that are stubbornly adversarial, a challenging transition for institutions to undergo a complete reinvention, and an implausible forfeiture of the never-ending human vices of greed and vanity.

So, should we surrender?

I say: Absolutely not! The opportunity to reach plenitude presents humanity with the ultimate test: As the confluence of magic-like technologies allows us to build almost anything for almost free, do we succumb to the meaningless temptation to accumulate wealth when there is nothing to spend the wealth on? Do we deliberately turn a blind eye to extreme poverty when there is enough for everyone? The answers are clear. We must find an economic model that is subordinate to human needs and not human greed. We face daunting challenges and overwhelming odds, but there are also unprecedented rewards. Never has the potential for human flourishing been higher, or the stakes of failure been greater.

WHAT COMES AFTER PLENITUDE—SINGULARITY?

At the outset of the book, I declared that I wanted to look toward a distant horizon: 2041. But now that we've come to the end of the book, let's consider what horizons may lie just beyond that. What could be beyond something as grandiose as plenitude? Some futurists have predicted that "the singularity" will come by 2045, which is right around the corner from 2041.

According to the singularity theory, due to the exponential growth of computing, self-directed AI will also grow exponentially and gain superintelligence faster than we can imagine, because humans cannot fathom exponential phenomena. In other words, singularity is the moment in which machine intelligence outpaces human intelligence—and when AI could snatch control of our world from humans. On the question of the singularity, futurists' minds gravitate to different extremes, resulting in sharply contrasting visions of what may come to pass—visions that have captured public attention and divided much of the technological community.

Singularity utopians believe that once AI far surpasses human intelligence, it will provide us with near-magical tools for alleviating suffering and realizing human potential. In this vision, superintelligent AI systems will so deeply understand the universe that they will act as omnipotent oracles, answering humanity's most vexing questions and conjuring brilliant solutions. Some believe that we must become cyborgs to become one with the omnipotent AI, lest we drift into oblivion.

But not everyone is so optimistic. In the singularity dystopian camp are people like Elon Musk, who has called superintelligent AI systems "the biggest risk we face as a civilization," comparing their creation to "summoning the demon." This group warns that when humans create self-improving AI programs whose intellect dwarfs our own, they will want to control us, or at least make us totally irrelevant.

Which singularity vision—cyborgs or machine overlords—might arrive by 2041? I'd say neither. Believers in singularity argue that exponentially improving technologies will lead to superintelligence. I agree that AI computational prowess has indeed increased exponentially, but exponentially faster computing power alone does not lead to qualitatively better AI. To deliver qualitatively better AI, new scientific breakthroughs like deep learning are also needed. Suppose we had all the computing power today but no deep learning; then the whole AI industry would be nonexistent.

To achieve superintelligence in the future, we absolutely need more scientific breakthroughs. For example, we will need to figure out: How can we effectively model creativity in the arts and sciences? Or strategic thinking, reasoning, and counterfactual thinking? Or compassion, empathy, and human trust? Or consciousness and its accompanying needs, wants, and emotions? Without these capabilities, AI cannot even become human, not to mention a god or a demon. Take consciousness as an example. Not only are we incapable of building a conscious AI, but we don't even understand the underlying physiological mechanisms behind human consciousness.

Are these scientific breakthroughs possible? Perhaps one day, but they will not come easily or quickly. In the sixty-five years of AI history, arguably there has been just a single breakthrough: deep learning. We will need at least a dozen more breakthroughs to reach superintelligence, which is unlikely to happen in just twenty years.

THE STORY OF AI: A HAPPY ENDING?

In *AI 2041*, we saw that AI will open the door to a radiant future for humanity. AI will create unbelievable wealth, amplify our capabilities through human-AI symbiosis, improve how we work, play, and communicate, liberate us from routine tasks, and, as we learned in this chapter, usher us into the age of plenitude.

At the same time, AI will bring about myriad challenges and perils: AI biases, security risks, deepfakes, privacy infringements, autonomous weapons, and job displacements. These problems were not inflicted by AI, but by humans who use AI maliciously or carelessly. In the ten stories collected here, these problems were overcome by human creativity, resourcefulness, tenacity, wisdom, courage, compassion, and love. In particular, our sense of justice, our capacity to learn, our audacity to dream, and our faith in human agency always saved the day.

We will not be passive spectators in the story of AI—we are the authors of it. The values underpinning our visions of an AI future will become self-fulfilling prophesies. If we believe we will become the "useless class" as AI's capabilities expand, then we will obliterate any chance of reinventing ourselves. If we become complacent with the gifts of plenitude and stop enriching our minds and bonds with one another, then we will terminate the evolution of our own species. If we feel hopeless and capitulate as singularity draws near, then we will usher in a winter of despair, whether singularity arrives or not.

On the other hand, if we are thankful for the liberation from routine work and from the fear of hunger and poverty, if we cherish our free will (which AI does not have), and if we have faith that human-AI symbiosis is much greater than the sum of the two parts, then we will work to mold AI into a perfect complement to help us "boldly go where no one has gone before." We will explore new worlds with AI, but, more important, we will explore ourselves. AI will give us a comfortable life and a sense of security, pushing us to pursue love and self-actualization. AI will reduce our fear, vanity, and greed, helping us to connect with more-noble human needs and wants. AI will take care of all that is routine, invigorating us to explore what makes us human and what our destiny should be. In the end, the story that we write is not just the story of AI, but the story of ourselves.

In the story of AI and humans, if we get the dance between artificial intelligence and human society right, it would unquestionably be the single greatest achievement in human history.

ACKNOWLEDGMENTS

This book was born out of serendipity. Lin Qiling and Anita Huang introduced the two of us and proposed the idea of a book of "scientific fiction," a fusion of fiction and popular science, as a way of exploring the future. Shortly thereafter, Kai-Fu serendipitously met Markus Dohle, the CEO of Penguin Random House, following an impromptu introduction by Laurie Erlam at a conference. Markus was immediately enthusiastic about the premise for this book and connected us to the fantastic David Drake and Paul Whitlatch at PRH's Crown division. The next serendipity came when Kai-Fu and Laurie managed to squeeze in a stop between trips to see David and Paul in New York, just before the pandemic travel ban. And finally, working from home in 2020 serendipitously gave us the quiet seclusion to finish the book, a rather mammoth undertaking, in just one year.

Serendipity creates opportunity, but it's always people who make things happen. We are deeply thankful to Qiling, Anita, Markus, David, and Paul for their creativity, enthusiasm, and diligence, which have made this book possible. We would also like to thank other members of the team at Crown and Currency, including Katie Berry, Gillian Blake, Annsley Rosner, Dyana Messina, Julie Cepler, Emily Hotaling, Sarah Breivogel, Robert Siek, Edwin Vazquez, Michelle Giuseffi, Jennifer Backe, Sally Franklin, and Allison Fox.

The creation of the book was a complex undertaking, with Stan's stories first composed in Chinese and Kai-Fu's commentary in English. Writing in two languages simultaneously was quite a challenge, and we are thankful to our team of skillful translators: Emily Jin, Andy Dudak, Blake Stone-Banks, and Benjamin Zhou. Paul Whitlatch is a fastidious and diligent editor, but equally important, he is an exceptional writer. These two skills enabled him to go well beyond the call of duty to help us maintain a consistent and readable style.

We are also thankful to the following technologists who answered questions and helped validate the feasibility of various technologies explored in this book: Professor Xiao Wei, Professor Ni Jianquan, Professor Ma Weixiong, Dr. Tony Han, Dr. He Xiaofei, Dr. Wang Jiaping, Dr. Shi Chengxi, Dr. Zhang Tong, Wang Yonggang, and other colleagues from Sinovation Ventures, including Melody Xu. We also want to thank the whole team at Erlam & Co—Mike Harvey, Amy Holmes, Reese Duerden, Daniel Orchard, Alexandra Schiel, and Helen Glover—for reading many drafts of this book and giving extensive suggestions. In particular, we are deeply indebted to Scott Meredith for his comprehensive edits, insightful comments, and excellent ideas for several stories.

Lastly, we thank all the science fiction writers, past and present, whose collective imagination became a blueprint for AI, and all the AI scientists who are building advanced technologies that are indistinguishable from magic.

KAI-FU LEE is the CEO of Sinovation Ventures and the *New York Times* bestselling author of *AI Superpowers.* Lee was formerly the president of Google China and a senior executive at Microsoft, SGI, and Apple. Co-chair of the Artificial Intelligence Council at the World Economic Forum, he has a bachelor's degree from Columbia and a PhD from Carnegie Mellon. Lee's numerous honors include being named to the *Time* 100 and *Wired* 25 Icons lists. He is based in Beijing.

CHEN QIUFAN (aka Stanley Chan) is an award-winning author, translator, creative producer, and curator. He is the president of the World Chinese Science Fiction Association. His works include *Waste Tide, Future Disease*, and *The Algorithms for Life.* The founder of Thema Mundi, a content development studio, he lives in Beijing and Shanghai.